Essentials of Measure Theory

Carlos S. Kubrusly

Essentials of Measure Theory

Springer

Carlos S. Kubrusly
Electrical Engineering Department
Catholic University of Rio de Janeiro
Rio de Janeiro, Brazil

ISBN 978-3-319-37200-6 ISBN 978-3-319-22506-7 (eBook)
DOI 10.1007/978-3-319-22506-7

Mathematics Subject Classification (2010): 28Axx, 28Bxx, 28Cxx, 46Gxx

Springer Cham Heidelberg New York Dordrecht London

Softcover re-print of the Hardcover 1st edition 2015

Printed on acid-free paper

Springer International Publishing AG Switzerland is part of Springer Science+Business Media (www.springer.com)

To J & A, ever

You can't be perfect,
but if you don't try,
you won't be good enough.

Paul Halmos

Preface

The book is split into two parts. Part I is a first course in measure theory with integration, consisting of a revised, corrected, enlarged, updated, and thoroughly rewritten text based on the author's 2007 *Measure Theory: A First Course* [25]. Part II is a second course, dealing with measure and integration on topological spaces.

Part I, *Introduction to Measure and Integration*, designed to be a textbook for a first course in measure theory, gives an abstract approach to measure and integration, in which the classical concrete cases of Lebesgue measure and Lebesgue integral are presented as an important particular case of the general theory. Part I contains nine chapters. Chapter 1 considers real-valued (and extended real-valued) measurable functions with respect to a σ-algebra, and Chapter 2 introduces the concepts of measure and signed measure. The integral of nonnegative measurable functions with respect to a given measure is addressed in Chapter 3. The notion of integral is extended to real-valued measurable functions in Chapter 4, and L^p spaces are constructed in Chapter 5. Convergence of sequences of measurable functions is discussed in Chapter 6, where several concepts are compared. Decomposition of measures is investigated in Chapter 7, and extension theorems are treated in Chapter 8, where the Lebesgue measure is built up and discussed in detail. Product measures and integrals with respect to product measures (in particular, iterated integrals) close Part I in Chapter 9.

Part II, *Measures on Topological Spaces*, extends the material of Part I by equipping a nonempty set with a topology and considering σ-algebras of subsets of it containing the topology. This second part investigates measures and integrals on such Borel σ-algebras. It contains four chapters. Chapter 10 is an introduction to Part II, examining fundamental properties of integrals with respect to positive, real, and complex measures. Measures on topological spaces are introduced in Chapter 11, whose central theme is the construction of Borel measures on a Borel σ-algebra of subsets of a locally compact Hausdorff space. Several forms of the Riesz Representation Theorem are considered in Chapter 12 after an introduction to continuous functions with compact support and bounded linear functionals. Invariant measures are focused in Section 13, where the main topic is the construction of Haar measure on a Borel σ-algebra of subsets of a locally compact Hausdorff group.

The final section of each chapter in Part I contains *Problems*, and is an integral part of the chapter, not only just a set of routine exercises. The majority of those problems consists of auxiliary results, extensions of the theory, examples, and mainly counterexamples. The reader is encouraged to look at these problems with the same care as expected for a conventional theory section. Indeed, part of the theory is sometimes shifted to the problems section, and when this happens those problems are accompanied by *Hints* (sometimes, by detailed hints). The intention is to motivate readers to take an active part in the development of the theory presented in Part I.

Part II evidently is more advanced than Part I, addressed to more experienced readers. Thus, unlike Part I, the last section of each chapter of Part II consists of *Additional Propositions*, containing auxiliary and complementary results. These are followed by a set of *Notes*, in which each proposition is briefly discussed, and references are provided indicating proofs for all of them. These additional propositions can be viewed as a set of more complex proposed problems, and the respective notes as hints for dealing with them.

Each chapter in both Parts I and II ends with a collection of *Suggested Readings*. This has a triple purpose: to offer a reasonable bibliography including most of the classics as well as some recent texts, to point out where different approaches and proofs can be found, and also to indicate alternate routes towards additional results (so that some of the references are suggested as a second or third reading on the subject).

The material in Part I was devised to be covered in a one-semester beginning graduate course. Although naturally addressed to graduate students, Part I certainly is accessible to advanced undergraduate students as well. In fact, it is self-contained, and the prerequisites are very modest, namely, conventional undergraduate introductory analysis and, just for Chapter 5, linear spaces as usually taught in standard linear algebra courses. No acquaintance with functional analysis is required in Part I, but elementary set theory is obviously required. In particular, it is assumed that the reader be familiar with the notions of cardinality, countable and uncountable sets, and also with the following basic results: the set of all rational numbers is countable, countable union of countable sets is countable, infinite sets (in particular, uncountable sets) have a countably infinite proper subset, and countably infinite families can be reenumerated into a sequence. Part II was prepared to be covered in a one-semester graduate course as well, subsequent to a first course based on the material of Part I. The prerequisites for Part II is just Part I.

All in all, the resulting text of the whole book is the outcome of attempts to meet the needs of a contemporary course in measure theory for mathematicians who will also be accessible to a wider audience of students in mathematics, statistics, economics, engineering, and physics, bearing a modest prerequisite. I tried to respond to the input from those students by presenting a text that contains complete proofs, including answers to several questions they have raised throughout the years. The logical dependence of the various sections and chapters is roughly linear and reflects approximately the minimum amount of material needed to proceed further.

I have been lecturing on this subject for a long time. Thus, I benefited from the help of many friends among students and colleagues and I am truly grateful to all of them, in particular to Renato A.A. da Costa, Sergio Franklin, Leonardo B. Gonçalves, Johnny Kwong, André L. Pulcherio, Luciano R. da Silveira, Alexandre Street, and João Zanni who helped with the quest for typos in Part I, and Lucas Freire and Joaquim D. Garcia who helped with the quest for typos in Part II. Special thanks are due to Jessica Q. Kubrusly from whom I stole the notes for Section 7.3, to Adrian H. Pizzinga who read the entire text of Part I and corrected a number of typos and inaccuracies, to Richard Delaware who (together with his students) has scanned Part I as well and contributed with many corrections, and to an anonymous reviewer who made sensible suggestions to improve the text. Thanks are also due to my friend and colleague Marcelo D. Fragoso who was indeed an accomplice in writing this book. Let me also thank Elizabeth Loew and Ann Kostant from Springer New York for a pleasant and lasting partnership. I am grateful to the Catholic University of Rio de Janeiro for providing the release time that made this project possible.

Rio de Janeiro, Brazil   Carlos S. Kubrusly
June 2015

Contents

Part I

Introduction to Measures and Integration

1

Measurable Functions

1.1 Measurable Space

The *power set* $\wp(X)$ of a given set X is the collection of all subsets of X. We will work with set functions (i.e., functions whose domains are sets of sets) from Chapter 2 onwards. To begin with, we might assume that a natural candidate for the domain of such functions of sets would be the power set $\wp(X)$ of a given set X. This indeed would be an admissible candidate. However, as we will see in subsequent chapters, there are instances where the power set is too large a set to be the domain of some set functions we wish to consider. This means that some functions may lose essential properties if their domain is too large or, in other words, some functions would not be well-behaved when defined on a domain that is as big as a power set (this will be discussed in detail in Chapter 8). Given an arbitrary set X, a collection of subsets of X (i.e., a subcollection of the power set $\wp(X)$) that will be appropriate to our purpose is a *σ-algebra*.

Definition 1.1. An *algebra* $\mathcal{A}$ (or a *field*, or a *Boolean algebra*) of sets is a collection of subsets of a set X (i.e., $\mathcal{A} \subseteq \wp(X)$) fulfilling the next axioms.

(a) The whole set X and the empty set $\varnothing$ belong to $\mathcal{A}$.

(b) The complement $X \backslash E$ of a set E in $\mathcal{A}$ belongs to $\mathcal{A}$.

(c) The union of a finite collection of sets in $\mathcal{A}$ belongs to $\mathcal{A}$.

C.S. Kubrusly, *Essentials of Measure Theory*,
DOI 10.1007/978-3-319-22506-7_1

If the restriction of finite union in axiom (c) is relaxed to allow countably infinite unions as well, then the algebra receives a special name and notation.

A *σ-algebra* (or a *σ-field*, or a *Borel field*) on X, denoted by $\mathcal{X}$, is an algebra of subsets of a set X for which axiom (c) is extended to axiom ($\widehat{\text{c}}$) below.

($\widehat{\text{c}}$) The union of a countable collection of sets in $\mathcal{X}$ belongs to $\mathcal{X}$.

A pair $(X, \mathcal{X})$ consisting of an arbitrary set X and a σ-algebra $\mathcal{X}$ of subsets of X is called a *measurable space*. Sets in $\mathcal{X}$ are referred to as *measurable sets* (measurable with respect to the σ-algebra $\mathcal{X}$) or as *$\mathcal{X}$-measurable sets*.

Axioms (c) and ($\widehat{\text{c}}$) in the preceding definition of algebra and σ-algebra can be replaced, respectively, by the following axioms.

(c′) The intersection of a finite collection of sets in $\mathcal{A}$ belongs to $\mathcal{A}$.

($\widehat{\text{c}}'$) The intersection of a countable collection of sets in $\mathcal{X}$ belongs to $\mathcal{X}$.

Indeed, (b) and the De Morgan laws (viz., $X\backslash\big(\bigcup_\alpha A_\alpha\big) = \bigcap_\alpha(X\backslash A_\alpha)$ and $X\backslash\big(\bigcap_\alpha A_\alpha\big) = \bigcup_\alpha(X\backslash A_\alpha)$ for any collection $\{A_\alpha\}$ of subsets of X) ensure that (b) and (c) $\big($(b) and ($\widehat{\text{c}}$)$\big)$ is equivalent to (b) and (c′) $\big($to (b) and ($\widehat{\text{c}}'$)$\big)$.

Remark: Since $A\backslash B = A \cap X\backslash B$ for every $A, B \subseteq X$, axioms (a), (b), (c) are equivalent to axioms (a′), (b′), (c), where (a′) and (b′) are given as follows.

(a′) X belongs to $\mathcal{A}$.

(b′) The difference $E\backslash F$ of sets E and F in $\mathcal{A}$ belongs to $\mathcal{A}$.

A nonempty collection of sets that satisfies axioms (b′) and (c) is called a *ring* (or a *Boolean ring*) of sets. If it satisfies axioms (b′) and ($\widehat{\text{c}}$), then it is a *σ-ring*. Every algebra (every σ-algebra) is a ring (a σ-ring). If a ring (a σ-ring) of subsets of a set X contains X, then it is an algebra (a σ-algebra).

Example 1A. Take a nonempty set X. The power set $\wp(X)$ is a σ-algebra of subsets of X (it is more than that since the union of an arbitrary collection of sets in $\wp(X)$ lies in $\wp(X)$), which is the largest σ-algebra of subsets of X. At the other end, the collection $\{\varnothing, X\}$ is the smallest σ-algebra of subsets of X. (The notions of large and small are defined in terms of the inclusion ordering). A *partition* of a set X is any collection of pairwise disjoint subsets of X that cover X. If $\{A, B\}$ is a partition of X (i.e., if $A \cup B = X$ and $A \cap B = \varnothing$), then $\mathcal{X} = \{\varnothing, A, B, X\}$ is also a σ-algebra of subsets of X.

The intersection of any collection of σ-algebras of subsets of a set X is again a σ-algebra of subsets of X. This is readily verified by the definition of σ-algebra. If $\mathcal{C}$ is a nonempty collection of subsets of a set X (i.e., if $\mathcal{C} \subseteq \wp(X)$), then there exists a smallest (inclusion ordering) σ-algebra $\mathcal{X}_\mathcal{C}$

of subsets of X that includes $\mathcal{C}$. Indeed, let $\boldsymbol{X}_{\mathcal{C}}$ denote the collection of all σ-algebras of subsets of X that include $\mathcal{C}$. Note that $\boldsymbol{X}_{\mathcal{C}}$ is nonempty because the power set $\wp(X)$ is an element of $\boldsymbol{X}_{\mathcal{C}}$ (i.e., $\wp(X) \in \boldsymbol{X}_{\mathcal{C}}$). Now set $\mathcal{X}_{\mathcal{C}} = \bigcap \boldsymbol{X}_{\mathcal{C}}$, the intersection of all σ-algebras of subsets of X that include $\mathcal{C}$, which is itself a σ-algebra of subsets of X that includes $\mathcal{C}$, included in any σ-algebra of subsets of X that includes $\mathcal{C}$ (i.e., if $\mathcal{X}$ is a σ-algebra of subsets of X such that $\mathcal{C} \subseteq \mathcal{X}$, then $\mathcal{X}_{\mathcal{C}} \subseteq \mathcal{X}$). This smallest σ-algebra $\mathcal{X}_{\mathcal{C}}$ is referred to as *the σ-algebra generated by $\mathcal{C}$*.

The *Borel σ-algebra of subsets of* $\mathbb{R}$ is the σ-algebra $\mathfrak{R}$ generated by the collection of all open (or closed) intervals of the real line $\mathbb{R}$, also called the *Borel algebra of subsets of* $\mathbb{R}$, for short. This coincides with the σ-algebra generated by the open subsets of $\mathbb{R}$ (see the remark following Problem 1.14). The elements of $\mathfrak{R}$ (i.e., the $\mathfrak{R}$-measurable sets) are called *Borel sets*. The notion of Borel σ-algebra will be extended in Chapter 11.

1.2 Real-Valued Measurable Functions

Definition 1.2. Consider a measurable space $(X, \mathcal{X})$. A function $f\colon X \to \mathbb{R}$ is *measurable* with respect to the σ-algebra $\mathcal{X}$, or $\mathcal{X}$*-measurable*, if the inverse image of (α, ∞) under f is a measurable set for any real number α:

$$f^{-1}((\alpha, \infty)) = \{x \in X\colon f(x) > \alpha\} \in \mathcal{X} \quad \text{for every } \alpha \in \mathbb{R}.$$

Remark: The sign $>$ in Definition 1.2 can be replaced with $\geq$, $<$, or $\leq$, yielding equivalent definitions of a measurable function (cf. Problem 1.1). Moreover, it can verified that $f\colon X \to \mathbb{R}$ is *measurable* if and only if the inverse image of open sets in $\mathbb{R}$ are measurable sets in $\mathcal{X}$ (Problem 1.7(b) — the notion of measurable function will be extended in Chapter 11).

Example 1B. Let $(X, \mathcal{X})$ be a measurable space, take any set $E \in \wp(X)$, and consider the *characteristic function* $\chi_E\colon X \to \{0, 1\} \subset \mathbb{R}$ of E; that is,

$$\chi_E(x) = \begin{cases} 1, & x \in E, \\ 0, & x \in X \backslash E. \end{cases}$$

The characteristic function χ_E of a set $E \subseteq X$ is an $\mathcal{X}$-measurable function if and only if E is an $\mathcal{X}$-measurable set (i.e., if and only if $E \in \mathcal{X}$). Indeed,

$$\chi_E^{-1}((\alpha, \infty)) = \{x \in X\colon \chi_E(x) > \alpha\} = \begin{cases} \varnothing, & 1 \leq \alpha, \\ E, & 0 \leq \alpha < 1, \\ X, & \alpha < 0. \end{cases}$$

Observe that the function $1: X \to \mathbb{R}$ such that $1(x) = 1$ for all $x \in X$ coincides with the characteristic function χ_X of the entire set X. Moreover, it is also readily verified that γf is measurable for every $\gamma \in \mathbb{R}$ whenever $f: X \to \mathbb{R}$ is a measurable function. Therefore, *every constant function is measurable* (since a constant function is precisely $\gamma\chi_X$ for some $\gamma \in \mathbb{R}$).

Example 1C. Take the set $\mathbb{N}$ of all positive integers, and let $\mathbb{N}_e$ and $\mathbb{N}_o$ be the subsets of $\mathbb{N}$ consisting of all even and odd numbers, respectively. Thus $\{\mathbb{N}_o, \mathbb{N}_e\}$ forms a partition of $\mathbb{N}$. Example 1A says that $\mathcal{N} = \{\varnothing, \mathbb{N}_o, \mathbb{N}_e, \mathbb{N}\}$ is a σ-algebra of subsets of $\mathbb{N}$. Consider the measurable space $(\mathbb{N}, \mathcal{N})$, and take the identity function $f: \mathbb{N} \to \mathbb{N}\mathbb{R}$ (i.e., $f(n) = n$ for every $n \in \mathbb{N}$). This function f is not measurable (i.e., it is not $\mathcal{N}$-measurable). In fact,

$$f^{-1}((1, \infty)) = \{n \in \mathbb{N}: f(n) > 1\} = \{2, 3, 4, \ldots\} \notin \mathcal{N}.$$

Take a σ-algebra $\mathcal{X}$ of subsets of a set X, and let $E \in \mathcal{X}$ be an arbitrary measurable set of $\mathcal{X}$. Recall that the power set $\wp(E)$ of E is a σ-algebra of subsets of $E \subseteq X$. Thus the collection of all $\mathcal{X}$-measurable subsets of E,

$$\mathcal{E} = \wp(E) \cap \mathcal{X},$$

is again a σ-algebra, now of subsets of $E \subseteq X$. Indeed, since $\mathcal{X}$ and $\wp(X)$ are σ-algebras, countable unions of sets in $\mathcal{E}$ lies in $\mathcal{E}$. Moreover, if $B \in \mathcal{E}$, then $B \subseteq E$ and $B \in \mathcal{X}$, and hence $E \backslash B \in \wp(E)$ and $E \backslash B = E \cap (X \backslash B) = X \backslash [(X \backslash E) \cup B] \in \mathcal{X}$. Observe that the smallest σ-algebra of subsets of E, namely $\{E, \varnothing\}$, is a subcollection of $\mathcal{E}$, and $\mathcal{E}$ itself is the largest σ-algebra of subsets of E included in $\mathcal{X}$.

Proposition 1.3. *Restrictions of $\mathcal{X}$-measurable functions to $\mathcal{X}$-measurable sets E are $\mathcal{E}$-measurable functions.*

Proof. Take a σ-algebra $\mathcal{X}$ of subsets of a set X and an $\mathcal{X}$-measurable function $f: X \to \mathbb{R}$. Let E be an $\mathcal{X}$-measurable set and consider the restriction $f|_E: E \to \mathbb{R}$ of f to E (i.e., $f|_E(e) = f(e)$ for every $e \in E$). Let $\alpha \in \mathbb{R}$ be an arbitrary real number and observe that

$$\{e \in E: f|_E(e) > \alpha\} = \{x \in X: f(x) > \alpha\} \cap E \quad \in \quad \mathcal{E} = \wp(E) \cap \mathcal{X}.$$

In fact, as a subset of E, this obviously lies in $\wp(E)$, and it lies in $\mathcal{X}$ since it is the intersection of two $\mathcal{X}$-measurable sets (because f is an $\mathcal{X}$-measurable function). Therefore, $f|_E$ is an $\mathcal{E}$-measurable function. □

Proposition 1.4. *Consider a measurable space $(X, \mathcal{X})$. Let A and B be $\mathcal{X}$-measurable sets, and take the σ-algebras $\mathcal{A} = \wp(A) \cap \mathcal{X}$ and $\mathcal{B} = \wp(B) \cap \mathcal{X}$*

of subsets of A and B, respectively. Take a function $f\colon X \to \mathbb{R}$ and consider its restrictions $f|_A\colon A \to \mathbb{R}$ and $f|_B\colon B \to \mathbb{R}$ to A and B, respectively. If $A \cup B = X$, then f is $\mathcal{X}$-measurable if and only if $f|_A$ is $\mathcal{A}$-measurable and $f|_B$ is $\mathcal{B}$-measurable.

Proof. Proposition 1.3 ensures the "only if" part. To prove the "if" part proceed as follows. Take an arbitrary $\alpha \in \mathbb{R}$ and note that since $A \cup B = X$,

$$\big\{x \in X\colon\, f(x) > \alpha\big\} = \big\{a \in A\colon\, f|_A(a) > \alpha\big\} \cup \big\{b \in B\colon\, f|_B(b) > \alpha\big\},$$

which is the union of an $\mathcal{A}$-measurable set and a $\mathcal{B}$-measurable set (because $f|_A$ is an $\mathcal{A}$-measurable function and $f|_B$ is an $\mathcal{B}$-measurable function). Since the σ-algebras $\mathcal{A}$ and $\mathcal{B}$ are included in $\mathcal{X}$, both sets lie in $\mathcal{X}$, and so their union $\{x \in X\colon\, f(x) > \alpha\}$ is an $\mathcal{X}$-measurable set. This means that f is an $\mathcal{X}$-measurable function. □

There is a huge supply of measurable functions as we will see in the rest of this chapter (see also Problem 1.2 for the particular case of real-valued functions on $\mathbb{R}$ with respect to the Borel algebra $\mathfrak{R}$). Let $\mathcal{X}$ be a σ-algebra of subsets of a nonempty set X and take a pair of $\mathcal{X}$-measurable functions, say, $f\colon X \to \mathbb{R}$ and $g\colon X \to \mathbb{R}$. A polynomial $p(f,g)$ of a pair of functions f and g is an arbitrary (finite) linear combination of products of powers of f and g; that is, $p(f,g) = \sum_{j,k=0}^{m,n} \gamma_{j,k}\, f^j\, g^k$ with real coefficients $\gamma_{j,k}$.

Proposition 1.5. *If f and g are measurable functions, then so is any $p(f,g)$.*

Proof. Consider a measurable space $(X, \mathcal{X})$. Let f and g be measurable functions and let γ be any real number. We have already seen that γ (a constant function) and γf (a multiple of a measurable function) are measurable functions. Thus it is enough to show that $f+g$ and fg are measurable functions in order to ensure that each (finite) linear combination of $\{f^j g^k\}_{j,k\ge 0}$ is a measurable function. Let $\mathbb{Q}$ denote the rational field, take an arbitrary $\rho \in \mathbb{Q}$, an arbitrary $\alpha \in \mathbb{R}$, and note that $f^{-1}((\rho,\infty)) \cap g^{-1}((\alpha-\rho,\infty))$ is a measurable set (it is the intersection of measurable sets since f and g are measurable functions). Let $E_{\alpha,\rho}$ denote such a measurable set, viz.,

$$E_{\alpha,\rho} = \big\{x \in X\colon\, f(x) > \rho \;\text{ and }\; g(x) > \alpha - \rho\big\}.$$

Recall: $(f+g)(x) = f(x) + g(x)$. Thus for each real $\alpha \in \mathbb{R}$, $(f+g)(x) > \alpha$ if and only if $f(x) > \rho$ and $g(x) > \alpha - \rho$ for some rational $\rho \in \mathbb{Q}$. That is, if and only if $x \in E_{\alpha,\rho}$ for some $\rho \in \mathbb{Q}$. Thus, for every $\alpha \in \mathbb{R}$,

$$(f+g)^{-1}((\alpha,\infty)) = \big\{x \in X\colon\, (f+g)(x) > \alpha\big\} = \bigcup\nolimits_{\rho} E_{\alpha,\rho}.$$

Since each $E_{\alpha,\rho}$ is a measurable set and since $\mathbb{Q}$ is a countable set, it follows that $\bigcup_{\rho\in\mathbb{Q}} E_{\alpha,\rho}$ is a measurable set for every $\alpha \in \mathbb{R}$, and hence $f+g$ is a measurable function (Definitions 1.1 and 1.2). Next note that f^2 (defined by $f^2(x) = f(x)^2$ for every $x \in X$) is also a measurable function. In fact, $\{x \in X\colon\, f^2(x) > \alpha\} = X$ if $\alpha < 0$ and $f(x)^2 > \alpha$ if and only if either $f(x) > \sqrt{\alpha}$ or $f(x) < -\sqrt{\alpha}$ whenever $\alpha \geq 0$, which implies that the set $(f^2)^{-1}((\alpha,\infty)) = \{x \in X\colon\, f^2(x) > \alpha\}$ is measurable:

$$(f^2)^{-1}((\alpha,\infty)) = \begin{cases} X, & \alpha < 0,\\ f^{-1}((\sqrt{\alpha},\infty)) \cup (-f)^{-1}((\sqrt{\alpha},\infty)), & \alpha \geq 0,\end{cases}$$

where, for $\alpha \geq 0$, we have a union of measurable sets. This ensures that f^2 is a measurable function. However, since

$$fg = \tfrac{1}{4}\big((f+g)^2 - (f-g)^2\big),$$

whose expression involves multiplication by a constant, addition, and squaring of measurable functions, it follows that fg is a measurable function. □

Take an arbitrary function $f\colon X \to \mathbb{R}$, set

$$F^+ = \{x \in X\colon\, f(x) \geq 0\} \quad \text{and} \quad F^- = \{x \in X\colon\, f(x) \leq 0\},$$

and let χ_{F^+} and χ_{F^-} be the characteristic functions of F^+ and F^-, respectively. Consider the following real-valued functions on X:

$$f^+ = f\chi_{F^+} \quad \text{and} \quad f^- = -f\chi_{F^-}.$$

These are the *positive part* of f and the *negative part* of f, respectively. Note that $f^+\colon X \to \mathbb{R}$ and $f^-\colon X \to \mathbb{R}$ are both nonnegative functions (i.e., $f^+(x) \geq 0$ and $f^-(x) \geq 0$ for every $x \in X$). Moreover, the functions f and its *absolute value* $|f|\colon X \to \mathbb{R}$ (given by $|f|(x) = |f(x)|$ for every $x \in X$) can be expressed in terms of the positive and negative parts of f as follows.

$$f = f^+ - f^- \quad \text{and} \quad |f| = f^+ + f^-.$$

The above identities can be reversed yielding

$$f^+ = \tfrac{1}{2}(|f| + f) \quad \text{and} \quad f^- = \tfrac{1}{2}(|f| - f).$$

Proposition 1.6. *The following assertions are pairwise equivalent.*

(a) *f is a measurable function.*

(b) *f^+ and f^- are measurable functions.*

(c) *F^+ and F^- are measurable sets and $|f|$ is a measurable function.*

Proof. Consider a measurable space $(X, \mathcal{X})$. Let f be a measurable function. Thus F^+ and F^- are both measurable sets (cf. Definition 1.2), and so χ_{F^+} and χ_{F^-} are measurable functions (cf. Example 1B). Then Proposition 1.5 ensures that $f^+ = f\chi_{F^+}$ and $f^- = -f\chi_{F^-}$ are measurable functions, and so is $|f| = f^+ + f^-$. Note that the identity $f = f^+ - f^-$ ensures the converse: if f^+ and f^- are measurable, then f is measurable. So far we have proved that (a) and (b) are equivalent, and (a) implies (c). Next we verify that (c) implies (a). Note that the restrictions of $|f|$ to F^+ and F^- coincide with the restrictions of f to F^+ and with the restriction of $-f$ to F^-; that is,

$$f|_{F^+} = |f|\big|_{F^+} : F^+ \to \mathbb{R} \quad \text{and} \quad f|_{F^-} = -|f|\big|_{F^-} : F^- \to \mathbb{R}.$$

Since $F^+ \cup F^- = X$, it follows by Proposition 1.4 that if F^+ and F^- are $\mathcal{X}$-measurable sets and $|f|$ is an $\mathcal{X}$-measurable function, then $f|_{F^+}$ is $\mathcal{F}^+$-measurable and $-f|_{F^-}$ (and so $f|_{F^-}$) is $\mathcal{F}^-$-measurable, with respect to the σ-algebras $\mathcal{F}^+ = \wp(F^+) \cap \mathcal{X}$ and $\mathcal{F}^- = \wp(F^-) \cap \mathcal{X}$, respectively. Another application of Proposition 1.4 ensures that f is $\mathcal{X}$-measurable. $\square$

Observe that "$|f|$ measurable" does not imply "f measurable" (i.e., (c) does not imply (a) without the assumption that F^+ and F^- are measurable). For instance, take a measurable space $(X, \mathcal{X})$ and suppose there exists a partition $\{A, A'\}$ of X, where A and A' are not $\mathcal{X}$-measurable sets, and consider the function $f: X \to \mathbb{R}$ given by $f(x) = 1$ for $x \in A$ and $f(x) = -1$ for $x \in A'$ so that $F^+ = A$ and $F^- = A'$. It is clear that f is not an $\mathcal{X}$-measurable function (A is not an $\mathcal{X}$-measurable set). But $|f|$ is a constant function, thus measurable (with respect to any σ-algebra — Example 1.B).

1.3 Extended Real-Valued Measurable Functions

When considering the notion of length of subsets of the real line $\mathbb{R}$ (as it will be done from Chapter 2 onwards), it emerges the need for dealing with the length of unbounded subsets of $\mathbb{R}$ (such as $\mathbb{R}$ itself), and also with the notion of inf and sup of unbounded subsets of $\mathbb{R}$. This lead us to introduce the *extended real number system* (or the *extended real line*), which is the collection $\overline{\mathbb{R}} = \mathbb{R} \cup \{-\infty, +\infty\}$ consisting of the real field $\mathbb{R}$ and a pair of symbols, namely, $-\infty$ and $+\infty$. These symbols are not numbers, certainly not real numbers. We might have chosen any other pair of symbols such as, for instance, (α, ω) or $(\triangleleft, \triangleright)$ instead of the "set-down-eights" $(-\infty, +\infty)$. However, we will stick with the standard notation. Regarding the natural ordering of the real line $\mathbb{R}$, we postulate that $-\infty < x < +\infty$ for all x in $\mathbb{R}$, and extend the natural order of $\mathbb{R}$ to $\overline{\mathbb{R}}$. Note that $\overline{\mathbb{R}}$ is not a field, even though arithmetics with the new symbols are partially defined in the usual fashion, with some exceptions (e.g., $-\infty$ and $+\infty$ cannot be added together in any order; equivalently, the subtraction of $+\infty$ with itself is not defined).

We transfer the definition of measurable function to extended real-valued functions exactly as in Definition 1.2, namely, if $(X, \mathcal{X})$ is a measurable space, then an extended real-valued function $f: X \to \overline{\mathbb{R}}$ is *measurable* (with respect to the σ-algebra $\mathcal{X}$) or $\mathcal{X}$-*measurable* if the inverse image of (α, ∞) under f is a measurable set for any real number (i.e., for any $\alpha \in \mathbb{R}$):

$$f^{-1}((\alpha, \infty)) = \{x \in X\colon f(x) > \alpha\} \in \mathcal{X} \quad \text{for every } \alpha \in \mathbb{R}.$$

As in Definition 2.1, the sign $>$ in the above expression can be replaced with $\geq$, $<$, or $\leq$, yielding equivalent definitions of an extended real-valued measurable function (cf. Problem 1.1). If f is an extended real-valued $\mathcal{X}$-measurable function, then it is easy to verify (cf. Problem 1.4) that

$$F_{+\infty} = \{x \in X\colon f(x) = +\infty\} \quad \text{and} \quad F_{-\infty} = \{x \in X\colon f(x) = -\infty\}$$

are $\mathcal{X}$-measurable sets. Let $\chi_{X \backslash (F_{+\infty} \cup F_{-\infty})}$ be the characteristic function of the complement of $F_{+\infty} \cup F_{-\infty}$ and consider the real-valued function

$$f_{\mathbb{R}} = f \chi_{X \backslash (F_{+\infty} \cup F_{-\infty})} \colon X \to \mathbb{R}$$

(i.e., $f_{\mathbb{R}}(x) = f(x)$ if $x \notin (F_{+\infty} \cup F_{-\infty})$ and $f_{\mathbb{R}}(x) = 0$ if $x \in (F_{+\infty} \cup F_{-\infty})$; in other words, $f_{\mathbb{R}}(x) = f(x)$ if $f(x) \neq \pm\infty$, and $f_{\mathbb{R}}(x) = 0$ if $f(x) = \pm\infty$).

Proposition 1.7. *Consider a measurable space $(X, \mathcal{X})$. An extended real-valued function $f: X \to \overline{\mathbb{R}}$ is measurable if and only if the real-valued function $f_{\mathbb{R}}: X \to \mathbb{R}$ is measurable and both sets $F_{+\infty}$ and $F_{-\infty}$ are measurable.*

Proof. Take a function $f: X \to \overline{\mathbb{R}}$, and an arbitrary $\alpha \in \mathbb{R}$. Observe that

$$\{x \in X\colon f(x) > \alpha\} = \begin{cases} \{x \in X\colon f_{\mathbb{R}}(x) > \alpha\} \cup F_{+\infty}, & \alpha \geq 0, \\ \{x \in X\colon f_{\mathbb{R}}(x) > \alpha\} \backslash F_{-\infty}, & \alpha < 0. \end{cases}$$

Then f is a measurable function whenever $f_{\mathbb{R}}$ is a measurable function and $F_{+\infty}$ and $F_{-\infty}$ are measurable sets (recall: $A \backslash B = X \backslash [(X \backslash A) \cup B]$). Since

$$\{x \in X\colon f_{\mathbb{R}}(x) > \alpha\} = \begin{cases} \{x \in X\colon f(x) > \alpha\} \backslash F_{+\infty}, & \alpha \geq 0, \\ \{x \in X\colon f(x) > \alpha\} \cup F_{-\infty}, & \alpha < 0, \end{cases}$$

and since $F_{+\infty}$ and $F_{-\infty}$ are measurable sets whenever f is a measurable function, it follows that $f_{\mathbb{R}}$ is measurable whenever f is. □

Let S be an arbitrary set. An *S-valued sequence* (or a sequence of elements in S) is just an S-valued function defined on $\mathbb{N}$, the positive integers (or on $\mathbb{N}_0 = \{0\} \cup \mathbb{N}$, the nonnegative integers). Take an arbitrary $\overline{\mathbb{R}}$-valued sequence $\{\alpha_n\}$. Let $\inf_n \alpha_n$ and $\sup_n \alpha_n$ denote the greatest lower bound and the least upper bound of $\{\alpha_n\}$, respectively, which exist and are unique in $\overline{\mathbb{R}}$ (but may not exist in $\mathbb{R}$). The *limit inferior* (notation: $\liminf_n \alpha_n$) and *limit superior* (notation: $\limsup_n \alpha_n$) of $\{\alpha_n\}$ are defined in $\overline{\mathbb{R}}$ by

$$\liminf_n \alpha_n = \sup_n \inf_{n \le k} \alpha_k \quad \text{and} \quad \limsup_n \alpha_n = \inf_n \sup_{n \le k} \alpha_k.$$

If $\liminf_n \alpha_n = \limsup_n \alpha_n = \alpha$, then we say that $\{\alpha_n\}$ *converges* to $\alpha \in \overline{\mathbb{R}}$, and write $\lim_n \alpha_n = \alpha$. If $\{\alpha_n\}$ is *increasing* (i.e., $\alpha_n \le \alpha_{n+1}$), or *decreasing* (i.e., $\alpha_{n+1} \le \alpha_n$), then it is a *monotone* sequence. An $\overline{\mathbb{R}}$-valued monotone sequence $\{\alpha_n\}$ converges to a limit α in $\overline{\mathbb{R}}$. For $\mathbb{R}$-valued sequences this definition of convergence is equivalent to the standard definition of convergence (restricted to $\mathbb{R}$); viz., a real-valued sequence $\{\alpha_n\}$ *converges* to $\alpha \in \mathbb{R}$ if for every $\varepsilon > 0$ there exists an integer $n_\varepsilon \ge 1$ such that $|\alpha_n - \alpha| < \varepsilon$ whenever $n \ge n_\varepsilon$. If a real-valued sequence $\{\alpha_n\}$ is *bounded* (i.e., if $\sup_n |\alpha_n|$ lies in $\mathbb{R}$), then the sequences $\{\inf_{n \le k} \alpha_k\}$ and $\{\sup_{n \le k} \alpha_k\}$ converge in $\mathbb{R}$ to $\liminf_n \alpha_n \in \mathbb{R}$ and $\limsup_n \alpha_n \in \mathbb{R}$, respectively. In this case,

$$\liminf_n \alpha_n = \lim_n \inf_{n \le k} \alpha_k \quad \text{and} \quad \limsup_n \alpha_n = \lim_n \sup_{n \le k} \alpha_k.$$

An $\mathbb{R}$-valued monotone and bounded sequence $\{\alpha_n\}$ converges to a limit α in $\mathbb{R}$. Let $\{f_n\}$ be a sequence of extended real-valued functions $f_n : X \to \overline{\mathbb{R}}$ on X. Since each $f_n(x)$ lies in $\overline{\mathbb{R}}$ for every x in X, set

$$\phi(x) = \inf_n f_n(x), \qquad \Phi(x) = \sup_n f_n(x),$$

$$\underline{f}(x) = \liminf_n f_n(x), \qquad \overline{f}(x) = \limsup_n f_n(x),$$

for every $x \in X$. These values always exist as elements of $\overline{\mathbb{R}}$, and so they define four functions, viz., $\phi: X \to \overline{\mathbb{R}}$, $\Phi: X \to \overline{\mathbb{R}}$, $\underline{f}: X \to \overline{\mathbb{R}}$, and $\overline{f}: X \to \overline{\mathbb{R}}$. If $\underline{f}(x) = \overline{f}(x)$ for every $x \in X$, then the $\overline{\mathbb{R}}$-valued sequence $\{f_n(x)\}$ converges in $\overline{\mathbb{R}}$ for every $x \in X$. The common value $\underline{f}(x) = \overline{f}(x) \in \overline{\mathbb{R}}$, denoted by $f(x)$ for each $x \in X$, is called the *limit* of the $\overline{\mathbb{R}}$-valued sequence $\{f_n(x)\}$. This defines a function $f: X \to \overline{\mathbb{R}}$. In this case (i.e., when $\underline{f} = \overline{f}$), we say that the sequence of functions $\{f_n\}$ *converges pointwise* to the *limit function* f and write $f = \lim_n f_n$, which means that for every $x \in X$,

$$f(x) = \lim_n f_n(x).$$

For a sequence $\{f_n\}$ of real-valued functions this coincides with the standard definition of pointwise convergence (restricted to $\mathbb{R}$); viz., a sequence of real-valued functions $f_n: X \to \mathbb{R}$ converges pointwise to a real-valued function $f: X \to \mathbb{R}$ if $|f_n(x) - f(x)| \to 0$ as $n \to \infty$ for every $x \in X$. In other words, a sequence $\{f_n\}$ of $\mathbb{R}$-valued functions on X converges pointwise if there exists an $\mathbb{R}$-valued function f on X such that, for every $\varepsilon > 0$ and each $x \in X$, there is an $n_{\varepsilon,x} \in \mathbb{N}$ such that $|f_n(x) - f(x)| \leq \varepsilon$ whenever $n \geq n_{\varepsilon,x}$.

Proposition 1.8. *If each function f_n is measurable, then so are the functions ϕ, Φ, $\underline{f}$ and $\overline{f}$, and also is the limit $f = \lim_n f_n$ if the limit exists.*

Proof. Recall that arbitrary intersections of closed sets are closed, and arbitrary unions of open sets are open. Take any $\alpha \in \mathbb{R}$ and note that

$$\{x \in X\colon \phi(x) \geq \alpha\} = \bigcap\nolimits_n \{x \in X\colon f_n(x) \geq \alpha\},$$
$$\{x \in X\colon \Phi(x) > \alpha\} = \bigcup\nolimits_n \{x \in X\colon f_n(x) > \alpha\}.$$

Suppose $\{f_n\}$ is a sequence of measurable functions. The preceding sets are measurable (because they consist of countable intersections and countable unions of measurable sets), and so ϕ and Φ are measurable functions. Set

$$\phi_n(x) = \inf_{n \leq k} f_k(x) \quad \text{and} \quad \Phi_n(x) = \sup_{n \leq k} f_k(x)$$

for each integer $n \in \mathbb{N}$ and every point $x \in X$. Since each function f_n is measurable, the same argument ensures that the functions ϕ_n and Φ_n are measurable too, and so are the functions $\underline{f}$ and $\overline{f}$. In fact,

$$\underline{f}(x) = \sup_n \inf_{n \leq k} f_k(x) = \sup_n \phi_n(x) \quad \text{and} \quad \overline{f}(x) = \inf_n \sup_{n \leq k} f_k(x) = \inf_n \Phi_n(x)$$

for every $x \in X$. Recall that f exists if and only if $\underline{f} = \overline{f}$. □

Appropriate versions of Propositions 1.5 and 1.6 still hold for extended real-valued functions as we will see in Proposition 1.9. Let us first recall some usual conventions. We declare that $0 \cdot (\pm\infty) = 0$ so that if $\gamma = 0$, then $\gamma f(x) = 0$ for all $x \in X$ (i.e., $\gamma f = 0$ if $\gamma = 0$) for every $\overline{\mathbb{R}}$-valued function f on X. If f and g are $\overline{\mathbb{R}}$-valued $\mathcal{X}$-measurable functions on X, then the sets

$$F_{+\infty} \cap G_{-\infty} = \{x \in X\colon f(x) = +\infty \text{ and } g(x) = -\infty\},$$
$$G_{+\infty} \cap F_{-\infty} = \{x \in X\colon g(x) = +\infty \text{ and } f(x) = -\infty\},$$

are $\mathcal{X}$-measurable (cf. Problem 1.4). However, the function $f+g$ is not pointwise defined on these sets. That is, $f+g$ is not defined by $(f+g)(x) = f(x)+g(x)$ if x lies in $F_{+\infty} \cap G_{-\infty}$ or in $G_{+\infty} \cap F_{-\infty}$. So we declare that

$$(f+g)(x) = 0 \quad \text{for all} \quad x \in (F_{+\infty} \cap G_{-\infty}) \cup (G_{+\infty} \cap F_{-\infty}).$$

Recall the definition of the $\overline{\mathbb{R}}$-valued functions $f \wedge g$ and $f \vee g$ on X: for each $x \in X$, $(f \wedge g)(x) = \min\{f(x), g(x)\}$ and $(f \vee g)(x) = \max\{f(x), g(x)\}$.

Proposition 1.9. *If f and g are $\overline{\mathbb{R}}$-valued measurable function on X, then*

$$\gamma f, \quad f+g, \quad f^+, \quad f^-, \quad |f|, \quad f \wedge g, \quad f \vee g, \quad fg,$$

are $\overline{\mathbb{R}}$-valued measurable functions on X as well.

Proof. Consider a measurable space $(X, \mathcal{X})$, take a pair of measurable $\overline{\mathbb{R}}$-valued functions f and g on X, and let γ be any *real* number. The previous conventions ensure that the functions γf and $f+g$ are well defined, and they are measurable as well (use the same argument in the proof of Proposition 1.5). The arguments in the first part of the proof of Proposition 1.6 still hold for $\overline{\mathbb{R}}$-valued functions, so f^+, f^-, and $|f|$ also are well-defined measurable functions. Similarly (cf. Problem 1.3), $f \wedge g$ and $f \vee g$ also are well defined and measurable. Note that fg is a well-defined $\overline{\mathbb{R}}$-valued function. To show that it is measurable proceed as follows. For each pair of integers $m, n \in \mathbb{N}$ take the truncated functions $f_n\colon X \to \mathbb{R}$ and $g_m\colon X \to \mathbb{R}$ given by

$$f_n(x) = \begin{cases} f(x), & |f(x)| \leq n, \\ n, & f(x) > n, \\ -n, & f(x) < -n, \end{cases} \qquad g_m(x) = \begin{cases} g(x), & |g(x)| \leq m, \\ m, & g(x) > m, \\ -m, & g(x) < -m, \end{cases}$$

for every $x \in X$. Since $\{f_n\}$ and $\{g_m\}$ are sequences of $\mathbb{R}$-valued $\mathcal{X}$-measurable functions (cf. Problem 1.5), it follows that $f_n\, g_m$ are $\mathbb{R}$-valued $\mathcal{X}$-measurable functions for each pair of integers $m, n \in \mathbb{N}$ according to Proposition 1.5. Note that the sequences $\{f_n\}$ and $\{g_m\}$ clearly converge pointwise to f and g, respectively. Since $\{f_n\, g_m\}$ is a sequence of $\mathcal{X}$-measurable functions for each $m \in \mathbb{N}$, and since

$$f g_m = \lim_n f_n\, g_m$$

for each $m \in \mathbb{N}$, it follows that $f g_m$ is $\mathcal{X}$-measurable for each $m \in \mathbb{N}$ by Proposition 1.8. Analogously, since each $f g_m$ is $\mathcal{X}$-measurable, and since

$$fg = \lim_m f g_m,$$

it follows by Proposition 1.8 that fg is $\mathcal{X}$-measurable. □

Remark: The argument in the proof of Proposition 1.5 can be used to show that the function f^2 is well defined and $\mathcal{X}$-measurable, but the above conventions are not enough to ensure the identity $fg = \frac{1}{4}((f+g)^2 - (f-g)^2)$.

Let $\mathcal{M}(X, \mathcal{X})$ denote the collection of all real-valued functions on X that are measurable with respect to a σ-algebra $\mathcal{X}$ of subsets of X. Write $\mathcal{M}$ for $\mathcal{M}(X, \mathcal{X})$ when the measurable space $(X, \mathcal{X})$ is clear in the context; that is,

$$\mathcal{M} = \mathcal{M}(X, \mathcal{X}) = \{f : X \to \mathbb{R} \colon f \text{ is } \mathcal{X}\text{-measurable}\}.$$

The collection consisting of all real-valued functions on a set X is usually denoted by $\mathbb{R}^X$. It is well-known (and readily verified) that $\mathbb{R}^X$ is a real *linear space*. If $\mathcal{X}$ is a σ-algebra of subsets of X, then Proposition 1.5 ensures that the collection $\mathcal{M}(X, \mathcal{X})$ of all $\mathcal{X}$-measurable real-valued functions on X forms a *linear manifold* of the linear space $\mathbb{R}^X$ (i.e., addition and product by a scalar of $\mathcal{X}$-measurable real-valued functions are again $\mathcal{X}$-measurable real-valued functions). Thus $\mathcal{M}(X, \mathcal{X})$ is itself a linear space.

The collection of all *extended* real-valued functions on a set X that are measurable with respect to a σ-algebra $\mathcal{X}$ of subsets of X will be denoted by $\mathcal{M}(X, \mathcal{X})$ (or simply by $\mathcal{M}$) as well — same notation. However, in this case, $\mathcal{M}(X, \mathcal{X})$ is not a linear space (since $\overline{\mathbb{R}}^X$ is not a *real* linear space).

1.4 Problems

Problem 1.1. Take a measurable space $(X, \mathcal{X})$. If $f : X \to \mathbb{R}$ is a real-valued function on X, then show that the next assertions are pairwise equivalent.

(a) $f^{-1}((\alpha, \infty)) = \{x \in X \colon f(x) > \alpha\} \in \mathcal{X}$ for every $\alpha \in \mathbb{R}$,

(b) $f^{-1}([\alpha, \infty)) = \{x \in X \colon f(x) \geq \alpha\} \in \mathcal{X}$ for every $\alpha \in \mathbb{R}$,

(c) $f^{-1}((-\infty, \alpha)) = \{x \in X \colon f(x) < \alpha\} \in \mathcal{X}$ for every $\alpha \in \mathbb{R}$,

(d) $f^{-1}((-\infty, \alpha]) = \{x \in X \colon f(x) \leq \alpha\} \in \mathcal{X}$ for every $\alpha \in \mathbb{R}$.

Also show that these four assertions can be equivalently stated if we replace the assumption "for every $\alpha \in \mathbb{R}$" with "for every $\alpha \in \mathbb{Q}$".

Hint: If $\alpha \in \mathbb{R}$, then $\alpha = \lim_n \alpha_n$ with $\alpha_{n+1} < \alpha_n \in \mathbb{Q}$ for each $n \geq 1$ (decreasing rational sequence) so that $f^{-1}((\alpha, \infty)) = \bigcup_n \{x \in X \colon f(x) > \alpha_n\}$.

Therefore, the preceding eight assertions are equivalent forms of defining a measurable function, and they still hold if $f : X \to \overline{\mathbb{R}}$ take values in $\overline{\mathbb{R}}$.

Problem 1.2. Consider the real line $\mathbb{R}$ and take the Borel algebra $\mathfrak{R}$. By a measurable real-valued function on $\mathbb{R}$ we mean an $\mathfrak{R}$-measurable (or *Borel measurable*) function. Prove the next three assertions.

(a) Every continuous function $f\colon \mathbb{R} \to \mathbb{R}$ is measurable.

(b) Every monotone function $f\colon \mathbb{R} \to \mathbb{R}$ is measurable.

Now consider the characteristic function $\chi_{\mathbb{Q}}$ of the rational set $\mathbb{Q}$, which is far from being continuous or monotone. This is called the *Dirichlet function*.

(c) The Dirichlet function $\chi_{\mathbb{Q}}\colon \mathbb{R} \to \mathbb{R}$ is measurable.

Problem 1.3. Let $f\colon X \to \mathbb{R}$ and $g\colon X \to \mathbb{R}$ be arbitrary measurable functions (with respect to a σ-algebra $\mathcal{X}$ of subsets of X). Define the functions $e\colon X \to \mathbb{R}$ and $h\colon X \to \mathbb{R}$ as follows. For each $x \in X$,

$$e(x) = \min\{f(x), g(x)\} \quad \text{and} \quad h(x) = \max\{f(x), g(x)\},$$

denoted by $e = f \wedge g = \inf\{f, g\}$ and $h = f \vee g = \sup\{f, g\}$. Show that

$$e = \tfrac{1}{2}(f + g - |f - g|) \quad \text{and} \quad h = \tfrac{1}{2}(f + g + |f - g|).$$

Also show that e and h are $\mathcal{X}$-measurable functions.

Problem 1.4. Take an extended real-valued function $f\colon X \to \overline{\mathbb{R}}$, and show that if f is measurable with respect to a σ-algebra $\mathcal{X}$ of subsets of X, then

$$F_{+\infty} = \{x \in X\colon\ f(x) = +\infty\} \quad \text{and} \quad F_{-\infty} = \{x \in X\colon\ f(x) = -\infty\}$$

are $\mathcal{X}$-measurable sets. *Hint:* With n ranging over $\mathbb{N}$,

$$F_{+\infty} = \bigcap\nolimits_{n\in\mathbb{N}}\{x \in X\colon\ f(x) > n\},$$

$$F_{-\infty} = \bigcap\nolimits_{n}\{x \in X\colon\ f(x) \le -n\} = X\backslash\bigcup\nolimits_{n}\{x \in X\colon\ f(x) > -n\};$$

countable intersection and complement of countable union.

Problem 1.5. Consider a measurable space $(X, \mathcal{X})$. Let $f\colon X \to \overline{\mathbb{R}}$ be an arbitrary extended real-valued measurable function on X. For each *positive real* number β define the *β-truncation* of f as the real-valued function $f_\beta\colon X \to \mathbb{R}$ on X defined for each $x \in X$ by

$$f_\beta(x) = \begin{cases} f(x), & |f(x)| \le \beta, \\ \beta, & f(x) > \beta, \\ -\beta, & f(x) < -\beta. \end{cases}$$

Show that f_β is a measurable function.

Problem 1.6. Let $\mathcal{X}$ be a σ-algebra of subsets of X, and take a *nonnegative* $\mathcal{X}$-measurable function $f\colon X \to \overline{\mathbb{R}}$ on X. Show that there exists a sequence $\{\varphi_n\}$ of $\mathcal{X}$-measurable functions $\varphi_n\colon X \to \mathbb{R}$ with the following properties.

(i) Each φ_n is nonnegative (i.e., $0 \le \varphi_n(x)$ for every $x \in X$ for each n).

(ii) $\{\varphi_n\}$ is increasing (i.e., $\varphi_n(x) \le \varphi_{n+1}(x)$ for every $x \in X$ for each n).

(iii) $\{\varphi_n\}$ converges pointwise to f (i.e., $f(x) = \lim_n \varphi_n(x)$ for every $x \in X$).

(iv) Each φ_n has a finite range (i.e., $\{\alpha \in \overline{\mathbb{R}} : \alpha = \varphi_n(x) \text{ for some } x \in X\}$ is a finite set for each n).

Now show that if f is *bounded* (i.e., if $\sup_{x\in X} |f(x)| < \infty$), then $\{\varphi_n\}$ *converges uniformly* to f (i.e., $\sup_{x\in X} |\varphi_n(x) - f(x)| \to 0$ as $n \to \infty$).

Hint: Take an arbitrary $n \in \mathbb{N}$ and, for each integer $0 \le k \le n2^n$, set

$$E_{n,k} = \begin{cases} \{x \in X\colon\, k2^{-n} \le f(x) < (k+1)2^{-n}\}, & k < n2^n, \\ \{x \in X\colon\, f(x) \ge n\}, & k = n2^n. \end{cases}$$

Verify that $\{E_{n,k}\}_{0\le k\le n2^n}$ is a partition of X made up of $\mathcal{X}$-measurable sets. Then, for each $n \in \mathbb{N}$, set $\varphi_n(x) = 2^{-n}\sum_{k=0}^{n2^n} k\chi_{E_{n,k}}(x)$ for every $x \in X$.

Problem 1.7. Take an arbitrary complex-valued function $f\colon X \to \mathbb{C}$ on X. Consider its Cartesian decomposition: $f = f_1 + i\, f_2$, where f_1 and f_2 are real-valued functions on X (called the *real* and *imaginary parts* of f, which are and defined by $f_1(x) = \operatorname{Re} f(x)$ and $f_2(x) = \operatorname{Im} f(x)$ for every $x \in X$). We say that a complex-valued function f is *measurable* (with respect to the σ-algebra $\mathcal{X}$) or $\mathcal{X}$*-measurable* if its real and imaginary parts f_1 and f_2 are both (real-valued) $\mathcal{X}$-measurable functions. (Compare with Remark 10.2.)

(a) Verify that $f\colon X \to \mathbb{C}$ is $\mathcal{X}$-measurable if and only if the inverse image of every open rectangle of the complex plane $\mathbb{C}$ is an $\mathcal{X}$-measurable set. That is, the set $\{x \in X\colon\, \alpha < f_1(x) < \beta \text{ and } \gamma < f_2(x) < \delta\}$ lies in $\mathcal{X}$ for all real numbers α, β, γ, and δ.

(b) Generalize the above characterization: $f\colon X \to \mathbb{C}$ is $\mathcal{X}$-measurable if and only if the inverse image of every open set of $\mathbb{C}$ is an $\mathcal{X}$-measurable set.

(c) Also verify that sums and products of complex-valued measurable functions are again measurable, as well as the limit of every (pointwise) convergent sequence of complex-valued measurable functions.

Problem 1.8. Consider two measurable spaces $(X, \mathcal{X})$ and $(Y, \mathcal{Y})$. Let F be a function of X into Y. We say that F is a *measurable transformation* (with respect to the σ-algebras $\mathcal{X}$ and $\mathcal{Y}$) if the inverse image under F of every $\mathcal{Y}$-measurable set is an $\mathcal{X}$-measurable set. That is, $F\colon X \to Y$ is measurable with respect to the σ-algebras $\mathcal{X}$ and $\mathcal{Y}$ of subsets of X and Y if

$$F^{-1}(E) = \{x \in X\colon Fx \in E\} \in \mathcal{X} \quad \text{for every} \quad E \in \mathcal{Y}.$$

Now set $Y = \mathbb{R}$, $\mathcal{Y} = \mathfrak{R}$ (the Borel algebra) and consider a real-valued function $f\colon X \to \mathbb{R}$. Show that f is $\mathcal{X}$-measurable in the sense of Definition 1.2 if and only if it is measurable in the above sense; that is, $f^{-1}(E) \in \mathcal{X}$ for every Borel set E. (Note the analogy with the definition of continuous function between topological spaces. Such an analogy and also an alternate definition of measurable transformation will be discussed in Theorem 11.4.)

Problem 1.9. Let F be an arbitrary function of a set X into a set Y. Let $\mathcal{Y}$ be a σ-algebra of subsets of Y. First show that the collection of the inverse images under F of each $\mathcal{Y}$-measurable set, $\mathcal{X} = \{F^{-1}(E)\colon E \in \mathcal{Y}\}$, forms a σ-algebra of subsets of X. Given any function $F\colon X \to Y$ and a σ-algebra $\mathcal{Y}$ of subsets of Y, also show that this $\mathcal{X}$ is the smallest σ-algebra of subsets of X that makes F measurable. This is called the σ-algebra of subsets of X *inversely induced* by F. (Note the similarity between this concept and that of the topology inversely induced on X by F, which is the weakest topology on X that makes F continuous.)

Hint: Let $F\colon X \to Y$ be any function X into a set Y, and recall that the inverse image of any subset B of Y under F is the subset of X given by

$$F^{-1}(B) = \{x \in X\colon F(x) \in B\}.$$

Show that $\varnothing = F^{-1}(\varnothing)$, $X = F^{-1}(Y)$, $X \backslash F^{-1}(B) = F^{-1}(Y \backslash B)$ for every $B \subseteq Y$, and that $\bigcup_\gamma F^{-1}(B_\gamma) = F^{-1}\big(\bigcup_\gamma B_\gamma\big)$ for every nonempty collection $\{B_\gamma\}$ of subsets of Y.

Problem 1.10. Consider three measurable spaces, namely, $(X, \mathcal{X})$, $(Y, \mathcal{Y})$, and $(Z, \mathcal{Z})$. Let $F\colon X \to Y$ and $G\colon Y \to Z$ be measurable functions (with respect to the σ-algebras $\mathcal{X}$ and $\mathcal{Y}$, and $\mathcal{Y}$ and $\mathcal{Z}$, respectively). Show that the composition $G \circ F\colon X \to Z$ is a measurable function (with respect to the σ-algebras $\mathcal{X}$ and $\mathcal{Z}$) — Compare with Theorem 11.4(d).

Problem 1.11. Let $f\colon X \to \mathbb{R}$ be a real-valued $\mathcal{X}$-measurable function on a set X, where $\mathcal{X}$ is a σ-algebra of subsets of the set X, and let $g\colon \mathbb{R} \to \mathbb{R}$ be a continuous function. Show that the composition $g \circ f\colon X \to \mathbb{R}$ is measurable. (*Hint:* Problems 1.2 and 1.10.) This will be extended in Theorem 11.4(e).

Problem 1.12. A collection $\mathcal{T} \subseteq \wp(X)$ of subsets of a set X is a *topology* on X if it satisfies the following three axioms.

(i) The whole set X and the empty set $\varnothing$ belong to $\mathcal{T}$.

(ii) The intersection of a finite collection of sets in $\mathcal{T}$ belongs to $\mathcal{T}$.

(iii) The union of an arbitrary collection of sets in $\mathcal{T}$ belongs to $\mathcal{T}$.

A *topological space*, denoted by $(X, \mathcal{T})$ or simply X (if $\mathcal{T}$ is clear or immaterial) is a set X equipped with a topology $\mathcal{T}$. The sets in $\mathcal{T}$ are called the *open* sets of X with respect to $\mathcal{T}$. Let X be equipped with a topology, and equip the real line $\mathbb{R}$ with its usual topology (the one induced by the usual metric on $\mathbb{R}$). Show that if a real-valued function $f: X \to \mathbb{R}$ is *continuous* (with respect to the above topologies, which means that the inverse image under f of open sets of $\mathbb{R}$ are open sets of X), then f is measurable with respect to the σ-algebra $\mathcal{X}_{\mathcal{T}}$ generated by a topology $\mathcal{T}$ on X. (In the jargon of Section 11.1 this means that *continuous functions are Borel measurable*).

Problem 1.13. This is a generalization of Problem 1.12. Let $\mathcal{X}_{\mathcal{T}}$ be the σ-algebra generated by a topology $\mathcal{T}_X$ on a set X and let $\mathcal{Y}_{\mathcal{T}}$ be the σ-algebra generated by a topology $\mathcal{T}_Y$ on a set Y. If $F: X \to Y$ is a *continuous mapping* (i.e., if $F^{-1}(U) \in \mathcal{T}_X$ for every $U \in \mathcal{T}_Y$), then F is measurable (in the sense of Problem 1.8). Note: "continuity" is with respect to the topologies $\mathcal{T}_X$ and $\mathcal{T}_Y$, and "measurability" is with respect to the σ-algebras $\mathcal{X}_{\mathcal{T}}$ and $\mathcal{Y}_{\mathcal{T}}$. The notion of measurable functions will be discussed again in Section 11.1.

Problem 1.14. Take a topological space X with a topology $\mathcal{T}$. A subcollection $\mathcal{B}$ of $\mathcal{T}$ is a *base* (or a *topological base*) for X if it covers every open subset of X (i.e., every $U \in \mathcal{T}$ is the union of some subcollection of $\mathcal{B}$ — we return to this notion in Section 11.1.) If X is a metric space equipped with the metric topology $\mathcal{T}$, then it has a base of open balls (every open set is the union of open balls). A topological space is *separable* if it has a countable dense subset. A metric space is separable if and only if it has a countable base of open balls. Show that the σ-algebra $\mathcal{X}_{\mathcal{T}}$ generated by the metric topology $\mathcal{T}$ on a separable metric space coincides with the σ-algebra $\mathcal{X}_{\mathcal{B}}$ generated by any countable base $\mathcal{B}$ of open balls.

Remark: It is well known that the real line $\mathbb{R}$ (equipped with its usual topology) is a separable metric space. (Indeed, the set $\mathbb{Q}$ of all rational numbers is countable and dense in $\mathbb{R}$.) So the Borel algebra $\mathfrak{R}$ (the σ-algebra generated by the open intervals) coincides with the σ-algebra generated by any countable base of open intervals, which coincides with the σ-algebra generated by the topology on $\mathbb{R}$ (the σ-algebra generated by the open sets). For this reason $\mathcal{X}_{\mathcal{T}}$ is called the *Borel σ-algebra* of subsets of a set X generated a topology $\mathcal{T}$ on X. (Borel σ-algebra will be revisited in Section 11.2.)

Problem 1.15. A nonempty class (i.e., a nonempty collection) $\mathcal{K}$ of subsets of a set X that contains the union of every increasing sequence in $\mathcal{K}$ and the intersection of every decreasing sequence in $\mathcal{K}$ is called a *monotone class*. That is, $\mathcal{K}$ is a *monotone class* if, whenever $\{E_n\}$ is an increasing sequence ($E_n \subseteq E_{n+1}$) of sets in $\mathcal{K}$ and $\{F_n\}$ is a decreasing sequence ($F_{n+1} \subseteq F_n$) of sets in $\mathcal{K}$, then $\bigcup_n E_n$ and $\bigcap_n F_n$ are sets in $\mathcal{K}$. The remaining problems are all about monotone classes, leading to the central result of Problem 1.18 (The Monotone Class Lemma), which will be required in the sequel. To begin with, prove the following assertions (where monotone classes, algebras, and σ-algebras are supposed to consist of subsets of the same fixed set X).

(a) Every σ-algebra is a monotone class.

(b) A monotone class is not necessarily a σ-algebra.

A nonempty collection of subsets of a set X that is both a monotone class and an algebra is called a *monotone algebra*.

(c) Every monotone algebra is a σ-algebra.

Hint: Let $\mathcal{K}$ be a monotone algebra and let $\{E_n\}$ be a sequence of sets in $\mathcal{K}$. Since $\mathcal{K}$ is an algebra, $\{\bigcup_{i=1}^n E_i\}$ is an increasing sequence of sets in $\mathcal{K}$. Since $\mathcal{K}$ is a monotone class, $\{\bigcup_{i=1}^\infty E_i\}$ lies in $\mathcal{K}$.

Problem 1.16. Let $\mathcal{C} \subseteq \wp(X)$ be a nonempty collection of subsets of a set X. Prove that there exists a smallest (in the inclusion ordering) monotone class $\mathcal{K}_\mathcal{C}$ of subsets of X that includes $\mathcal{C}$. That is, there exists a monotone class $\mathcal{K}_\mathcal{C}$ included in any monotone class that includes $\mathcal{C}$.

Hint: The intersection of any collection of monotone classes of subsets of X is again a monotone class of subsets of X.

This smallest monotone class $\mathcal{K}_\mathcal{C}$ is the *monotone class generated by* $\mathcal{C}$.

Problem 1.17. Let $\mathcal{C} \subseteq \wp(X)$ be a arbitrary nonempty collection of subsets of a given set X. Show that the monotone class $\mathcal{K}_\mathcal{C}$ generated by $\mathcal{C}$ is included in the σ-algebra $\mathcal{X}_\mathcal{C}$ generated by $\mathcal{C}$:

$$\mathcal{C} \subseteq \mathcal{K}_\mathcal{C} \subseteq \mathcal{X}_\mathcal{C}.$$

Give an example where the above inclusions are all proper.

Problem 1.18. Now prove the *Monotone Class Lemma*, which says that if the preceding collection is an algebra $\mathcal{A}$, then

$$\mathcal{K}_\mathcal{A} = \mathcal{X}_\mathcal{A}.$$

In other words, *the monotone class $\mathcal{K}_{\mathcal{A}}$ generated by an algebra $\mathcal{A}$ of subsets of a set X coincides with the σ-algebra $\mathcal{X}_{\mathcal{A}}$ generated by $\mathcal{A}$.* This is the Monotone Class Lemma, which plays a major role in Chapter 9.

Hint: Consider an algebra $\mathcal{A}$ of subsets of a set X, and let $\mathcal{K}_{\mathcal{A}}$ be the monotone class generated by $\mathcal{A}$. For each $E \in \mathcal{K}_{\mathcal{A}}$, set

$$\mathcal{E}_{\mathcal{A}} = \{F \in \mathcal{K}_{\mathcal{A}} \colon E \backslash F \in \mathcal{K}_{\mathcal{A}},\ E \cap F \in \mathcal{K}_{\mathcal{A}} \text{ and } F \backslash E \in \mathcal{K}_{\mathcal{A}}\} \subseteq \mathcal{K}_{\mathcal{A}}.$$

Show that

(i) $\mathcal{E}_{\mathcal{A}}$ is a monotone class,

(ii) $F \in \mathcal{E}_{\mathcal{A}}$ if and only if $E \in \mathcal{F}_{\mathcal{A}}$,

where the definition of $\mathcal{F}_{\mathcal{A}}$ is analogous to that of $\mathcal{E}_{\mathcal{A}}$ (swapping E for F). Since $\mathcal{A}$ is an algebra and $\mathcal{A} \subseteq \mathcal{K}_{\mathcal{A}}$ (cf. Problem 1.17), also show that

(iii) $\varnothing$, E, $X \backslash E$, and X all lie in $\mathcal{E}_{\mathcal{A}}$,

(iv) $E \in \mathcal{A}$ implies $\mathcal{A} \subseteq \mathcal{E}_{\mathcal{A}}$.

(Recall that $E \cap F = X \backslash ((X \backslash E) \cup (X \backslash F))$ and $E \backslash F = E \cap (X \backslash F)$, and hence $E \backslash F$, $E \cap F$, and $F \backslash E$ all lie in $\mathcal{A}$ if E and F lie in $\mathcal{A}$.) Now show from (i) and (iv) that (cf. Problem 1.16)

$$\mathcal{E}_{\mathcal{A}} = \mathcal{K}_{\mathcal{A}} \quad \text{for every} \quad E \in \mathcal{A}.$$

Thus, if $E \in \mathcal{A}$ and $F \in \mathcal{K}_{\mathcal{A}}$, then $F \in \mathcal{E}_{\mathcal{A}}$, and so $E \in \mathcal{F}_{\mathcal{A}}$ by (ii). Hence $\mathcal{A} \subseteq \mathcal{F}_{\mathcal{A}}$ for every $F \in \mathcal{K}_{\mathcal{A}}$. Then show from (i) that (cf. Problem 1.16)

$$\mathcal{F}_{\mathcal{A}} = \mathcal{K}_{\mathcal{A}} \quad \text{for every} \quad F \in \mathcal{K}_{\mathcal{A}}.$$

So, if $E, F \in \mathcal{K}_{\mathcal{A}}$, then $E \cap F$, $E \backslash F$, and $F \backslash E$ lie in $\mathcal{K}_{\mathcal{A}}$. Furthermore, property (iii) ensures that $\varnothing$ and X also lie in $\mathcal{K}_{\mathcal{A}}$. Thus verify that the intersection and the complement of sets in $\mathcal{K}_{\mathcal{A}}$ remain in $\mathcal{K}_{\mathcal{A}}$, and so a finite union of sets in $\mathcal{K}_{\mathcal{A}}$ remain in $\mathcal{K}_{\mathcal{A}}$. (Recall that $F \cup E = X \backslash ((X \backslash E) \cap (X \backslash F))$.) Therefore, according to Definition 1.1,

$$\mathcal{K}_{\mathcal{A}} \text{ is an algebra.}$$

Finally, use Problem 1.15(c) to show that $\mathcal{K}_{\mathcal{A}}$ is a σ-algebra, and then conclude that $\mathcal{X}_{\mathcal{A}} \subseteq \mathcal{K}_{\mathcal{A}}$, and therefore, by using Problem 1.17, infer that

$$\mathcal{K}_{\mathcal{A}} = \mathcal{X}_{\mathcal{A}}.$$

Problem 1.19. Prove that if a monotone class includes an algebra $\mathcal{A}$, then it also includes the σ-algebra $\mathcal{X}_{\mathcal{A}}$ generated by $\mathcal{A}$.

Remark: Problems 1.15 to 1.19 (and their hints) give a rather complete account of monotone classes which will be enough for our purposes. In particular, the Monotone Class Theorem as in Problem 1.18 will play an important role in Lemma 9.7, which is crucial to proving Tonelli and Fubini Theorems (Theorems 9.7 and 9.8) in Chapter 9.

Suggested Reading

Bartle [4], Berberian [7], Halmos [18], Royden [35], Rudin [36]. For introductory set theory, see [19], [39], [40] and the first chapter of [9], [12], [18], [21], [24], [26], [29], [35]. For general topology, see [9], [12], [16, Chapter 4], [21], [24, Chapters 2 and 3], [26, Chapter 3], [29, Chapters 7 & 8], [35, Part Two].

2

Measure on a σ-Algebra

2.1 Measure and Measure Space

A function whose domain is a collection of sets is called a *set function.* A *measure* is a nonnegative extended real-valued set function satisfying some further conditions. The domain of a measure is a subcollection of the power set $\wp(X)$ of a given set X. It is advisable to require that the empty set $\varnothing$ and the whole set X itself belong to the domain, and to assign the minimum (zero) for the value of the function at the empty set. It is also convenient to require *additivity* in the following sense. Assume that every finite union of sets in the domain is again a set in the domain. This indicates that the domain might be an algebra. Then assume that the value of the function at any finite union of disjoint sets in the domain equals the sum of the values of the function at each set. Actually, this leads to a possible definition of a concept of measure (measures defined on an algebra will be considered in Chapter 8). However, such an approach lacks an important feature that is needed to build up a useful theory, namely, *countable additivity*. That is, it is required that the notion of additivity also holds for countably infinite unions of disjoint sets, and so countably infinite unions of sets are supposed to be in the domain of a measure. This compels the domain to be a σ-algebra.

Definition 2.1. A *measure* is an extended real-valued set function μ on a σ-algebra $\mathcal{X}$ of subsets of a set X,

$$\mu\colon \mathcal{X} \to \overline{\mathbb{R}},$$

C.S. Kubrusly, *Essentials of Measure Theory*,
DOI 10.1007/978-3-319-22506-7_2

that fulfills the following axioms (referred to as the *measure axioms*).

(a) $\mu(\varnothing) = 0$,

(b) $\mu(E) \geq 0$ for every $E \in \mathcal{X}$,

(c) $\mu\big(\bigcup_n E_n\big) = \sum_n \mu(E_n)$
for every countable family $\{E_n\}$ of pairwise disjoint ($E_m \cap E_n = \varnothing$ whenever $m \neq n$) sets in $\mathcal{X}$ (i.e., μ is *countably additive*).

A triple $(X, \mathcal{X}, \mu)$ consisting of an arbitrary set X, a σ-algebra $\mathcal{X}$ of subsets of X, and a measure μ on $\mathcal{X}$ is called a *measure space*.

Remarks: Since μ takes values in $\overline{\mathbb{R}}$, it is possible that $\mu(E) = +\infty$ for some $\mathcal{X}$-measurable sets E. If the countable family $\{E_n\}$ in (c) is infinite, and if $\mu(E_n) \in \mathbb{R}$ for every n (i.e., if all the sets E_n have a finite measure), then we get an infinite sum of nonnegative real numbers in (c), and therefore either a convergent (in fact, unconditionally convergent), or a divergent series of nonnegative real numbers — in the latter case, $\mu\big(\bigcup_n E_n\big) = +\infty$.

A real-valued measure $\mu\colon \mathcal{X} \to \mathbb{R}$ is called a *finite measure*, which means that $\mu(E) < \infty$ for all $E \in \mathcal{X}$. However, as we will see in Proposition 2.2(a), this is equivalent to saying that a measure μ is *finite* if $\mu(X) < \infty$. In particular, if $\mu(X) = 1$, then μ is called a *probability measure*, and $(X, \mathcal{X}, \mu)$ is called a *probability space.* If there exists a countable covering of X consisting of $\mathcal{X}$-measurable sets of finite measure, then μ is referred to as a *σ-finite measure*. Equivalently, a measure $\mu\colon \mathcal{X} \to \overline{\mathbb{R}}$ is *σ-finite* if there exists an $\mathcal{X}$-valued sequence $\{E_n\}$ such that $\mu(E_n) < \infty$ for every n and $X = \bigcup_n E_n$.

An *atom* of a measure $\mu\colon \mathcal{X} \to \overline{\mathbb{R}}$ on a σ-algebra $\mathcal{X}$ of subsets of a nonempty set X is a measurable set $A \in \mathcal{X}$ such that (i) $\mu(A) > 0$, and (ii) either $\mu(E) = 0$ or $\mu(E) = \mu(A)$ for every measurable subset E of A. In other words, a measurable set $A \in \mathcal{X}$ is an *atom* of a measure $\mu\colon \mathcal{X} \to \overline{\mathbb{R}}$ if $\mu(A) > 0$ and $\mu(E) = \mu(A)$ whenever $E \in \wp(A) \cap \mathcal{X}$ is such that $\mu(E) \neq 0$).

Example 2A. Consider any σ-algebra $\mathcal{X}$ of subsets of a nonempty set X. Associated to any point $x \in X$, define a set function $\delta_x\colon \mathcal{X} \to \mathbb{R}$ as follows.

$$\delta_x(E) = \begin{cases} 1, & x \in E, \\ 0, & x \in X \backslash E, \end{cases}$$

for every measurable set $E \in \mathcal{X}$. According to Definition 2.1, δ_x is a measure, actually, a probability measure, called the *Dirac measure* at x (or the *unit point measure* concentrated at x). Observe that any measurable set containing x is an atom of δ_x. A *singleton* is a set containing just one element. If singletons are measurable sets with respect to the σ-algebra $\mathcal{X}$ (i.e., if $\{x\} \in \mathcal{X}$ for every $x \in X$), then $\{x\}$ is an atom of each δ_x; and $\delta_x(\{y\})$ is 1 if $y = x$ and 0 if $y \neq x$ for every singleton $\{y\}$ in $\mathcal{X}$.

Example 2B. Consider the σ-algebra $\wp(\mathbb{N})$ of all subsets of the positive integers $\mathbb{N}$. Let the symbol $\#$ stand for cardinality (for a *finite* set S, $\#(S)$ is the number of elements in S). Take the function $\mu \colon \wp(\mathbb{N}) \to \overline{\mathbb{R}}$ defined by

$$\mu(E) = \begin{cases} \#(E), & E \text{ is finite}, \\ +\infty, & E \text{ is infinite}, \end{cases}$$

for each $E \in \wp(\mathbb{N})$. It is readily verified by Definition 2.1 that μ is a measure. Set $E_n = \{n\} \in \wp(\mathbb{N})$ for each $n \in \mathbb{N}$ so that $\mathbb{N} = \bigcup_n E_n$ and $\mu(E_n) = 1$. Then the measure μ is σ-finite, but it is not finite (there are infinite sets in $\wp(\mathbb{N})$; e.g., $\mathbb{N}$ itself). This is called the *counting measure* on $\mathbb{N}$.

Example 2C. Consider the σ-algebra $\mathfrak{R}$ of subsets of $\mathbb{R}$ generated by the collection of all open intervals. That is, consider the Borel algebra $\mathfrak{R}$. We will prove in Chapter 8 the existence and uniqueness of a measure $\lambda \colon \mathfrak{R} \to \overline{\mathbb{R}}$ such that $\lambda((\alpha, \beta)) = \beta - \alpha$ for every open interval (α, β) of $\mathbb{R}$. In other words, this is the only measure on $\mathfrak{R}$ that has the property of assigning to each open interval its own length. It is referred to as the *Lebesgue measure* on $\mathfrak{R}$, which is not finite (e.g., $\lambda(\mathbb{R}) = +\infty$) but is σ-finite. In fact, for an arbitrary $\varepsilon > 0$ and for each integer $k \in \mathbb{Z}$, let $E_k = (q_k - \varepsilon, q_k + \varepsilon)$ be the open interval of radius ε centered at q_k, where $\{q_k\}_{k \in \mathbb{Z}}$ is an enumeration of the rational numbers $\mathbb{Q}$, so that $\mathbb{R} = \bigcup_k E_k$ and $\mu(E_k) = 2\varepsilon$ for all $k \in \mathbb{Z}$.

Example 2D. Scaling the Lebesgue measure λ of Example 2C leads to another measure on $\mathfrak{R}$. Indeed, for any real $\gamma > 0$, the function $\lambda_\gamma \colon \mathfrak{R} \to \overline{\mathbb{R}}$ defined by $\lambda_\gamma = \gamma\lambda$, so that $\lambda_\gamma((\alpha, \beta)) = \gamma\lambda((\alpha, \beta)) = \gamma(\beta - \alpha) = \gamma\beta - \gamma\alpha$ for every open interval (α, β) of $\mathbb{R}$, is again a measure on $\mathfrak{R}$. This is a homogeneous scaling. Inhomogeneous scaling also yields new measures. For instance, let $F \colon \mathbb{R} \to \mathbb{R}$ be a nondecreasing function (i.e., $F(x) \leq F(y)$ for $x \leq y$), which ensures that F has a left and a right limit at each point $x \in \mathbb{R}$, denoted by $F(x^-) = \lim_{\varepsilon \to 0} F(x - |\varepsilon|)$ and $F(x^+) = \lim_{\varepsilon \to 0} F(x + |\varepsilon|)$ — if F is continuous, then $F(x^+) = F(x^-) = F(x)$. The same argument used to construct the Lebesgue measure λ in Chapter 8 can be readily modified to show that there is a unique measure $\lambda_F \colon \mathfrak{R} \to \overline{\mathbb{R}}$ such that $\lambda_F((\alpha, \beta)) = F(\beta^-) - F(\alpha^+)$ for every open interval (α, β) of $\mathbb{R}$. This measure λ_F is the *Borel–Stieltjes measure* generated by F, which again is σ-finite. Particular cases: (i) if $F = \chi_{[0,\infty)} \colon \mathbb{R} \to \{0, 1\}$ is the characteristic function of $[0, \infty)$, then $\lambda_F = \delta_0 \colon \mathfrak{R} \to \{0, 1\}$ is the Dirac measure at 0 of Example 2.A; (ii) if F is continuously differentiable, then $\lambda_F((\alpha, \beta)) = F(\beta) - F(\alpha) = \int_\alpha^\beta \frac{dF}{dx}\, dx$.

Recall the definition of increasing and decreasing sequences of sets: a sequence $\{A_n\}$ of subsets of X is *increasing* if $A_n \subseteq A_{n+1}$, and *decreasing* if $A_{n+1} \subseteq A_n$, for every n. It is *monotone* if it is either increasing or decreasing. The next proposition presents some basic properties of measures.

Proposition 2.2. *Let $\mathcal{X}$ be a σ-algebra of subsets of a set X. Take arbitrary (measurable) sets A and B in $\mathcal{X}$, and an arbitrary sequence $\{E_n\}$ of sets in $\mathcal{X}$. The following properties hold true for every measure $\mu\colon \mathcal{X} \to \overline{\mathbb{R}}$.*

(a) $\mu(A) \le \mu(B)$ *if* $A \subseteq B$.

(b) $\mu(B\backslash A) = \mu(B) - \mu(A)$ *if* $A \subseteq B$ *and* $\mu(A) \ne +\infty$.

(c) $\mu\big(\bigcup_n E_n\big) = \lim_n \mu(E_n)$ *if* $\{E_n\}$ *is increasing.*

(d) $\mu\big(\bigcap_n E_n\big) = \lim_n \mu(E_n)$ *if* $\{E_n\}$ *is decreasing and* $\mu(E_1) \ne +\infty$.

Proof. If $A \subseteq B \subseteq X$, then $B = A \cup (B\backslash A)$ and $A \cap (B\backslash A) = \varnothing$. If both A and B are $\mathcal{X}$-measurable, then so is $B\backslash A = B \cap (X\backslash A)$ according to Definition 1.1(b,c)). Then, by Definition 2.1(b,c),

$$\mu(A) \le \mu(A) + \mu(B\backslash A) = \mu(A \cup (B\backslash A)) = \mu(B),$$

proving (a). If, in addition, $\mu(A) \ne +\infty$, then $\mu(B) - \mu(A)$ is in $\overline{\mathbb{R}}$ so that

$$\mu(B) - \mu(A) = \mu(B\backslash A),$$

which proves (b). Next let $\{E_n\}$ be an increasing sequence of $\mathcal{X}$-measurable sets. If one of E_n has an infinite measure, then assertion (c) follows from assertion (a). Thus suppose $\mu(E_n) \in \mathbb{R}$ for every $n \in \mathbb{N}$, and take a sequence $\{E'_n\}$ of $\mathcal{X}$-measurable sets recursively defined as follows: $E'_1 = E_1$ and $E'_{n+1} = E_{n+1}\backslash E_n$ for each $n \in \mathbb{N}$, which is a sequence of pairwise disjoint sets, and so (cf. Definition 2.1(c) and also recall that $\bigcup_n E_n = \bigcup_n E'_n$)

$$\mu\Big(\bigcup\nolimits_n E_n\Big) = \mu\Big(\bigcup\nolimits_n E'_n\Big) = \sum\nolimits_n \mu(E'_n) = \mu(E'_1) + \lim_m \sum_{n=1}^{m} \mu(E'_{n+1}).$$

It then follows from (b) that for an arbitrary integer $m > 1$,

$$\begin{aligned}\sum_{n=1}^{m} \mu(E'_{n+1}) &= \sum_{n=1}^{m} \mu(E_{n+1}\backslash E_n)\\ &= \sum_{n=1}^{m} \big(\mu(E_{n+1}) - \mu(E_n)\big) = \mu(E_{m+1}) - \mu(E_1),\end{aligned}$$

and hence $\mu\big(\bigcup_n E_n\big) = \lim_m \mu(E_{m+1})$, proving (c). Next suppose $\{E_n\}$ is a decreasing sequence of $\mathcal{X}$-measurable sets and set $E''_n = E_1\backslash E_n$ for each

n so that $\{E''_n\}$ is an increasing sequence. Thus we can apply (c), but first note that since $E_n \subseteq E_1$ for all n and $\mu(E_1) \neq +\infty$, we get from (a) that $\mu(E_n) \in \mathbb{R}$ for all n (and hence $\mu(\bigcap_n E_n) \in \mathbb{R}$). Then by (b), for each n,

$$\mu(E_1 \backslash E_n) = \mu(E_1) - \mu(E_n) \quad \text{and} \quad \mu\Big(E_1 \backslash \bigcap\nolimits_n E_n\Big) = \mu(E_1) - \mu\Big(\bigcap\nolimits_n E_n\Big).$$

Recalling De Morgan laws and applying (c), it then follows that

$$\begin{aligned}\mu(E_1) - \mu\Big(\bigcap\nolimits_n E_n\Big) &= \mu\Big(E_1 \backslash \bigcap\nolimits_n E_n\Big) = \mu\Big(\bigcup\nolimits_n (E_1 \backslash E_n)\Big) = \mu\Big(\bigcup\nolimits_n E''_n\Big) \\ &= \lim_n \mu(E''_n) = \lim_n \mu(E_1 \backslash E_n) = \mu(E_1) - \lim_n \mu(E_n),\end{aligned}$$

completing the proof of (d) since $\mu(E_1) \neq +\infty$. □

Consider a measure space $(X, \mathcal{X}, \mu)$. If a statement (or a proposition) $P(x)$ holds for every $x \in X \backslash N$ for some $N \in \mathcal{X}$ such that $\mu(N) = 0$ (i.e., if it holds up to a set of measure zero), then we say that $P(x)$ holds *μ-almost everywhere* (or *almost everywhere with respect to* μ, or simply *almost everywhere* if the measure μ is clear in the context, or still *almost sure* if $(X, \mathcal{X}, \mu)$ is a probability space). Summing up: a proposition $P(x)$ holds *μ-almost everywhere* if $P(x)$ is true up to a set of measure zero, which means that there exists an $\mathcal{X}$-measurable set N with $\mu(N) = 0$ such that $P(x)$ holds true for all x in the complement $X \backslash N$ of N.

Example 2E. Let $f: X \to Z$ and $g: X \to Z$ be functions from a set X to a set Z. The standard definition of equality between functions is pointwise interpreted: $f = g$ if $f(x) = g(x)$ for every $x \in X$. That is, equality is pointwise defined everywhere in X. Now suppose $\mu: \mathcal{X} \to \overline{\mathbb{R}}$ is a measure defined on a σ-algebra $\mathcal{X}$ of subsets of X. The functions f and g are *equal almost everywhere with respect to* μ (or *equal μ-almost everywhere*), denoted by

$$f = g \quad \mu\text{-a.e.},$$

if $f(x) = g(x)$ for every $x \in X \backslash N$ for some $N \in \mathcal{X}$ such that $\mu(N) = 0$. That is, $f(x) = g(x)$ for every x in the complement of a set of measure zero. For instance, take the measure space $(\mathbb{R}, \Re, \lambda)$ of Example 2C, and let f and g be real-valued be functions on $\mathbb{R}$. If $f = 0$ (i.e., $f(x) = 0$ for all $x \in \mathbb{R}$), and $g(x) = 0$ for every $x \in \mathbb{R} \backslash \{0\}$ and $g(0) = 1$, then $f = g$ λ-a.e., but $f \neq g$ pointwise. (If both are continuous, then $f = g$ λ-a.e. if and only if $f = g$.)

Example 2F. Consider a sequence $\{f_n\}$ of real-valued functions $f_n: X \to \mathbb{R}$ on a set X. The sequence of functions $\{f_n\}$ is said to *converge pointwise* to a function $f: X \to \mathbb{R}$ if the sequence of real numbers $\{f_n(x)\}$ converges to the real number $f(x)$ for every $x \in X$, which means convergence of $\{f_n(x)\}$ everywhere in X. Let $\mu: \mathcal{X} \to \overline{\mathbb{R}}$ be a measure on a σ-algebra $\mathcal{X}$ of subsets of X. The sequence of functions $\{f_n\}$ *converges to* f *almost everywhere with respect to* μ (or *converges* μ*-almost everywhere to* f), denoted by

$$f_n \to f \quad \mu\text{-a.e.} \qquad \text{or} \qquad \lim_n f_n = f \quad \mu\text{-a.e.},$$

if $f_n(x) \to f(x)$ for every $x \in X \backslash N$ for some $N \in \mathcal{X}$ such that $\mu(N) = 0$. That is, $f_n(x) \to f(x)$ for every x except perhaps in a set of measure zero.

2.2 Signed Measure

Consider a measurable space $(X, \mathcal{X})$, let μ and λ be two measures on $\mathcal{X}$, and let α be a nonnegative real number. It is readily verified that $\alpha\mu$ and $\mu + \lambda$ are again measures on $\mathcal{X}$ (pointwise defined; that is, $(\alpha\mu)(E) = \alpha\mu(E)$ and $(\mu + \lambda)(E) = \mu(E) + \lambda(E)$ for every $E \in \mathcal{X}$)). This is easily extended (by a trivial induction) so that every (finite) linear combination $\sum_{i=1}^n \alpha_i\mu_i$ with nonnegative coefficients α_i of measures μ_i on $\mathcal{X}$ is again a measure on $\mathcal{X}$. The assumption of nonnegative coefficients is imposed to ensure nonnegativeness for the resulting measure (Definition 2.1(b)). If we ignore this nonnegativeness requirement, then we might consider real coefficients. For instance, we might consider the set function $\mu - \lambda$ on $\mathcal{X}$ (pointwise defined: $(\mu - \lambda)(E) = \mu(E) - \lambda(E)$ for every $E \in \mathcal{X}$). However, if there exists a set E in $\mathcal{X}$ such that $\mu(E) = \lambda(E) = +\infty$, then $\mu(E) - \lambda(E)$ is not well defined. An obvious way to avoid this problem is to assume that at least one of μ or λ does not take on the value $+\infty$. Another (and simpler) way consists in assuming, symmetrically, that μ and λ are both real-valued measures.

Definition 2.3. A *signed measure* is a real-valued set function ν on a σ-algebra $\mathcal{X}$ of subsets of a set X,

$$\nu\colon \mathcal{X} \to \mathbb{R},$$

fulfilling the next two axioms.

(a) $\nu(\varnothing) = 0$, and

(b) $\nu\left(\bigcup_n E_n\right) = \sum_n \nu(E_n)$
for every countable family $\{E_n\}$ of pairwise disjoint sets in $\mathcal{X}$ for which the series $\sum_n \nu(E_n)$ is unconditionally convergent.

Remark: A signed measure is a real-valued function on a σ-algebra that may fail to be a measure (a finite measure, actually) because it may not satisfy nonnegativeness in the second axiom of Definition 2.1(b). We are now dealing with real-valued functions. Thus the series in (b) must converge (otherwise the left-hand side of (b) is not well defined). In Definition 2.1(c) we had a series of nonnegative terms, where convergence coincides with absolute convergence, which (in $\mathbb{R}$) means unconditional convergence. Now we have a series of real numbers, where unconditional convergence is not a consequence of plain convergence. But the identity in (b) requires unconditional convergence (the union in the left-hand side is order invariant, and so is the series on the right-hand side) or, equivalently, absolute convergence.

The properties in Proposition 2.2(b,c,d) still hold for a signed measure (with essentially the same proof). Observe that any (finite) linear combination of signed measures is again a signed measure. If μ and λ are finite measures, then the function $\nu = \mu - \lambda$ is a signed measure. Such a setup was our motivation for defining signed measures. There are other ways to get signed measures from measures (cf. Lemma 4.6). Another important question is how to get measures from signed measures (cf. Section 7.1). The next example shows that what might seem the obvious way simply fails. The following two examples exhibit measures generated by a signed measure.

Example 2G. Take a measurable space $(X,\mathcal{X})$, let $\nu\colon\mathcal{X}\to\mathbb{R}$ be a signed measure, and consider a set function $\pi\colon\mathcal{X}\to\mathbb{R}$ defined for each $E\in\mathcal{X}$ by

$$\pi(E) = |\nu(E)|.$$

This $\pi\colon\mathcal{X}\to\mathbb{R}$ may *not* be a measure. Actually, if the signed measure ν is not a measure itself, then there is a set $B\in\mathcal{X}$ for which $\nu(B)<0$. If there is a set $A\in\mathcal{X}$ such that $A\cap B=\varnothing$ and $\nu(A)=-\nu(B)$, then $\nu(A\cup B)=0$. Thus $A\subset A\cup B$ and $\pi(A)\not\leq\pi(A\cup B)$, and so π is not a measure (Proposition 2.2(a) — i.e., π is not increasing). Another way to see this:

$$\pi(A\cup B) = |\nu(A\cup B)| = 0 < |\nu(A)| + |\nu(B)| = \pi(A) + \pi(B).$$

Hence π is not additive, not even finite additive, and so π is not a measure (Definition 2.1(c) — a concrete example is exhibited in Problem 2.12).

Example 2H. Let $(X,\mathcal{X})$ be a measurable space. Consider the σ-algebra $\mathcal{E}=\wp(E)\cap\mathcal{X}$ for each $\mathcal{X}$-measurable set E. Let $\nu\colon\mathcal{X}\to\mathbb{R}$ be a signed measure. We show that the function $\mu\colon\mathcal{X}\to\mathbb{R}$ defined for each $E\in\mathcal{X}$ by

$$\mu(E) = \sup_{A\in\mathcal{E}} \nu(A),$$

is a finite measure on $\mathcal{X}$. In fact, let $\{E_n\}$ be a arbitrary countable family of pairwise disjoint sets in $\mathcal{X}$, set $E=\bigcup E_n\in\mathcal{X}$, and take any $n\in\mathbb{N}$.

Claim. For each $\varepsilon > 0$ there is an $\mathcal{X}$-measurable set $A_n \subseteq E_n$ such that

$$\nu(A_n) \le \mu(E_n) \le \nu(A_n) + \tfrac{\varepsilon}{2^n}.$$

Proof. Observe that $\nu(A_n) \le \mu(E_n)$ according to the definition of μ. Take an arbitrary $\varepsilon > 0$. If $\nu(A) + \varepsilon < \mu(E)$ for every $A \in \mathcal{E}$, then $\mu(E) + \varepsilon = \sup_{A\in\mathcal{E}} \nu(A) + \varepsilon < \mu(E)$, which is a contradiction. Thus, for every $\varepsilon > 0$ and every $E \in \mathcal{X}$ there is an $A_\varepsilon \in \mathcal{E}$ such that $\mu(E) \le \nu(A_\varepsilon) + \varepsilon$. Therefore, for any integer $n \in \mathbb{N}$ and any $\varepsilon > 0$, set $\varepsilon_n = \frac{\varepsilon}{2^n}$ so that there exists an $\mathcal{X}$-measurable $A_n \subseteq E_n$ for which $\mu(E_n) \le \nu(A_n) + \frac{\varepsilon}{2^n}$. □

Thus,

$$\sum\nolimits_n \nu(A_n) \le \sum\nolimits_n \mu(E_n) \le \sum\nolimits_n \nu(A_n) + \varepsilon$$

because $\sum_{n\in\mathbb{N}} \frac{1}{2^n} = 1$. Since $\{A_n\}$ is a disjoint sequence (reason: $\{E_n\}$ is a disjoint sequence) of sets in $\mathcal{X}$, and since ν is a signed measure, we get

$$\nu\Big(\bigcup\nolimits_n A_n\Big) = \sum\nolimits_n \nu(A_n).$$

Set $A = \bigcup_n A_n$ in $\mathcal{X}$ so that $A \subseteq E = \bigcup_n E_n$. Hence $A \in \mathcal{E}$ and

$$\nu(A) \le \sum\nolimits_n \mu(E_n) \le \nu(A) + \varepsilon.$$

This implies that

$$\mu(E) \le \sum\nolimits_n \mu(E_n) \le \mu(E) + \varepsilon$$

for every $\varepsilon > 0$ because $\mu(E) = \sup_{A\in\mathcal{E}} \nu(A)$. Therefore, since $\bigcup_n E_n = E$,

$$\mu\Big(\bigcup\nolimits_n E_n\Big) = \mu(E) = \sum\nolimits_n \mu(E_n),$$

so that axiom (c) of Definition 2.2 (countable additivity) is satisfied. Since axioms (a) and (b) are trivially satisfied ($\wp(\varnothing) \cap \mathcal{X} = \{\varnothing\}$ and $\nu(\varnothing) = 0$, so $\mu(\varnothing) = 0$ and $\mu(E) \ge 0$ for every $E \in \mathcal{X}$), μ is a measure on $\mathcal{X}$. Moreover, it is readily verified that μ is indeed a finite measure (actually, it is dominated by the measure in the next example, which will also be shown to be finite).

Example 2I. A *covering* of a subset A of a set X is a collection of subsets of X that cover A (i.e., whose union includes A). A *partition* of A is then a covering of it consisting of disjoint subsets of it (i.e., a disjoint covering). Take a σ-algebra $\mathcal{X}$ of subsets of X and let E be any $\mathcal{X}$-measurable set. A *measurable covering* of E is a covering of it made up of $\mathcal{X}$-measurable sets. A *measurable partition* of E is a disjoint covering of it consisting of sets in $\mathcal{E} = \wp(E) \cap \mathcal{X}$. For each integer $n \in \mathbb{N}$ let $\boldsymbol{E}(n)$ be the collection of all measurable partitions of E containing precisely n sets, so that $\bigcup \boldsymbol{E}(n)$ is the collection of all *finite* measurable partitions of E. Take any signed measure $\nu\colon \mathcal{X} \to \mathbb{R}$ and consider the function $\mu\colon \mathcal{X} \to \mathbb{R}$ defined for each $E \in \mathcal{X}$ by

$$\mu(E) \;=\; \sup_{\{E_i\}\in \bigcup \boldsymbol{E}(n)} \sum\nolimits_i |\nu(E_i)|.$$

Every finite partition $\{E_i\} \in \bigcup \boldsymbol{E}(n)$ can be written as $\{E_i\} = \{E_j\} \cup \{E_k\}$, where $\{E_j\}$ and $\{E_k\}$ are disjoint collections such that $\nu(E_j) \geq 0$ for each j and $\nu(E_k) \leq 0$ for each k. Set $E^+ = \bigcup_j E_j$ and $E^- = \bigcup_k E_k$ in $\mathcal{E}$ so that

$$\nu(E^+) = \nu\Big(\bigcup\nolimits_j E_j\Big) = \sum\nolimits_j \nu(E_j) \geq 0,$$

$$\nu(E^-) = \nu\Big(\bigcup\nolimits_k E_k\Big) = \sum\nolimits_k \nu(E_k) \leq 0.$$

Therefore, $\sum_i |\nu(E_i)| = \nu(E^+) - \nu(E^-)$, where $\{E^+, E^-\} \in \boldsymbol{E}(2)$. Thus

$$\mu(E) \;=\; \sup_{\{E^+,\,E^-\}\in \boldsymbol{E}(2)} \big(\nu(E^+) - \nu(E^-)\big) \quad \text{for every} \quad E \in \mathcal{X},$$

where the supremum is taken over all measurable partitions $\{E^+, E^-\}$ of E consisting of two sets such that $\nu(E^+) \geq 0$ and $\nu(E^-) \leq 0$. This μ is a finite measure on $\mathcal{X}$, called the *variation* of the signed measure ν. Indeed, μ coincides with the "total variation" of ν, which is a finite measure $|\nu|\colon \mathcal{X} \to \mathbb{R}$ that will be discussed in Proposition 7.4 and Example 7A of Section 7.1.

2.3 Completion of Measure Spaces

Consider a measure space $(X, \mathcal{X}, \mu)$. It is said to be *complete* if the σ-algebra contains all subsets of sets of measure zero. That is, $(X, \mathcal{X}, \mu)$ is *complete* if

$$N \in \mathcal{X}, \;\; \mu(N) = 0 \;\text{ and }\; A \subseteq N \quad \text{imply} \quad A \in \mathcal{X}.$$

If a measure space $(X, \mathcal{X}, \mu)$ is complete, then $\mathcal{X}$ is said to be a *complete σ-algebra* (with respect to a measure μ) and μ is said to be a *complete measure* on $\mathcal{X}$. Each measure space can be completed by adding up enough subsets of measure zero to the σ-algebra, as we will see next. Consider a measure space $(X, \mathcal{X}, \mu)$, and take the collection

$$\mathcal{N} = \{N \in \mathcal{X}\colon \mu(N) = 0\}$$

of all sets in $\mathcal{X}$ of measure zero. Let $\overline{\mathcal{X}}$ be the collection of all sets of the form $E \cup A$, where E is a set in $\mathcal{X}$ and A is a subset of some set in $\mathcal{N}$:

$$\overline{\mathcal{X}} = \big\{\overline{E} \subseteq X\colon \overline{E} = E \cup A \text{ with } E \in \mathcal{X} \text{ and } A \subseteq N \text{ for some } N \in \mathcal{N}\big\}.$$

Consider a set function $\overline{\mu}: \overline{\mathcal{X}} \to \overline{\mathbb{R}}$ defined by

$$\overline{\mu}(\overline{E}) = \mu(E)$$

for each set $\overline{E} \in \overline{\mathcal{X}}$, where E is any set in $\mathcal{X}$ for which $\overline{E} = E \cup A$ for some subset A of some N in $\mathcal{N}$. This function $\overline{\mu}$ is well defined. In fact, take an arbitrary $\overline{E} \in \overline{\mathcal{X}}$ and consider any pair of possible representations of it, say, $\overline{E} = E_1 \cup A_1 = E_2 \cup A_2$, where E_1, E_2 are sets in $\mathcal{X}$ and A_1, A_2 are subsets of some sets N_1, N_2 in $\mathcal{N}$, respectively. Since $E_1 \subseteq E_1 \cup A_1 = E_2 \cup A_2 \subseteq E_2 \cup N_2$, and since $E_2 \cup N_2$ lies in $\mathcal{X}$ (Definition 1.1(c$'$)), it follows from Proposition 2.2(a) and Definition 2.1(c) that $\mu(E_1) \leq \mu(E_2 \cup N_2) = \mu(E_2)$. Applying the same argument, $\mu(E_2) \leq \mu(E_1 \cup N_1) = \mu(E_1)$. So $\mu(E_1) = \mu(E_2)$. Thus the function $\overline{\mu}$ is well defined: it assigns to each $\overline{E}$ the value $\mu(E)$, which is invariant for all representations of $\overline{E} = E \cup A$. Moreover, as it will be shown in the forthcoming Proposition 2.4,

(i) $\overline{\mathcal{X}}$ is a σ-algebra of subsets of X that includes the σ-algebra $\mathcal{X}$ of subsets of X (which means that $\mathcal{X}$ is a *sub-σ-algebra* of $\overline{\mathcal{X}}$), and

(ii) $\overline{\mu}$ is a measure on $\overline{\mathcal{X}}$ that agrees with μ on $\mathcal{X}$ (that is, $\overline{\mu}$ is an *extension of μ over $\overline{\mathcal{X}}$* or, equivalently, μ is a *restriction of $\overline{\mu}$ to $\mathcal{X}$*).

Proposition 2.4. *Take an arbitrary measure space* $(X, \mathcal{X}, \mu)$.

(a) $\overline{\mathcal{X}}$ *is a σ-algebra of subsets of* X *such that* $\mathcal{X} \subseteq \overline{\mathcal{X}}$.

(b) $\overline{\mu}: \overline{\mathcal{X}} \to \overline{\mathbb{R}}$ *is a measure on* $\overline{\mathcal{X}}$ *such that* $\overline{\mu}(E) = \mu(E)$ *for every* $E \in \mathcal{X}$.

(c) $(X, \overline{\mathcal{X}}, \overline{\mu})$ *is a complete measure space.*

Proof. First note that $\mathcal{X} \subseteq \overline{\mathcal{X}}$, since $E = E \cup \varnothing$ for every $E \in \mathcal{X}$. In particular, the empty set and the whole set trivially lie in $\overline{\mathcal{X}}$. Take an arbitrary $\overline{E} \in \overline{\mathcal{X}}$. Next we show that $X \backslash \overline{E} \in \overline{\mathcal{X}}$. Indeed, if $\overline{E} = E \cup A$, then $X \backslash \overline{E} = X \backslash (E \cup A) = (X \backslash E) \cap (X \backslash A) = E' \backslash A$. Here $E' = X \backslash E$, which lies in $\mathcal{X}$ because $E \in \mathcal{X}$. Let N be any set in $\mathcal{N} \subseteq \mathcal{X}$ such that $A \subseteq N$. Thus $E' \backslash A = (E' \backslash N) \cup (N \backslash A) = E_1 \cup A_1$, where $E_1 = E' \backslash N$ lies in $\mathcal{X}$ since both E' and N lie in $\mathcal{X}$, and $A_1 = N \backslash A \subseteq N$. Thus,

$$X \backslash \overline{E} = E' \backslash A = E_1 \cup A_1 \quad \text{lies in} \quad \overline{\mathcal{X}}.$$

Let $\{\overline{E}_n\}$ be an arbitrary sequence of sets in $\overline{\mathcal{X}}$. Thus $\overline{E}_n = E_n \cup A_n$, where $E_n \in \mathcal{X}$ and $A_n \subseteq N_n$ with $N_n \in \mathcal{N}$, for each n. Set $E = \bigcup_n E_n$ in $\mathcal{X}$ (Definition 1.1.(c$'$)), and $A = \bigcup_n A_n \subseteq N = \bigcup_n N_n$. Note that $N \in \mathcal{N}$ (Definition 2.1(c)). Then $\overline{\mathcal{X}}$ is a σ-algebra (cf. Definition 1.1) since

$$\bigcup_n \overline{E}_n = \bigcup_n (E_n \cup A_n) = \Big(\bigcup_n E_n\Big) \cup \Big(\bigcup_n A_n\Big) = E \cup A \quad \text{lies in} \quad \overline{\mathcal{X}},$$

completing the proof of (a). Note that the set function $\overline{\mu}$ agrees with the measure μ on $\mathcal{X}$ (i.e., $\overline{\mu}(E)=\mu(E)$ for every $E\in\mathcal{X}$) by the definition of $\overline{\mu}$ (because $E=E\cup\varnothing$). In particular, $\overline{\mu}(\varnothing)=0$ and $\overline{\mu}(\overline{E})=\mu(E)\geq 0$ for every $\overline{E}\in\overline{\mathcal{X}}$. Moreover, suppose the arbitrary sequence $\{\overline{E}_n\}$ of sets in $\overline{\mathcal{X}}$ (i.e., $\overline{E}_n=E_n\cup A_n$, where $E_n\in\mathcal{X}$ and $A_n\subseteq N_n$ with $N_n\in\mathcal{N}$) is made up of pairwise disjoint set. Thus $\{E_n\}$ is a sequence of pairwise disjoint sets in $\mathcal{X}$, and so it follows from the above displayed identity that

$$\overline{\mu}\Big(\bigcup_n\overline{E}_n\Big)=\overline{\mu}(E\cup A)=\mu(E)=\mu\Big(\bigcup_n E_n\Big)=\sum_n\mu(E_n)=\sum_n\overline{\mu}(\overline{E}_n)$$

by the definition of $\overline{\mu}$, since μ is a measure on $\mathcal{X}$ (Definition 2.1(c)). Then $\overline{\mu}$ is a measure on $\overline{\mathcal{X}}$ (cf. Definition 2.1). This proves (b). Finally, set

$$\overline{\mathcal{N}}=\{\overline{N}\in\overline{\mathcal{X}}\colon\overline{\mu}(\overline{N})=0\}.$$

If $\overline{N}\in\overline{\mathcal{N}}$, then $\overline{N}\in\overline{\mathcal{X}}$. Thus the set $\overline{N}$ is of the form $\overline{N}=N'\cup A$ with $N'\in\mathcal{X}$ and $A\subseteq N$ for some $N\in\mathcal{N}$ (by the definition of $\overline{\mathcal{X}}$). However, $\mu(N')=\overline{\mu}(\overline{N})=0$ (cf. definition of $\overline{\mu}$), and so $N'\in\mathcal{N}$. Outcome:

$$\overline{N}\in\overline{\mathcal{N}}\quad\text{implies}\quad\overline{N}=N'\cup A\ \text{with}\ A\subseteq N\ \text{for some}\ N,N'\in\mathcal{N}.\quad(*)$$

Also, if $A\subseteq N\in\mathcal{N}\subseteq\mathcal{X}$, then $A=\varnothing\cup A$ must lie in $\overline{\mathcal{X}}$ since $\varnothing\in\mathcal{X}$:

$$A\subseteq N\in\mathcal{N}\quad\text{implies}\quad A\in\overline{\mathcal{X}}.\qquad(**)$$

Hence, if $\overline{A}$ is any subset of an arbitrary set $\overline{N}$ in $\overline{\mathcal{N}}$, then $\overline{A}\subseteq\overline{N}=N'\cup A$ with $A\subseteq N$ for some pair $N,N'\in\mathcal{N}$ by $(*)$. Therefore, $\overline{A}=A'\cup A''$, where $A'\subseteq N'$ and $A''\subseteq A\subseteq N$. But both A' and A'' lie in $\overline{\mathcal{X}}$ by $(**)$, and so $\overline{A}\in\overline{\mathcal{X}}$ (Definition 1.1(c')). That is, $\overline{\mathcal{X}}$ is a complete σ-algebra with respect to the measure $\overline{\mu}$, which completes the proof of (c). □

The complete measure space $(X,\overline{\mathcal{X}},\overline{\mu})$ is referred to as the *completion of the measure space* $(X,\mathcal{X},\mu)$. Accordingly, we say that $\overline{\mathcal{X}}$ is the *completion of the σ-algebra* $\mathcal{X}$ (with respect to the measure μ) and $\overline{\mu}$ is the *completion of the measure* μ on $\mathcal{X}$.

Remark: Consider the Lebesgue measure λ on the σ-algebra $\mathfrak{R}$ of Borel sets as in Example 2C. The measure space $(\mathbb{R},\mathfrak{R},\lambda)$ is not complete. That is, on the Borel algebra $\mathfrak{R}$ the Lebesgue measure λ is not complete. We will see this in Chapter 8, and also how to build its completion, where the Lebesgue measure $\overline{\lambda}$ on the *Lebesgue algebra* $\overline{\mathfrak{R}}$ (the completion of the Borel algebra $\mathfrak{R}$ with respect to λ) makes a complete measure space $(\mathbb{R},\overline{\mathfrak{R}},\overline{\lambda})$. It is worth noticing that the notation $\overline{\mathfrak{R}}$ is tricky; it does not mean the σ-algebra generated by the collection of all open intervals of the extended real line $\overline{\mathbb{R}}$.

2.4 Problems

Problem 2.1. Consider a measurable space $(X, \mathcal{X})$, and let $\mu\colon \mathcal{X} \to \overline{\mathbb{R}}$ be a measure on the σ-algebra $\mathcal{X}$.

(a) Show that μ is a finite measure if and only if there exists $0 \le \alpha \in \mathbb{R}$ such that $\mu(E) < \alpha$ for all $E \in \mathcal{X}$.

Actually, it is readily verified by Proposition 2.2(a) that μ is finite if and only if $\mu(X) < \infty$. Consider the definition of a σ-finite measure and assume that there exists an $\mathcal{X}$-valued sequence $\{E_n\}$ and a real (nonnegative) number α such that $\mu(E_n) \le \alpha$ for all n and $X = \bigcup_n E_n$. In this case μ is said to be *uniformly σ-finite.*

(b) Verify that the counting measure of Example 2B is uniformly σ-finite.

(c) Verify that the Lebesgue measure of Example 2C is uniformly σ-finite.

(d) When (i.e., for which class of functions F) is the Borel–Stieltjes measure λ_F of Example 2D finite? When is it uniformly σ-finite?

Problem 2.2. Every finite measure is uniformly σ-finite, and every uniformly σ-finite measure is σ-finite. (This is clear, isn't it?) However, the converses fail. Indeed, consider the σ-algebra $\wp(\mathbb{N})$ of all subsets of the natural numbers $\mathbb{N}$. According to the previous problem the counting measure on $\wp(\mathbb{N})$ is uniformly σ-finite, and it is clearly not finite (the value of it at $\mathbb{N}$ is not finite). Now consider the function $\mu\colon \wp(\mathbb{N}) \to \overline{\mathbb{R}}$ defined by

$$\mu(E) = \sum_{k \in E} k$$

for every $E \in \wp(\mathbb{N})$ (by convention, the empty sum is null). Verify that this is a measure on $\wp(\mathbb{N})$, which is σ-finite but not uniformly σ-finite.

Problem 2.3. Prove that a measure $\mu\colon \mathcal{X} \to \overline{\mathbb{R}}$ on a σ-algebra $\mathcal{X}$ of subsets of a set X is σ-finite if and only if there exists a countable family $\{E_k\}$ of *disjoint* sets in $\mathcal{X}$ such that $\mu(E_k) < \infty$ for every k and $X = \bigcup_k E_k$.

Hint: Every sequence of sets $\{X_n\}_{n\in\mathbb{N}}$ has a *disjointification* $\{Y_n\}_{n\in\mathbb{N}}$ (i.e., $\{Y_n\}_{n\in\mathbb{N}}$ is a sequence of pairwise disjoint sets and $\bigcup_n Y_n = \bigcup_n X_n$).

Problem 2.4. Let $\mathcal{X}$ be a σ-algebra of subsets of an arbitrary set X. A measure $\mu\colon \mathcal{X} \to \overline{\mathbb{R}}$ is called *semifinite* if every measurable set of infinite measure includes a measurable set of arbitrarily large finite measure.

(a) Prove that every σ-finite measure is semifinite.

Hint: Suppose a measure is not semifinite but σ-finite (Problem 2.3).

Now suppose that X is an uncountable set. Let $\#$ stand for cardinality, and take two set functions $\mu\colon \mathcal{X} \to \overline{\mathbb{R}}$ and $\lambda\colon \mathcal{X} \to \overline{\mathbb{R}}$ defined for each $E \in \mathcal{X}$ by

$$\mu(E) = \begin{cases} \#(E), & E \text{ is finite}, \\ +\infty, & E \text{ is infinite}, \end{cases} \qquad \lambda(E) = \begin{cases} 0, & E \text{ is countable}, \\ +\infty, & E \text{ is uncountable}. \end{cases}$$

(b) Verify that μ is a measure on $\mathcal{X}$ (the counting measure on an *uncountable* set) that is semifinite but not σ-finite.

(c) Verify that λ is a measure on $\mathcal{X}$ that is not semifinite (thus not σ-finite according to (a)).

Hint: Infinite sets have countably infinite (proper) subsets. Countable unions of countable sets are countable.

Problem 2.5. Suppose $\mathcal{X}$ is a σ-algebra of subsets of an infinite set X for which all singletons are measurable sets. Let $\pi\colon \mathcal{X} \to \overline{\mathbb{R}}$ and $\rho\colon \mathcal{X} \to \overline{\mathbb{R}}$ be set functions defined for each $E \in \mathcal{X}$ by

$$\pi(E) = \begin{cases} 0, & E \text{ is finite}, \\ +\infty, & E \text{ is infinite}, \end{cases} \qquad \rho(E) = \begin{cases} 0, & E \text{ is finite}, \\ 1, & E \text{ is infinite}. \end{cases}$$

Question: Why are these functions not measures?

Problem 2.6. This is a continuation of the previous problem for an uncountable set. Suppose X is an uncountable set, and let $\mathcal{X}$ be a collection of those subsets of X that either are countable or are the complement of a countable subset of X. First show that $\mathcal{X}$ is a σ-algebra of subsets of X. Next consider the set function $\mu\colon \mathcal{X} \to \mathbb{R}$ defined for each $E \in \mathcal{X}$ by

$$\mu(E) = \begin{cases} 0, & E \text{ is countable}, \\ 1, & X\backslash E \text{ is countable}. \end{cases}$$

Verify that μ is a measure (actually, a probability measure) on $\mathcal{X}$.

Problem 2.7. Consider the Lebesgue measure $\lambda\colon \mathfrak{R} \to \overline{\mathbb{R}}$ on the Borel algebra $\mathfrak{R}$ of subsets of $\mathbb{R}$ (cf. Example 2C). Prove the following assertions.

(a) Every singleton of $\mathbb{R}$ is $\mathfrak{R}$-measurable and has measure zero.

(b) Every countable subset of $\mathbb{R}$ is $\mathfrak{R}$-measurable and has measure zero.

(c) If $\alpha < \beta$, then the intervals (α, β), $[\alpha, \beta)$, $(\alpha, \beta]$, $[\alpha, \beta]$ are $\mathfrak{R}$-measurable and $\lambda((\alpha, \beta)) = \lambda([\alpha, \beta)) = \lambda((\alpha, \beta]) = \lambda([\alpha, \beta]) = \beta - \alpha$.

(d) Every nonempty open subset U of $\mathbb{R}$ is $\mathfrak{R}$-measurable and $\lambda(U) > 0$. (*Hint:* $\mathbb{R}$ has a countable base of open intervals — see Problem 1.14.)

(e) Every bounded $\mathfrak{R}$-measurable subset of $\mathbb{R}$ has a finite measure.

(f) A closed and bounded subset K of $\mathbb{R}$ is $\mathfrak{R}$-measurable and $\lambda(K) < \infty$.

Also show that λ is uniformly σ-finite by exhibiting a countably infinite family $\{E_k\}$ of *disjoint sets* in $\mathfrak{R}$ such that $\lambda(E_k) = 1$ for all k and $\mathbb{R} = \bigcup_k E_k$.

Problem 2.8. Let $\mu\colon \mathcal{X} \to \overline{\mathbb{R}}$ be a measure on a σ-algebra $\mathcal{X}$, and let $\{E_n\}$ be a sequence of $\mathcal{X}$-measurable sets. Apply Proposition 2.2 to show that

(a) $\mu\big(\bigcup_n E_n\big) = \lim_m \mu\big(\bigcup_{i=1}^m E_i\big)$,

(b) $\mu\big(\bigcup_n E_n\big) \le \sum_n \mu(E_n)$.

Hint: Set $A_m = \bigcup_{i=1}^m E_i$ to prove (a), and set $B_{m+1} = E_{m+1} \backslash \big(\bigcup_{i=1}^m E_i\big)$ so that $B_m \subseteq E_m$ and $\{B_m\}$ is pairwise disjoint to prove (b).

Problem 2.9. The *Cantor set* is a rather important well-known subset of the interval $[0,1]$ of the real line $\mathbb{R}$, possessing striking properties, which make it a significant source of counterexamples. The reader is referred to the bibliography mentioned in the Suggested Reading section for many of its aspects. Roughly speaking, the Cantor set C is the intersection of a decreasing sequence $\{C_n\}$ of closed subsets of $C_0 = [0,1]$ obtained by successive removal of the central open third. Among the main properties of the Cantor set $C = \bigcap_n C_n \subset [0,1] \subset \mathbb{R}$ we point out the following. The set C is nonempty, closed, and bounded; it has an empty interior and has no isolated point; it is uncountable and totally disconnected. Consider the Lebesgue measure $\lambda\colon \mathfrak{R} \to \overline{\mathbb{R}}$ as in Example 2C (see Problem 2.7 as well) and show that the Cantor set has measure zero. In other words, C lies in $\mathfrak{R}$ and $\lambda(C) = 0$. (*Hint:* Each C_n consists of 2^n disjoint intervals of length $\frac{1}{3^n}$.)

Problem 2.10. Now we build up a Cantor-like set S whose Lebesgue measure is not null. Consider the setup and the construction of the previous problem, where each set C_n (for $n \in \mathbb{N}$) is obtained from C_{n-1} by removing 2^{n-1} central open subintervals, each of length $\frac{1}{3^n}$. Now, instead of removing 2^{n-1} central open subintervals of length $\frac{1}{3^n}$ at each iteration, remove 2^{n-1} central open subintervals of length $\frac{1}{4^n}$ at each iteration. Let $\{S_n\}$ be the resulting decreasing sequence of closed subsets of the unit interval $S_0 = [0,1]$. Show that the length of each S_n is $\lambda(S_n) = \frac{1}{2} + \frac{1}{2^{n+1}}$ and conclude that $S = \bigcap_n S_n \subset [0,1] \subset \mathbb{R}$ lies in $\mathfrak{R}$ and $\lambda(S) = \frac{1}{2}$. (*Hint:* Proposition 2.2(c).)

Problem 2.11. Take a measure $\mu\colon \mathcal{X} \to \overline{\mathbb{R}}$ on a σ-algebra $\mathcal{X}$ of subsets of a set X. Let A be an arbitrary $\mathcal{X}$-measurable set, and consider the σ-algebra $\mathcal{A} = \wp(A) \cap \mathcal{X}$ of subsets of A, so that $\mathcal{A} \subseteq \mathcal{X}$. Define a pair of set functions $\mu_A\colon \mathcal{X} \to \overline{\mathbb{R}}$ and $\mu|_A\colon \mathcal{A} \to \overline{\mathbb{R}}$ as follows:

$$\mu_A(E) = \mu(E \cap A) \quad \text{for every} \quad E \in \mathcal{X},$$

$$\mu|_A(E) = \mu(E) \quad \text{for every} \quad E \in \mathcal{A}.$$

Verify that the set functions $\mu_A\colon \mathcal{X} \to \overline{\mathbb{R}}$ and $\mu|_A\colon \mathcal{A} \to \overline{\mathbb{R}}$ are measures on $\mathcal{X}$ and on $\mathcal{A}$, respectively. (*Hint:* $\big(\bigcup_n E_n\big) \cap A = \bigcup_n (E_n \cap A)$.) The measure $\mu|_A$ is *the restriction of both μ and μ_A to $\mathcal{A}$*, and so μ and μ_A are (different) *extensions of $\mu|_A$ over $\mathcal{X}$* — all these measures coincide on $\mathcal{A}$.

Problem 2.12. Consider the Lebesgue measure $\lambda\colon \mathfrak{R} \to \overline{\mathbb{R}}$ on the Borel algebra $\mathfrak{R}$ of subsets of the real line $\mathbb{R}$ (cf. Example 2C and Problem 2.7). Set $A = [1, 2]$, $B = [-2, -1]$, and take the measures $\lambda_A\colon \mathfrak{R} \to \overline{\mathbb{R}}$ and $\lambda_B\colon \mathfrak{R} \to \overline{\mathbb{R}}$ defined in Problem 2.11. Show that λ_A and λ_B are finite measures so that their difference $\nu = \lambda_A - \lambda_B\colon \mathfrak{R} \to \mathbb{R}$ is a signed measure. Also show that

$$\nu(A \cup B) = \lambda_A(A \cup B) - \lambda_B(A \cup B) = 0,$$

$$|\nu(A)| + |\nu(B)| = |\lambda_A(A) - \lambda_B(A)| + |\lambda_A(B) - \lambda_B(B)| = 2.$$

Then conclude that the set function $\pi\colon \mathfrak{R} \to \mathbb{R}$, defined for each $E \in \mathfrak{R}$ by $\pi(E) = |\nu(E)| = |\lambda_A(E) - \lambda_B(E)|$, is not a measure (cf. Example 2G).

Problem 2.13. Consider the measure space $(\mathbb{R}, \mathfrak{R}, \mu)$, where $\mu\colon \mathfrak{R} \to \overline{\mathbb{R}}$ is a measure on the Borel algebra $\mathfrak{R}$ of subsets of $\mathbb{R}$ such that $\mu(K) < \infty$ for every closed and bounded subset K of $\mathbb{R}$. This is referred to as a *Borel measure* (recall that all open, and so all closed, sets lie in $\mathfrak{R}$; that is, they are Borel sets). The general notion of Borel measure will be the subject of a whole chapter — Chapter 11. Verify that the Lebesgue measure of Example 2C is a Borel measure, and that every Borel measure is σ-finite. If μ is a Borel measure, then its *support* is the set $[\mu] = \mathbb{R}\backslash U$, where U is the union of all open sets of measure zero. Show that $[\mu]$ is a closed set (so $[\mu] \in \mathfrak{R}$), and $\mathbb{R}\backslash[\mu]$ is the largest (in the inclusion ordering) open set of measure zero. Show that a point $\alpha \in \mathbb{R}$ is not in the support of μ if and only if there exists an open subset of measure zero that contains α. Let A be a closed set with $0 < \mu(A) < \infty$, take the σ-algebra $\mathcal{A} = \wp(A) \cap \mathfrak{R}$, and take the restriction $\lambda = \mu|_A\colon \mathcal{A} \to \mathbb{R}$ of μ to $\mathcal{A}$, which is a finite measure. Examples: if μ is a finite measure, then A may be any closed subset of $\mathbb{R}$ of nonzero measure (e.g., $A = \mathbb{R}$ and $\mathcal{A} = \mathfrak{R}$); if μ is the Lebesgue measure, then A may be any closed and bounded *nondegenerate* interval (i.e., one that is not a singleton). Show that the support $[\lambda]$ of λ is the smallest (in the inclusion ordering) closed subset of A such that $\lambda([\lambda]) = \lambda(A)$.

Problem 2.14. Take an arbitrary measurable space $(X, \mathcal{X})$. Prove the assertion: the sum of σ-finite measures on $\mathcal{X}$ is a σ-finite measure on $\mathcal{X}$.

Hint: Suppose μ and λ are σ-finite measure on $\mathcal{X}$. Let $\{E_n\}$ and $\{F_n\}$ be sequences in $\mathcal{X}$ such that $\mu(E_n) < \infty$, $\lambda(F_n) < \infty$, and $\bigcup_n E_n = \bigcup_n F_n = X$. If $\lambda(E_i) < \infty$, then take E_i. If $\lambda(E_k) = \infty$, then take $\{F_{n_j}\}$ such that $E_k = \bigcup_j F_{n_j}$. Show that the collection of all those E_i and $\{F_{n_j}\}$ consists of a countable collection, and so conclude that $(\mu + \lambda)$ is σ-finite.

Problem 2.15. Take a measure space $(X, \mathcal{X}, \mu)$ and let $(X, \overline{\mathcal{X}}, \overline{\mu})$ be a completion of it. Suppose $\overline{f}\colon X \to \overline{\mathbb{R}}$ is $\overline{\mathcal{X}}$-measurable. Show that there is an $\mathcal{X}$-measurable function $f\colon X \to \overline{\mathbb{R}}$ such that $f = \overline{f}$ μ-almost everywhere.

Hint: Let $q \in \mathbb{Q}$ be an arbitrary rational number. Set $\overline{E}_q = \overline{f}^{-1}((q, \infty))$ in $\overline{\mathcal{X}}$, write $\overline{E}_q = E_q \cup A_q$ with $E_q \in \mathcal{X}$ and $A_q \subseteq N_q \in \mathcal{N}$, and set $N = \bigcup N_q$, which is a countable union, so that $N \in \mathcal{N}$ with $A_q \subseteq N$. Consider the function $f\colon X \to \overline{\mathbb{R}}$ such that $f(x) = \overline{f}(x)$ if $x \in X \backslash N$ and $f(x) = 0$ if $x \in N$. Show that $f^{-1}((q, \infty))$ is either in $\overline{E}_q \backslash N$ or in $\overline{E}_q \cup N$, and so $f^{-1}((q, \infty))$ lies in $\mathcal{X}$ for all $q \in \mathbb{Q}$, and hence f is $\mathcal{X}$-measurable (cf. Problem 1.1).

Problem 2.16. Let $\mathcal{X}$ be a σ-algebra of subsets of a set X. A complex-valued set function $\eta\colon \mathcal{X} \to \mathbb{C}$ satisfying axioms (a) and (b) of Definition 2.3, with absolute convergence on the right-hand side of (b), is a *complex measure.* Show that every complex measure η on $\mathcal{X}$ can be expressed as $\eta = \nu_1 + i\nu_2$, where ν_1 and ν_2 are (real-valued) signed measures on $\mathcal{X}$. Complex measures will be considered again in Chapter 10.

Problem 2.17. Show that for a real-valued measure (equivalently, for a finite measure) it is unnecessary to assume condition (a) in Definition 2.1 since this follows from countable additivity in condition (c); and verify that condition (a) cannot be dismissed for extended real-valued measures. Similarly, since a signed measure was defined as a real-valued set function satisfying Definition 2.3, also show that for a signed measure (and so for a complex measure) it is unnecessary to assume condition (a) in Definition 2.3 since this follows from countable additivity in condition (b).

Problem 2.18. Let $(X, \mathcal{X}, \mu)$ be a measure space, let $f\colon X \to \mathbb{R}$ be an $\mathcal{X}$-measurable real-valued function, let $\mathfrak{R}$ be the Borel algebra, and set

$$\mu^f(E) = \mu(f^{-1}(E)) = \mu(\{x \in X\colon f(x) \in E\}) \quad \text{for every} \quad E \in \mathfrak{R}.$$

Prove that this defines a measure μ^f on $\mathfrak{R}$ (see Problem 1.8). If $(X, \mathcal{X}, \mu)$ is a probability space, then show that $(\mathbb{R}, \mathfrak{R}, \mu^f)$ is also a probability space (i.e., μ^f is a probability measure whenever μ is). In this case, the $\mathcal{X}$-measurable real-valued function f is referred to as a real *random variable*.

Problem 2.19. Consider the definition of an atom (preceding Example 2A). Take any measure space $(X, \mathcal{X}, \mu)$. A set $E \in \mathcal{X}$ is *atomic* (or *purely atomic*) if every measurable subset of it is the union of a set of measure zero and a disjoint collection of atoms. A set $E \in \mathcal{X}$ is *atom-free* if it has no atom as a subset. Now take the measure space $(\mathbb{R}, \mathfrak{R}, \delta)$, where δ is any countable sum of Dirac measures on $\mathfrak{R}$ (Example 2A), and verify that every measurable set is purely atomic. Next take the measure space $(\mathbb{R}, \mathfrak{R}, \lambda)$, where λ is the Lebesgue measure on $\mathfrak{R}$, and show that every measurable set is atom-free.

Suggested Reading

Bartle [4], Berberian [7], Brown and Pearcy [8], Halmos [18], Royden [35]. For a discussion on unconditionally convergent series (as in Definition 2.3) see [26, Section 5.7]. For the construction and properties of the Cantor set (as in Problems 2.9 and 2.10) see, for instance, [1], [3], [9], [26], [32], [37].

3

Integral of Nonnegative Functions

3.1 Simple and Nonnegative Functions

Let X be an arbitrary set. A *simple function* on X is a real-valued function $\varphi\colon X \to \mathbb{R}$ with a finite range (i.e., a function that takes on only a finite number of distinct values). It is clear that φ is a simple function if and only if it can be represented as a linear combination of characteristic functions,

$$\varphi = \sum_{i=1}^{n} \alpha_i \chi_{E_i},$$

where $\{E_i\}_{i=1}^n$ is a finite collection of subsets of X and $\{\alpha_i\}_{i=1}^n$ is a finite set of real numbers. The above *representation* is not unique, but it becomes unique if it is assumed that $\{\alpha_i\}_{i=1}^n$ is a set of *distinct* coefficients and $\{E_i\}_{i=1}^n$ is a partition of X (i.e., a collection of *disjoint* sets that cover X). This is the representation of φ for which the set $\{\alpha_i\}_{i=1}^n$ is the range of φ, and for each index $i = 1, \dots, n$ the set E_i is the *inverse image* of the singleton $\{\alpha_i\}$, viz., $E_i = \varphi^{-1}(\{\alpha_i\}) = \{x \in X\colon\ \varphi(x) = \alpha_i\}$. This unique representation is referred to as the *canonical* (or *standard*) *representation* of φ.

Let $\mathcal{X}$ be σ-algebra of subsets of X. Take any $E \subseteq X$. Its characteristic function $\chi_E\colon X \to \{0,1\}$ is $\mathcal{X}$-measurable if and only if the set E is $\mathcal{X}$-measurable (Example 1B). Then a simple function $\sum_{i=1}^n \alpha_i \chi_{E_i}\colon X \to \mathbb{R}$ is measurable if $\{E_i\}_{i=1}^n$ is a collection of measurable sets (Proposition 1.5). The converse fails: take a partition $\{A, B\}$ of X made up of nonmeasurable sets $A, B \in \wp(X)\backslash\mathcal{X}$ and note that $\chi_X = \chi_A + \chi_B$. However, a canonical representation is a measurable function if and only if $\{E_i\}_{i=1}^n$ is a collection

C.S. Kubrusly, *Essentials of Measure Theory*,
DOI 10.1007/978-3-319-22506-7_3

of measurable sets, and a simple function is a *measurable simple function* if and only if its canonical representation is measurable or, equivalently, if it has a representation such that all sets E_i are measurable. Whenever we refer to a measurable simple function, we consider only representations of it for which all sets E_i are measurable. Since sum and (real) scalar multiple of measurable simple functions are again measurable simple functions, the collection of all $\mathcal{X}$-measurable simple functions forms a linear manifold of the real linear space of all $\mathcal{X}$-measurable functions, and so it is a linear space itself (cf. notes that close Section 1.3). A simple function is nonnegative if and only if all coefficients α_i of any representation are nonnegative numbers.

Definition 3.1. (The *integral* of a simple function). Consider a measure space $(X, \mathcal{X}, \mu)$. Let $\varphi: X \to \mathbb{R}$ be a nonnegative measurable simple function,

$$\varphi = \sum_{i=1}^{n} \alpha_i \chi_{E_i}.$$

The *integral* of φ with respect to μ is the nonnegative extended real number

$$\int \varphi \, d\mu = \sum_{i=1}^{n} \alpha_i \, \mu(E_i).$$

It is clear that φ must be measurable, since all E_i must be measurable; otherwise the definition of the integral $\int \varphi \, d\mu$ (with respect to any measure μ on $\mathcal{X}$) would not make sense. In particular, $\int \chi_E \, d\mu = \mu(E)$ for every $E \in \mathcal{X}$. It is readily verified that the integral of a nonnegative measurable simple function is independent of its representation. So the notion of integral of a simple function is unambiguously defined and we may assume the canonical representation of φ without loss of generality. To ensure that the integral of the null function ($\varphi = 0$) is well defined and equal to zero for every measure, including nonfinite measures, we declare again that $0 \cdot +\infty = 0$. The next proposition considers three fundamental properties, which will survive as long as the notion of integral is extended. The first two point out that the integral is nonnegative homogeneous and additive. The third one shows how the integral with respect to a measure yields a new measure. The reader is invited to prove Proposition 3.2 in Problem 3.2.

Proposition 3.2. *Consider any measure space* $(X, \mathcal{X}, \mu)$*. If* φ *and* ψ *are nonnegative measurable simple functions, and* γ *is a nonnegative real number, then* $\gamma\varphi$ *and* $\varphi + \psi$ *are nonnegative measurable simple functions and*

(a) $\int \gamma\varphi\, d\mu = \gamma \int \varphi\, d\mu$,

(b) $\int (\varphi + \psi)\, d\mu = \int \varphi\, d\mu + \int \psi\, d\mu$,

(c) $\lambda(E) = \int \varphi\chi_E\, d\mu$ *for every* $E \in \mathcal{X}$ *defines a measure* $\lambda\colon \mathcal{X} \to \overline{\mathbb{R}}$.

Let $(X, \mathcal{X})$ be a measurable space. In Chapter 1 we adopted the notation $\mathcal{M}(X, \mathcal{X})$, or simply $\mathcal{M}$, for the collection of all $\mathcal{X}$-measurable functions. Similarly, set $\mathcal{M}(X, \mathcal{X})^+$, or simply $\mathcal{M}^+$ if the measurable space is clear in the context, for the collection of all nonnegative functions from $\mathcal{M}(X, \mathcal{X})$:

$$\mathcal{M}^+ = \mathcal{M}(X, \mathcal{X})^+ = \big\{ f\colon X \to \overline{\mathbb{R}}\colon f \text{ is } \mathcal{X}\text{-measurable and } f(x) \geq 0\ \forall x \in X \big\}.$$

Extended real-valued functions are allowed in $\mathcal{M}$ and $\mathcal{M}^+$, but note that these collections also contain real-valued functions. In particular, nonnegative $\mathcal{X}$-measurable simple functions lie in $\mathcal{M}(X, \mathcal{X})^+$. Given an arbitrary (extended real-valued) function f in $\mathcal{M}(X, \mathcal{X})^+$, consider the set Φ_f^+ of all simple functions φ in $\mathcal{M}(X, \mathcal{X})^+$ that are dominated by f,

$$\Phi_f^+ = \big\{ \varphi \in \mathcal{M}\colon \varphi \text{ is simple and } 0 \leq \varphi(x) \leq f(x)\ \forall x \in X \big\} \subseteq \mathcal{M}^+.$$

Definition 3.3. (The *integral* of a nonnegative measurable function). Consider a measure space $(X, \mathcal{X}, \mu)$. The *integral* of a function $f \in \mathcal{M}(X, \mathcal{X})^+$ with respect to μ is the extended real number

$$\int f\, d\mu = \sup_{\varphi \in \Phi_f^+} \int \varphi\, d\mu.$$

The *integral* of f *over* a measurable set E with respect to μ is defined by

$$\int_E f\, d\mu = \int f\chi_E\, d\mu \quad \text{in} \quad \overline{\mathbb{R}}.$$

The function f in the definition of the integral must indeed be measurable (recall that the supremum of measurable functions is measurable; and $f^{-1}(\alpha, \infty) = \{x \in X\colon \alpha < f(x)\} = \{x \in X\colon \alpha < \sup_{\varphi \in \Phi^+_f} \varphi(x)\}$). Moreover, since $\chi_E \in \mathcal{M}(X, \mathcal{X})^+$, we get $f\chi_E \in \mathcal{M}(X, \mathcal{X})^+$, whenever $E \in \mathcal{X}$.

3.2 The Monotone Convergence Theorem

The Monotone Convergence Theorem is a fundamental result in the theory of integration, which is due to Beppo Levi. It is indeed a basic tool for almost everything that follows. A sequence $\{f_n\}$ of functions $f_n\colon X \to \overline{\mathbb{R}}$ is *increasing* if $f_n \leq f_{n+1}$ (i.e., $f_n(x) \leq f_{n+1}(x)$ for every $x \in X$) and *decreasing* if $f_{n+1} \leq f_n$ (i.e., $f_{n+1}(x) \leq f_n(x)$ for every $x \in X$) for each n. If it is either increasing or decreasing, then it is a *monotone* sequence. Observe, according to Section 1.3, that a monotone sequence of *extended* real-valued functions converges pointwise (to an *extended* real-valued function).

Theorem 3.4. (Monotone Convergence Theorem). *Let $(X, \mathcal{X}, \mu)$ be a measure space. If $\{f_n\}$ is an increasing sequence of functions in $\mathcal{M}(X, \mathcal{X})^+$, then it converges pointwise to a function $f: X \to \overline{\mathbb{R}}$ in $\mathcal{M}(X, \mathcal{X})^+$, and*

$$\int f\, d\mu = \lim_n \int f_n\, d\mu.$$

Proof. Recall that $\{f_n\}$ converges pointwise. Let f be its limit. Take any x in X. Since each $f_n(x) \geq 0$, it follows that $f(x) = \lim_n f_n(x) \geq 0$, and so f is in $\mathcal{M}(X, \mathcal{X})^+$ by Proposition 1.8. Since $f_n \leq f_{n+1}$, we get $f_n \leq f_{n+1} \leq f = \lim_n f_n$, and so $\int f_n\, d\mu \leq \int f_{n+1}\, d\mu \leq \int f\, d\mu$ (cf. Problem 3.3), for every n. Then the extended real-valued increasing sequence $\{\int f_n\, d\mu\}$ converges and

$$\lim_n \int f_n\, d\mu \leq \int f\, d\mu. \qquad (*)$$

To verify the reverse inequality, take a simple function φ in $\mathcal{M}(X, \mathcal{X})^+$ such that $0 \leq \varphi \leq f$ (i.e., any $\varphi \in \Phi_f^+$). Let α be any real number in $(0, 1)$ and set $\psi = \alpha\varphi$, which is a simple function in $\mathcal{M}(X, \mathcal{X})^+$ with the property that $\psi(x) = 0$ if $f(x) = 0$ and $0 \leq \psi(x) < f(x)$ if $f(x) \neq 0$. For every n set

$$E_n = \{x \in X \colon \psi(x) \leq f_n(x)\} = \{x \in X \colon f_n(x) - \psi(x) \geq 0\}.$$

Since f_n and ψ are measurable functions, $f_n - \psi$ is measurable (Proposition 1.9), and so each E_n is a measurable set. Thus, for every n,

$$\alpha \int_{E_n} \varphi\, d\mu = \int_{E_n} \psi\, d\mu \leq \int_{E_n} f_n\, d\mu \leq \int f_n\, d\mu$$

(Proposition 3.2(a,c)). Since $\{f_n\}$ is increasing, it follows that $\{E_n\}$ is increasing. Since $f_n \nearrow f$ (i.e., $\{f_n\}$ is increasing and converges to f) and $0 \leq \psi(x) < f(x)$ if $f(x) \neq 0$, it follows that for every $x \in X$ there is an m for which $\psi(x) \leq f_m(x) \leq f(x)$, and so x lies in E_m. Hence $X = \bigcup_n E_n$ so that $\{E_n\}$ is an increasing sequence of sets in $\mathcal{X}$ that cover X. Then

$$\int \varphi\, d\mu = \lambda(X) = \lim_n \lambda(E_n) = \lim_n \int_{E_n} \varphi\, d\mu$$

according to Proposition and 2.2(c), where λ is the measure of Proposition 3.2(c). So, by the previous two displayed expressions,

$$\alpha \int \varphi\, d\mu \leq \lim_n \int f_n\, d\mu,$$

which implies that

$$\int \varphi \, d\mu = \sup_{\alpha \in (0,1)} \alpha \int \varphi \, d\mu \le \lim_n \int f_n \, d\mu,$$

and therefore

$$\int f \, d\mu = \sup_{\varphi \in \Phi_f^+} \int \varphi \, d\mu \le \lim_n \int f_n \, d\mu. \qquad (**)$$

The inequalities $(*)$ and $(**)$ ensure the identity: $\int f \, d\mu = \lim_n \int f_n \, d\mu$. $\square$

The Monotone Convergence Theorem gives us the first evidence of continuity for the integral transformation. Theorem 3.4 will also give us the first hint of linearity (in Proposition 3.5(a,b) below). Of course, all this would make sense only if the domain $\mathcal{M}^+$ and codomain $\overline{\mathbb{R}}$ of the transformation

$$\int (\cdot) \, d\mu \colon \mathcal{M}^+ \to \overline{\mathbb{R}}$$

(that assigns to each function f in $\mathcal{M}^+$ the extended real number $\int f d\mu$) might be equipped with a suitable algebraic and topological structure. We will modify the domain $\mathcal{M}^+$ in order to endow the new domain with the proper algebraic structure in Chapter 4 (that makes the integral transformation a linear one), and with a proper topological structure in Chapter 5 (that makes the integral transformation a continuous one as well).

As a first application of the Monotone Convergence Theorem (among many to come) we extend in Proposition 3.5 below the results of Proposition 3.2, from measurable nonnegative simple functions to arbitrary measurable nonnegative (extended real-valued) functions. In particular, Theorem 3.4 shows (as stated in Proposition 3.5(c)) how the integral of a *nonnegative* function with respect to a measure yields a new measure, viz., the set function $\lambda \colon \mathcal{X} \to \overline{\mathbb{R}}$ given for each set $E \in \mathcal{X}$ by $\lambda(E) = \int_E f \, d\mu$.

Proposition 3.5. *Consider a measure space $(X, \mathcal{X}, \mu)$. If f and g are functions in $\mathcal{M}(X, \mathcal{X})^+$, and if γ is a nonnegative real number, then the functions γf and $f + g$ lie in $\mathcal{M}(X, \mathcal{X})^+$ and*

(a) $\int \gamma f \, d\mu = \gamma \int f \, d\mu$,

(b) $\int (f + g) \, d\mu = \int f \, d\mu + \int g \, d\mu$,

(c) $\lambda(E) = \int_E f \, d\mu$ *for every $E \in \mathcal{X}$ defines a measure $\lambda \colon \mathcal{X} \to \overline{\mathbb{R}}$.*

Proof. If f and g are in $\mathcal{M}(X, \mathcal{X})^+$, then it follows by Proposition 1.9 that γf and $f + g$ also are in $\mathcal{M}(X, \mathcal{X})^+$ and, according to Problem 1.6, there are increasing sequences $\{\varphi_n\}$ and $\{\psi_n\}$ of simple functions in $\mathcal{M}(X, \mathcal{X})^+$ for which $f = \lim_n \varphi_n$ and $g = \lim_n \psi_n$.

(a) Take $\gamma \geq 0$ so that $\{\gamma\varphi_n\}$ is an increasing sequence of simple functions in $\mathcal{M}(X, \mathcal{X})^+$ that converges to γf. Thus Proposition 3.2(a) and Theorem 3.4 (the Monotone Convergence Theorem) ensure that

$$\int \gamma f \, d\mu = \lim_n \int \gamma\varphi_n \, d\mu = \gamma \lim_n \int \varphi_n \, d\mu = \gamma \int f \, d\mu.$$

(b) Note that $\{\varphi_n + \psi_n\}$ is an increasing sequence of simple functions in $\mathcal{M}(X, \mathcal{X})^+$ that converges to $f + g$. Thus, again, Proposition 3.2(b) and the Monotone Convergence Theorem ensure that

$$\begin{aligned}\int (f+g) \, d\mu &= \lim_n \int (\varphi_n + \psi_n) \, d\mu = \lim_n \left(\int \varphi_n \, d\mu + \int \psi_n \, d\mu \right) \\ &= \lim_n \int \varphi_n \, d\mu + \lim_n \int \psi_n \, d\mu = \int f \, d\mu + \int g \, d\mu.\end{aligned}$$

(c) Observe that $\lambda(\varnothing) = 0$ since $\chi_\varnothing = 0$ so that $\int_\varnothing f \, d\mu = 0$, and $\lambda(E) \geq 0$ for all E in $\mathcal{X}$ by the definition of the integral of f in $\mathcal{M}(X, \mathcal{X})^+$. To verify countable additivity (Definition 2.1(c)), take any sequence $\{E_n\}$ of pairwise disjoint sets in $\mathcal{X}$. Since $\sum_{n=1}^m f\chi_{E_n} = f\chi_{\bigcup_{n=1}^m E_n}$, set for every integer $m \geq 1$

$$f_m = \sum_{n=1}^m f\chi_{E_n} = f\chi_{\bigcup_{n=1}^m E_n}.$$

Now observe that $\{f_m\}$ is an increasing sequence of functions in $\mathcal{M}(X, \mathcal{X})^+$ (Proposition 1.9) that converges pointwise to the function $f\chi_E$, with $E = \bigcup_n E_n$ in $\mathcal{X}$. Then the Monotone Convergence Theorem ensures that

$$\lambda\Big(\bigcup_n E_n\Big) = \int_E f \, d\mu = \int f\chi_E \, d\mu = \lim_m \int f_m \, d\mu = \lim_m \int \sum_{n=1}^m f\chi_{E_n} \, d\mu.$$

Thus a trivial induction, using additivity as in item (b), ensures that the integral of a finite sum coincides with the finite sum of each integral, so

$$\lim_m \int \sum_{n=1}^m f\chi_{E_n} \, d\mu = \lim_m \sum_{n=1}^m \int f\chi_{E_n} \, d\mu,$$

which completes the proof of (c). That is, $\lambda\big(\bigcup_n E_n\big) = \sum_n \lambda(E_n)$. In fact,

$$\lim_m \sum_{n=1}^m \int f\chi_{E_n} \, d\mu = \lim_m \sum_{n=1}^m \int_{E_n} f \, d\mu = \lim_m \sum_{n=1}^m \lambda(E_n) = \sum_n \lambda(E_n). \quad \square$$

Remarks: Theorem 3.4 deals with functions that are possibly extended real-valued and with measures that are not necessarily finite (not even σ-finiteness is assumed). Thus infinite integrals and infinite limits are allowed. For example, if λ is the Lebesgue measure on $\Re$ and $f_n = \chi_{[0,n)}$ for each integer $n \geq 1$, then $\{f_n\}$ is an increasing sequence of functions in $\mathcal{M}(\mathbb{R},\Re)^+$ with finite integral ($\int f_n \, d\lambda = n$ for each n) converging pointwise to the function $f = \chi_{[0,\infty)}$ that has an infinite integral. However, the real-valued sequence $\{\int f_n \, d\lambda\}$ is unbounded (and so it does not converge in $\mathbb{R}$) but it has the limit $+\infty$ in $\overline{\mathbb{R}}$. Hence, $\int f \, d\lambda = \lim_n \int f_n \, d\lambda = +\infty$. We will see in the next section that the Monotone Convergence Theorem still holds if pointwise convergence is weakened to almost everywhere convergence. Moreover, it holds without monotonicity by assuming convergence from below. This is Corollary 3.10. (Convergence from below can be dismissed if we impose uniform convergence and *finite* measure — see Problems 3.6 and 3.12.)

3.3 Monotone Convergence Corollaries

The next result, Fatou's Lemma, can be viewed as an important consequence of the Monotone Convergence Theorem, which will be applied to prove an extension of the Monotone Convergence Theorem that does not require monotonicity (as in the forthcoming Corollary 3.10). Recall from Chapter 1 that $\underline{f} = \liminf_n f_n$ is a measurable function for any sequence $\{f_n\}$ of extended real-valued measurable functions (Proposition 1.8), and that if $\{f_n\}$ converges pointwise to a function f (i.e., if $f = \lim_n f_n$), then $f = \underline{f}$.

Lemma 3.6. (Fatou's Lemma). *Take a measure space* $(X, \mathcal{X}, \mu)$. *If* $\{f_n\}$ *is a sequence of functions in* $\mathcal{M}(X,\mathcal{X})^+$, *then*

$$\int \underline{f} \, d\mu \leq \liminf_n \int f_n \, d\mu.$$

Proof. Let $\{f_n\}$ be a sequence of functions in $\mathcal{M}(X,\mathcal{X})^+$. For every n set

$$\phi_n = \inf_{n \leq k} f_k,$$

meaning that each $\phi_n \colon X \to \overline{\mathbb{R}}$ is a function defined by $\phi_n(x) = \inf_{n\leq k} f_k(x)$ for every $x \in X$. Proposition 1.8 ensures that each ϕ_n is measurable, and so each ϕ_n lies in $\mathcal{M}(X,\mathcal{X})^+$. By definition $\{\phi_n\}$ is an increasing sequence, and it converges pointwise to $\underline{f}$. Indeed,

$$\lim_n \phi_n(x) = \lim_n \inf_{n\leq k} f_k(x) = \liminf_n f_n(x) = \underline{f}(x)$$

for each $x \in X$. Then the Monotone Convergence Theorem ensures that

$$\int \underline{f}\, d\mu = \lim_n \int \phi_n\, d\mu.$$

But $\phi_n \leq f_k$ for every $k \geq n$, which implies that $\int \phi_n\, d\mu \leq \int f_k\, d\mu$ for every $k \geq n$, and so $\int \phi_n\, d\mu \leq \inf_{n\leq k} \int f_k\, d\mu$. Hence

$$\lim_n \int \phi_n\, d\mu \leq \lim_n \inf_{n\leq k} \int f_k\, d\mu = \liminf_n \int f_n\, d\mu. \qquad \square$$

Item (a) in Proposition 3.7 below is an application of Fatou's Lemma. It gives the first hint of what will make the basis for defining a new concept of equality in the spaces L^p (based on the notion of equivalence classes as introduced in Chapter 5), which reads as follows: if $f \in \mathcal{M}(X, \mathcal{X})^+$, then

$$f = 0 \quad \mu\text{-a.e.} \qquad \Longleftrightarrow \qquad \int f\, d\mu = 0.$$

Proposition 3.7. *Take a measure space* $(X, \mathcal{X}, \mu)$. *If* f *is a function in* $\mathcal{M}(X, \mathcal{X})^+$ *and* λ *is the measure on* $\mathcal{X}$ *defined in* Proposition 3.5(c), *then*

(a) $f = 0$ *μ-almost everywhere if and only if* $\int f\, d\mu = 0$,

(b) $\lambda(E) = 0$ *for every* $E \in \mathcal{X}$ *such that* $\mu(E) = 0$.

Proof. Take a measure space $(X, \mathcal{X}, \mu)$ and let f be a function in $\mathcal{M}(X, \mathcal{X})^+$.

(a) Since f is a measurable function, the set

$$E_n = \left\{x \in X \colon \tfrac{1}{n} < f(x)\right\}$$

is a measurable one such that $\frac{1}{n}\chi_{E_n} \leq f$, and so (cf. Proposition 3.5(a))

$$0 \leq \tfrac{1}{n}\,\mu(E_n) = \tfrac{1}{n}\int_{E_n} d\mu = \int \tfrac{1}{n}\chi_{E_n}\, d\mu \leq \int f\, d\mu$$

for every n. If $\int f\, d\mu = 0$, then $\mu(E_n) = 0$ for all n, and hence (since $0 \leq f$)

$$\mu(\{x \in X \colon f(x) \neq 0\}) = \mu(\{x \in X \colon 0 < f(x)\}) = \mu\Big(\bigcup_n E_n\Big) = 0.$$

This means that $f = 0$ μ-almost everywhere. Conversely, set $E = \bigcup_n E_n$ in $\mathcal{X}$, and set $f_n = n\,\chi_E$ for each n. It is clear that $\{f_n\}$ is a sequence of functions in $\mathcal{M}(X, \mathcal{X})^+$ converging pointwise to $f_\infty = +\infty\,\chi_E \colon X \to \overline{\mathbb{R}}$, and so $\underline{f} = \liminf_n f_n = \lim_n f_n = f_\infty$. Also, $0 \leq f \leq f_\infty = \underline{f}$. Thus

$$0 \leq \int f\, d\mu \leq \int \underline{f}\, d\mu \leq \liminf_n \int f_n\, d\mu$$

by Problem 3.3 and Lemma 3.6 (Fatou's Lemma). If $f=0$ μ-almost everywhere, then $\mu(E)=0$ so that $\int f_n\,d\mu = n\int \chi_E\,d\mu = n\mu(E) = 0$ for all n. Hence $\liminf_n \int f_n\,d\mu = 0$, and so $\int f\,d\mu = 0$ by the preceding inequality.

(b) If $\mu(E)=0$ for some $E\in\mathcal{X}$, and since $\int \chi_E\,d\mu = \mu(E)$, it follows by item (a) that $\chi_E = 0$, and so $f\chi_E = 0$, μ-almost everywhere. Let $\lambda\colon \mathcal{X}\to\overline{\mathbb{R}}$ be the measure of Proposition 3.5(c). Another application of item (a) yields

$$\lambda(E) = \int_E f\,d\mu = \int f\chi_E\,d\mu = 0. \qquad \square$$

The implication $\{\mu(E)=0 \Longrightarrow \lambda(E)=0\}$ in Proposition 3.7(b) is referred to by saying that the measure λ is *absolutely continuous* with respect to the measure μ. (Absolutely continuity will be discussed in Chapter 7.) Thus, by Propositions 3.5(c) and 3.7(b), *if μ is a measure on $\mathcal{X}$ and f is a function in $\mathcal{M}(X,\mathcal{X})^+$, then the set function λ on $\mathcal{X}$ defined for each $E\in\mathcal{X}$ by*

$$\lambda(E) = \int_E f\,d\mu$$

is a measure that is absolutely continuous with respect to μ. This is another important consequence of the Monotone Convergence Theorem. A crucial result of Chapter 7 (the Radon–Nikodým Theorem) asserts the converse: *if μ and λ are σ-finite measures and λ is absolutely continuous with respect to μ, then there exists a function f in $\mathcal{M}(X,\mathcal{X})^+$ such that for each $E\in\mathcal{X}$,*

$$\lambda(E) = \int_E f\,d\mu.$$

Corollary 3.8. *Take a measure space $(X,\mathcal{X},\mu)$. If $\{f_n\}$ is an increasing sequence of functions in $\mathcal{M}(X,\mathcal{X})^+$ that converges almost everywhere to a function $f\colon X\to\overline{\mathbb{R}}$ in $\mathcal{M}(X,\mathcal{X})^+$, then*

$$\int f\,d\mu = \lim_n \int f_n\,d\mu.$$

Proof. Suppose $\{f_n\}$ converges μ-almost everywhere to f, so that $\{f_n\}$ converges pointwise to f on $E = X\backslash N$ for some $N\in\mathcal{X}$ with $\mu(N)=0$. Thus $\{f_n\chi_E\}$ is an increasing sequence of functions in $\mathcal{M}(X,\mathcal{X})^+$ converging pointwise to $f\chi_E\in\mathcal{M}(X,\mathcal{X})^+$. By the Monotone Convergence Theorem,

$$\int f\chi_E\,d\mu = \lim_n \int f_n\chi_E\,d\mu.$$

However, Propositions 3.5(b) and 3.7(a) ensure (see Problem 3.8) that

$$\int f\chi_E \, d\mu = \int_E f \, d\mu = \int f \, d\mu \quad \text{and} \quad \int f_n\chi_E \, d\mu = \int_E f_n \, d\mu = \int f_n \, d\mu. \quad \square$$

Corollary 3.8 is Theorem 3.4 for almost everywhere convergence, which will also be referred to as the Monotone Convergence Theorem. It leads to the following version of Lemma 3.6 for almost everywhere convergence, again referred to as Fatou's Lemma.

Lemma 3.9. *Take a measure space* $(X, \mathcal{X}, \mu)$*. If* $\{f_n\}$ *is a sequence in* $\mathcal{M}(X, \mathcal{X})^+$ *that converges almost everywhere to* $f \in \mathcal{M}(X, \mathcal{X})^+$*, then*

$$\int f \, d\mu \le \liminf_n \int f_n \, d\mu.$$

Proof. Consider the setup of the previous proof, where the sequence $\{f_n\chi_E\}$ converges pointwise to $f\chi_E \in \mathcal{M}(X, \mathcal{X})^+$. By Lemma 3.6 (Fatou's Lemma),

$$\int f\chi_E \, d\mu \le \liminf_n \int f_n\chi_E \, d\mu.$$

Again, Propositions 3.5(b) and 3.7(a) ensure that (cf. Problem 3.8)

$$\int f\chi_E \, d\mu = \int f \, d\mu \quad \text{and} \quad \int f_n\chi_E \, d\mu = \int f_n \, d\mu. \qquad \square$$

A sequence $\{f_n\}$ *converges from below* to f if $f_n \le f$ for all n and if it converges to f in some sense. The following extension in Corollary 3.10 yields an ultimate version of the Monotone Convergence Theorem that assumes just almost everywhere convergence from below.

Corollary 3.10. *Let* $(X, \mathcal{X}, \mu)$ *be a measure space. If a sequence* $\{f_n\}$ *in* $\mathcal{M}(X, \mathcal{X})^+$ *converges almost everywhere to* $f \in \mathcal{M}(X, \mathcal{X})^+$ *from below, then*

$$\int f \, d\mu = \lim_n \int f_n \, d\mu.$$

Proof. Since $f_n \le f$, we get $\int f_n \, d\mu \le \int f \, d\mu$ for all n. Thus, by Lemma 3.9,

$$\int f \, d\mu \le \liminf_n \int f_n \, d\mu \le \limsup_n \int f_n \, d\mu \le \int f \, d\mu. \qquad \square$$

3.4 Problems

Problem 3.1. Consider a measurable space $(X, \mathcal{X})$. Thus, in this context, "measurable" means $\mathcal{X}$-measurable. Prove that sum, scalar multiplication, and product of measurable simple functions are measurable simple functions.

Next conclude that every polynomial $p(\varphi, \psi)$ of measurable simple functions φ and ψ is a measurable simple function. Moreover, also show that $\phi = \varphi \wedge \psi = \min\{\varphi, \psi\}$ and $\theta = \varphi \vee \psi = \max\{\varphi, \psi\}$ also are measurable simple functions. (*Hint:* Proposition 1.5 and Problem 1.3.)

Problem 3.2. Show that the definition of integral of a nonnegative measurable simple function does not depend on the representation for the simple function. Then, prove Proposition 3.2.

Hint: Verify homogeneity. To prove additivity proceed as follows. Show that

$$\varphi + \psi = \sum\nolimits_i \sum\nolimits_j (\alpha_i + \beta_j) \chi_{E_i \cap F_j},$$

where $\varphi = \sum_i \alpha_i \chi_{E_i}$ and $\psi = \sum_j \beta_j \chi_{F_j}$ are canonical representations. So

$$\begin{aligned}\int (\varphi + \psi)\, d\mu &= \sum\nolimits_i \sum\nolimits_j (\alpha_i + \beta_j)\, \mu(E_i \cap F_j)\\ &= \sum\nolimits_i \alpha_i \sum\nolimits_j \mu(E_i \cap F_j) + \sum\nolimits_j \beta_j \sum\nolimits_i \mu(E_i \cap F_j)\\ &= \sum\nolimits_i \alpha_i\, \mu(E_i) + \sum\nolimits_j \beta_j\, \mu(F_j) = \int \varphi\, d\mu + \int \psi\, d\mu\end{aligned}$$

since $\mu(E_i) = \sum_j \mu(E_i \cap F_j)$ and $\mu(F_j) = \sum_i \mu(E_i \cap F_j)$ because $\{E_i\}$ and $\{F_j\}$ are partitions of X. To prove (c) note that $\varphi \chi_E = \sum_i \alpha_i \chi_{E_i \cap E}$. Hence

$$\lambda(E) = \int \varphi \chi_E \, d\mu = \sum\nolimits_i \alpha_i\, \mu(E_i \cap E) = \sum\nolimits_i \alpha_i\, \mu_{E_i}(E) \quad \text{for every} \quad E \in \mathcal{X},$$

a (finite) linear combination with nonnegative coefficients α_i of measures μ_{E_i} on $\mathcal{X}$ (Problem 2.11), and so it is itself a measure on $\mathcal{X}$.

Problem 3.3. Let μ and λ be measures on a σ-algebra $\mathcal{X}$ of subsets of a set X, let E and F be sets in $\mathcal{X}$, and let f and g be functions in $\mathcal{M}(X, \mathcal{X})^+$. Recall that $\int_E f\, d\mu = \int f \chi_E \, d\mu$; in particular, $\int f\, d\mu = \int f \chi_X \, d\mu = \int_X f\, d\mu$ (cf. Definition 3.3). Prove the following assertions.

(a) $0 \le \int_E d\mu = \int \chi_E \, d\mu = \mu(E)$,

(b) $f \le g \quad \Longrightarrow \quad 0 \le \int_E f\, d\mu \le \int_E g\, d\mu$,

(c) $E \subseteq F \quad \Longrightarrow \quad 0 \le \int_E f\, d\mu \le \int_F f\, d\mu$,

(d) $\mu \le \lambda \quad \Longrightarrow \quad 0 \le \int_E f\, d\mu \le \int_E f\, d\lambda$.

Problem 3.4. Let $\mathbb{N}$ be the set of all natural numbers (i.e., of all positive integers), and consider the measurable space $(\mathbb{N}, \wp(\mathbb{N}))$.

(a) Verify that every nonnegative function $f\colon \mathbb{N} \to \overline{\mathbb{R}}$ lies in $\mathcal{M}(\mathbb{N}, \wp(\mathbb{N}))^+$.

Let μ be the counting measure of Example 2B. Apply Definition 3.3 to show that for every nonnegative function $f\colon \mathbb{N} \to \overline{\mathbb{R}}$,

(b) $\int_E f\,d\mu = \sum_{n\in E} f(n)$ for every $E \in \wp(\mathbb{N})$. Thus $\int f\,d\mu = \sum_{n=1}^{\infty} f(n)$.

Problem 3.5. Let x be an arbitrary point in $\mathbb{R}$, consider the Borel algebra $\mathfrak{R}$ of subsets of $\mathbb{R}$, and take the Dirac measure $\delta_x\colon \mathfrak{R} \to \mathbb{R}$ at x of Example 2A. If $f \in \mathcal{M}(\mathbb{R}, \mathfrak{R})^+$, then use Definition 3.3 to prove the following statements.

(a) $\int_E f\,d\delta_x = f(x)$ if $x \in E \in \mathfrak{R}$ and $\int_E f\,d\delta_x = 0$ if $x \notin E \in \mathfrak{R}$.

(b) $\int f\,d\delta_x = f(x)$ for every $x \in \mathbb{R}$.

If $\mu = \sum_{n=1}^{m} \alpha_n \delta_n$ with each $\alpha_n \geq 0$, then μ is a measure on $\mathfrak{R}$ for which

(c) $\int f\,d\mu = \sum_{n=1}^{m} \alpha_n \int f\,d\delta_n = \sum_{n=1}^{m} \alpha_n f(n)$.

We will return to Dirac measures in Problem 7.15(a).

Problem 3.6. A sequence $\{f_n\}$ of real-valued functions on a set X *converges uniformly* to a real-valued function f on X if, for each $\varepsilon > 0$, there is a positive integer n_ε such that $\sup_{x\in X} |f_n(x) - f(x)| < \varepsilon$ for all $n \geq n_\varepsilon$. Use Problem 3.3 and Proposition 3.5 to prove the assertion.

∘ If $(X, \mathcal{X}, \mu)$ is a *finite* measure space and $\{f_n\}$ is a sequence of real-valued functions in $\mathcal{M}(X, \mathcal{X})^+$ that converges uniformly to a real-valued function f, then f lies in $\mathcal{M}(X, \mathcal{X})^+$ and

$$\int f\,d\mu = \lim_n \int f_n\,d\mu.$$

Hint: For $k \geq 1$, set $\varepsilon = \frac{1}{k}$, and take $n \geq n_\varepsilon$. Set $E_k = \{x \in X : \frac{1}{k} \leq f(x)\}$ in $\mathcal{X}$. Uniform convergence implies $(f - \frac{1}{k})\chi_{E_k} \leq f_n \leq f + \frac{1}{k}$. Hence,

$$\int (f - \tfrac{1}{k})\chi_{E_k}\,d\mu \leq \liminf_n \int f_n\,d\mu \leq \limsup_n \int f_n\,d\mu \leq \int f\,d\mu + \tfrac{1}{k}\mu(X).$$

Show that $\int (f - \frac{1}{k})\chi_{E_k}\,d\mu = \int f\chi_{E_k}\,d\mu - \frac{1}{k}\mu(E_k)$ because $\mu(X) < \infty$. Since $\int f\chi_{E_k}\,d\mu = \lambda(E_k)$, where λ is the measure of Proposition 3.5(c), and since $\{E_k\}$ is increasing, verify that $\lim_k \int f\chi_{E_k}\,d\mu = \lambda\big(\bigcup_k E_k\big) = \lambda(X) = \int f\,d\mu$ (see Proposition 2.2). Since $\mu(X) < \infty$, it follows that $\lim_k \frac{1}{k}\mu(E_k) = 0$.

Problem 3.7. Prove the *Beppo Levi Theorem*, which says that if $(X, \mathcal{X}, \mu)$ is a measure space and $\{f_n\}$ is a sequence of functions in $\mathcal{M}(X, \mathcal{X})^+$, then

$$\int \sum_{n=1}^{m} f_n\,d\mu = \sum_{n=1}^{m} \int f_n\,d\mu$$

for each positive integer m. (*Hint:* Use Proposition 3.5(b).) Now use the Monotone Convergence Theorem (Theorem 3.4) to show that

$$\int \sum\nolimits_n f_n \, d\mu = \sum\nolimits_n \int f_n \, d\mu.$$

Problem 3.8. Consider a measure space $(X, \mathcal{X}, \mu)$ and take two functions $f, g \in \mathcal{M}(X, \mathcal{X})^+$. Let $\{E, F\}$ be a pair measurable partition of X. Apply Proposition 3.5(b) to show that

(a) $\int f \, d\mu = \int_E f \, d\mu + \int_F f \, d\mu$.

Let N be a set in $\mathcal{X}$. Use Problem 3.3(a) and Proposition 3.7(a) to show that the following propositions hold true.

(b) $\mu(N) = 0$ implies $\int_N f \, d\mu = 0$,

(c) $E = X \backslash N$ and $\mu(N) = 0$ imply $\int_E f \, d\mu = \int f \, d\mu$.

If the integrals are finite, apply Propositions 3.5(b) and 3.7(a) to show that

(d) $\int_E f \, d\mu = \int_E g \, d\mu$ for every $E \in \mathcal{X}$ implies $f = g$ μ-almost everywhere.

Hint: Set $A = \{x \in X : f(x) < g(x)\}$, $B = \{x \in X : f(x) > g(x)\}$, and $C = \{x \in X : f(x) = g(x)\}$. Thus $\{A, B, C\}$ is a measurable partition of X. Since $\int_A f \, d\mu = \int_A g \, d\mu$, and since $(g - f)\chi_A \in \mathcal{M}(X, \mathcal{X})^+$, verify that $\int_A (g - f) \, d\mu = 0$, and so $(g - f) = 0$ μ-a.e. on A. Similarly, $(f - g) = 0$ μ-a.e. on B. Moreover, $f = g$ on C trivially.

Problem 3.9. Consider a measure space $(X, \mathcal{X}, \mu)$. A (measurable) set E in $\mathcal{X}$ is *σ-finite* (with respect to the measure μ) if there exists a countable covering of E made up of measurable sets of finite measure (i.e., $E \subseteq \bigcup_n E_n$ with $\mu(E_n) < \infty$ for all n). Now take $f \in \mathcal{M}(X, \mathcal{X})^+$ and use Problem 3.3 and Proposition 3.5 to show that if $\int f \, d\mu < \infty$, then

(a) $\mu(\{x \in X : f(x) \geq \varepsilon\}) < \infty$ for each $\varepsilon > 0$,

(b) $\mu(\{x \in X : f(x) = +\infty\}) = 0$,

(c) $\{x \in X : f(x) \neq 0\}$ is a σ-finite set.

Hints: (a) If $F_\varepsilon = \{x \in X : \varepsilon \leq f(x)\}$, then $\varepsilon \chi_{F_\varepsilon} \leq f$. (b) Use Proposition 2.2(d) to $\{F_n\}$. (c) If $E_n = \{x \in X : \frac{1}{n} \leq f(x)\}$, then $\frac{1}{n} \chi_{E_n} \leq f$.

Problem 3.10. Let $(X, \mathcal{X}, \mu)$ be a measure space, take an arbitrary function f in $\mathcal{M}(X, \mathcal{X})^+$, and prove the following assertion.

∘ If $\int f \, d\mu < \infty$, then for every $\varepsilon > 0$ there exists a set $E_\varepsilon \in \mathcal{X}$ such that $\mu(E_\varepsilon) < \infty$ and $\int f \, d\mu \leq \int_{E_\varepsilon} f \, d\mu + \varepsilon$.

Hint: Set $E = \{x \in X\colon f(x) \neq 0\}$. Show that $\int f\,d\mu = \int_E f\,d\mu$ (Problem 3.8(a)). Set $E_n = \{x \in X\colon \frac{1}{n} \leq f(x)\}$. Show that $\{E_n\}$ is increasing, that $E = \bigcup_n E_n$, and that $\mu(E_n) = n\int f\,d\mu$ (Problem 3.9(c)). Use Theorem 3.4 to verify that $\lim_n \int f\chi_{E_n} d\mu = \int f\,d\mu$. Conclude: for each $\varepsilon > 0$ there is an n_ε for which, with $E_\varepsilon = E_{n_\varepsilon}$, it follows that $\int f\,d\mu - \int f\chi_{E_\varepsilon} d\mu < \varepsilon$.

Problem 3.11. Consider a measure space $(X, \mathcal{X}, \mu)$, take f in $\mathcal{M}(X,\mathcal{X})^+$, and let λ be the measure on $\mathcal{X}$ defined by

$$\lambda(E) = \int_E f\,d\mu \quad \text{for every} \quad E \in \mathcal{X} \tag{$*$}$$

as in Proposition 3.5(c). Show that if $(*)$ holds, then

$$\int g\,d\lambda = \int gf\,d\mu. \quad \text{for every} \quad g \in \mathcal{M}(X,\mathcal{X})^+.$$

This identity is sometimes abbreviated by writing

$$d\lambda = f\,d\mu,$$

where no independent meaning is assigned to the symbols $d\lambda$ and $d\mu$. In this case, the function f in $(*)$ is sometimes denoted by

$$f = \frac{d\lambda}{d\mu},$$

which again is mere notation (with no independent meaning). We will return to this point in Section 7.2.

Hint: If $\varphi \in \Phi_g^+$, that is, if $\varphi = \sum_{i=1}^n \alpha_i \chi_{E_i}$ is a measurable simple function such that $0 \leq \varphi \leq g$, then verify that $\int \varphi\,d\lambda = \int \varphi f\,d\mu = \sum_{i=1}^n \alpha_i \int_{E_i} f\,d\mu$. Now use Problem 1.6 and apply Corollary 3.10.

Problem 3.12. Monotone (increasing) convergence in Theorem 3.4 was weakened to convergence from below in Corollary 3.10, and such a version of the Monotone Convergence Theorem cannot be improved further (even under the assumption of uniform convergence — cf. Problem 3.6, which requires finite measure). In fact, let $(\mathbb{R}, \mathfrak{R}, \lambda)$ be the *Lebesgue measure space*, take the functions $f_n = \frac{1}{n}\chi_{[n,\infty)}$ and $g_n = \frac{1}{n}\chi_{[0,n]}$ for each positive integer n, and set $f = g = 0$, which are all functions in $\mathcal{M}(\mathbb{R},\mathfrak{R})^+$. Now show that

(a) $\{f_n\}$ decreases and converges uniformly to f but $\int f_n\,d\lambda = +\infty$ for all n, and so $0 = \int f\,d\lambda \neq \lim_n \int f_n\,d\lambda = +\infty$;

(b) $g \leq g_n$ for all n and $\{g_n\}$, which is not monotone, converges uniformly to g but $\int g_n\,d\lambda = 1$ for all n, and so $0 = \int g\,d\lambda \neq \lim_n \int g_n\,d\lambda = 1$.

Problem 3.13. Take a measure space $(X, \mathcal{X}, \mu)$. Let $\{f_n\}$ be a sequence of functions in $\mathcal{M}(X, \mathcal{X})^+$ converging pointwise to $f \in \mathcal{M}(X, \mathcal{X})^+$. If $\int f\, d\mu = \lim_n \int f_n\, d\mu < \infty$, then show that for every measurable set $E \in \mathcal{X}$

$$\int_E f\, d\mu = \lim_n \int_E f_n\, d\mu,$$

and verify that this may fail without the assumption $\lim_n \int f_n\, d\mu < \infty$.

Suggested Reading

Bartle [4], Berberian [7], Halmos [18], Royden [35], Rudin [36] (see also [2]).

4

Integral of Real-Valued Functions

4.1 Integrable Functions

A real-valued function $f: X \to \mathbb{R}$ on X can be expressed as $f = f^+ - f^-$, where the nonnegative functions $f^+: X \to \mathbb{R}$ and $f^-: X \to \mathbb{R}$ are the positive and negative parts of f. If f is measurable, then so are f^+ and f^- (Proposition 1.6). Integration of measurable real-valued functions, leading to real-valued integrals, are considered by using the above decomposition.

Consider a measure space $(X, \mathcal{X}, \mu)$, and let $\mathcal{L}(X, \mathcal{X}, \mu)$ — or simply $\mathcal{L}$ if the measure space is clear in the context — be the collection of all *real-valued* $\mathcal{X}$-measurable functions such that *both* positive and negative parts have a *finite* integral with respect to the measure μ. That is,

$$\mathcal{L} = \mathcal{L}(X, \mathcal{X}, \mu) = \big\{f: X \to \mathbb{R}\colon f \in \mathcal{M}(X, \mathcal{X}),\ \textstyle\int f^+ d\mu < \infty,\ \int f^- d\mu < \infty\big\}.$$

Definition 4.1. (The *integral* of a real-valued measurable function). Take a measure space $(X, \mathcal{X}, \mu)$. The *integral* of a function $f \in \mathcal{L}(X, \mathcal{X}, \mu)$ with respect to μ is the real number

$$\int f\, d\mu = \int f^+ d\mu - \int f^- d\mu.$$

C.S. Kubrusly, *Essentials of Measure Theory*,
DOI 10.1007/978-3-319-22506-7_4

The *integral* of f *over* a measurable set E with respect to μ is defined by

$$\int_E f\,d\mu = \int_E f^+ d\mu - \int_E f^- d\mu.$$

A function in $\mathcal{L}(X, \mathcal{X}, \mu)$ is called an *integrable* (or *μ-integrable*) *function*.

The integral $\int f\,d\mu$ of a *real-valued* function f in $\mathcal{M}(X, \mathcal{X})$ is then defined in terms of the integrals $\int f^{\pm}\,d\mu$ of their positive and negative parts $f^{\pm}$ in $\mathcal{M}(X, \mathcal{X})^+$ if these integrals (as in Chapter 3) are *finite*. Additional common notations: $\int f(x)\,d\mu$, $\int f(x)\,d\mu(x)$, or $\int f(x)\,\mu(dx)$. Consider the Lebesgue measure space $(\mathbb{R}, \mathfrak{R}, \mu)$ or $(\mathbb{R}, \overline{\mathfrak{R}}, \overline{\mu})$. The *Lebesgue integral* of a measurable function $f\colon \mathbb{R} \to \mathbb{R}$ is defined as the integral of f with respect to Lebesgue measure (λ on $\mathfrak{R}$, or $\overline{\lambda}$ on $\overline{\mathfrak{R}}$; see the remark that closes Chapter 2). If the Lebesgue integral of a real-valued function f exists in $\mathbb{R}$, then f is *Lebesgue integrable*. Another notation for the Lebesgue integral: $\int f(x)\,dx$.

Proposition 4.2. *Let $(X, \mathcal{X}, \mu)$ be a measure space.*

(a) *If f is a real-valued function in $\mathcal{M}(X, \mathcal{X})$ such that $f = 0$ μ-almost everywhere* (i.e., $f = 0$ *μ-a.e.*), *then f lies in $\mathcal{L}(X, \mathcal{X}, \mu)$ and $\int f\,d\mu = 0$.*

(b) *If $f \in \mathcal{L}(X, \mathcal{X}, \mu)$ and $g \in \mathcal{M}(X, \mathcal{X})$ is bounded, then $fg \in \mathcal{L}(X, \mathcal{X}, \mu)$.*

Proof.

(a) For any $f\colon \mathbb{R} \to \mathbb{R}$, consider the sets $F_+ = \{x \in X\colon f(x) > 0\}$, $F_- = \{x \in X\colon f(x) < 0\}$, and $F_0 = \{x \in X\colon f(x) = 0\}$, and set

$$F^+ = F_+ \cup F_0 = \{x \in X\colon f(x) \geq 0\} \quad\text{and}\quad F^- = F_- \cup F_0 = \{x \in X\colon f(x) \leq 0\}.$$

If $f \in M(X, \mathcal{X})$, then F^+ and F^- lie in $\mathcal{X}$ (Proposition 1.6) and so F_+, F_-, and F_0 also lie in $\mathcal{X}$. If $f = 0$ μ-a.e., then we have already seen that $\mu(F_+) = 0$. Recall that $f^+ = f\chi_{F^+}$, and hence $f^+ = f^+\chi_{F^+}$. Thus

$$\int f^+ d\mu = \int f^+ \chi_{F^+}\,d\mu = \int_{F^+} f^+ d\mu = \int_{F_+} f^+ d\mu + \int_{F_0} f^+ d\mu = \int_{F_0} f^+ d\mu = 0$$

by Problem 3.8. Similarly, $\int f^- d\mu = 0$. Hence $f \in \mathcal{L}(X, \mathcal{X}, \mu)$ if it is real-valued, and $\int f\,d\mu = 0$ by Definition 4.1.

(b) Note that $fg = (f^+ - f^-)(g^+ - g^-) = f^+g^+ + f^-g^- - f^+g^- - f^-g^+$. So

$$(fg)^+ = f^+g^+ + f^-g^- \quad\text{and}\quad (fg)^- = f^-g^+ + f^+g^-.$$

If g is bounded, set $\beta = \sup_{x\in X} |g(x)|$ so that $g^+ \leq \beta$ and $g^- \leq \beta$. Then $f^+g^+ \leq \beta f^+$ and so $\int f^+g^+ d\mu \leq \beta \int f^+ d\mu < \infty$ by Proposition 3.5(a) and Problem 3.3(b). Similarly, we get $\int f^-g^- d\mu < \infty$, $\int f^+g^- d\mu < \infty$, and

$\int f^- g^+ d\mu < \infty$. Hence, $\int (fg)^+ d\mu < \infty$ and $\int (fg)^- d\mu < \infty$ according to Proposition 3.5(b), and therefore $fg \in \mathcal{L}(X, \mathcal{X}, \mu)$. □

If f_1 and f_2 are real-valued functions in $\mathcal{M}(X, \mathcal{X})^+$ with $f_2 \leq f_1$ and $\int f_2 \, d\mu < \infty$, then $\int (f_1 - f_2) \, d\mu = \int f_1 d\mu - \int f_2 \, d\mu$. In fact, write $f_1 = (f_1 - f_2) + f_2$ and apply Propositions 1.5 and 3.5(b). In the next proposition we replace the assumption $f_2 \leq f_1$ by $\int f_1 \, d\mu < \infty$.

Proposition 4.3. *Take a measure space $(X, \mathcal{X}, \mu)$. If f_1 and f_2 are real-valued functions in $\mathcal{M}(X, \mathcal{X})^+$ with $\int f_1 \, d\mu < \infty$ and $\int f_2 \, d\mu < \infty$, then*

$$f_1 - f_2 \in \mathcal{L}(X, \mathcal{X}, \mu) \quad \text{and} \quad \int (f_1 - f_2) \, d\mu = \int f_1 \, d\mu - \int f_2 \, d\mu.$$

Proof. If f_1 and f_2 are real-valued functions in $\mathcal{M}(X, \mathcal{X})^+$, then $f = f_1 - f_2$ lies in $\mathcal{M}(X, \mathcal{X})$ (cf. Proposition 1.5). Since f_1, f_2, f^+, and f^- are functions in $\mathcal{M}(X, \mathcal{X})^+$ and $f^+ - f^- = f = f_1 - f_2$, it follows that $f^+ + f_2 = f_1 + f^-$ is in $\mathcal{M}(X, \mathcal{X})^+$, and so (cf. Proposition 3.5(b))

$$\int f^+ d\mu + \int f_2 \, d\mu = \int f_1 \, d\mu + \int f^- d\mu.$$

Note that $f^+ \leq f_1$ and $f^- \leq f_2$. Since $\int f_1 \, d\mu < \infty$ and $\int f_2 \, d\mu < \infty$, it follows by Problem 3.3(b) that $\int f^+ d\mu < \infty$ and $\int f^- d\mu < \infty$. Therefore, $(f_1 - f_2) = f$ lies in $\mathcal{L}(X, \mathcal{X}, \mu)$ and, according to Definition 4.1,

$$\int (f_1 - f_2) \, d\mu = \int f \, d\mu = \int f^+ d\mu - \int f^- d\mu = \int f_1 \, d\mu - \int f_2 \, d\mu. \quad \square$$

4.2 Three Fundamental Lemmas

Absolute integrability (of measurable functions), the first property in this section, is of crucial importance. It says that $|f|$ is integrable if and only if f is. This holds for every (measure-theoretic) integral, as defined in Definitions 3.3 and 4.1. In particular, absolute integrability holds for the Lebesgue integral (but it does not hold for the Riemann integral).

Actually, the well-known result stated in Problem 4.1 ensures that *if a bounded function on a closed and bounded* (i.e., on a compact) *interval of* $\mathbb{R}$ *has a Riemann integral, then it is $\overline{\mathfrak{R}}$-measurable and Lebesgue integrable, and its Riemann and Lebesgue integrals coincide*. However, there exist bounded functions defined on closed and bounded intervals that are Lebesgue but not Riemann integrable. For instance, $f_1(x) = 1$ for $x \in \mathbb{Q}$ and $f_1(x) = -1$ for $x \in \mathbb{R} \backslash \mathbb{Q}$ define a function $f_1 = (2\chi_{\mathbb{Q}} - 1)$ on $[0,1]$ for which the Riemann integral does not exist but the Lebesgue integral does exist (and is equal to -1). But $|f_1| = 1$, as a constant function on $[0,1]$, is trivially Riemann integrable and so it is Lebesgue integrable. On the other hand, the situation is different for improper Riemann integrals. If a function either is defined on an unbounded interval or is itself unbounded, then it may have a Riemann integral and not a Lebesgue integral. Example: $f_2(x) = \frac{\sin x}{x}$ on $[1, \infty)$ defines a Riemann integrable function f_2 (its improper Riemann integral exists and is finite) but $|f_2|$ is not integrable (i.e., it has no finite integral, in any sense) and so f_2 is not Lebesgue integrable.

Lemma 4.4. *If* $f \in \mathcal{M}(X, \mathcal{X})$, *then*

(a) $$f \in \mathcal{L}(X, \mathcal{X}, \mu) \quad \text{if and only if} \quad |f| \in \mathcal{L}(X, \mathcal{X}, \mu).$$

If $f \in \mathcal{L}(X, \mathcal{X}, \mu)$, *then*

(b) $$\left| \int f \, d\mu \right| \leq \int |f| \, d\mu.$$

Proof. (a) First note that for any function $f: X \to \mathbb{R}$,

$$|f|^+ = |f| = f^+ + f^- \quad \text{and} \quad |f|^- = 0, \tag{$*$}$$

and then recall from Proposition 1.6 that if f is measurable, then so are the functions f^+, f^-, and $|f|$. If $f \in \mathcal{L}(X, \mathcal{X}, \mu)$, then $\int f^+ d\mu < \infty$ and $\int f^- d\mu < \infty$. Hence, from $(*)$ and Proposition 3.5(b), $\int |f|^+ d\mu < \infty$ and $\int |f|^- d\mu < \infty$. So $|f| \in \mathcal{L}(X, \mathcal{X}, \mu)$. Conversely, if $|f| \in \mathcal{L}(X, \mathcal{X}, \mu)$, then $\int |f|^+ d\mu < \infty$. But $f^+ \leq |f|^+$ and $f^- \leq |f|^+$ by $(*)$. Then $\int f^+ d\mu < \infty$ and $\int f^- d\mu < \infty$ according to Problem 3.3(b), and therefore $f \in \mathcal{L}(X, \mathcal{X}, \mu)$.

(b) If $f \in \mathcal{L}(X, \mathcal{X}, \mu)$, then $f \in \mathcal{M}(X, \mathcal{X})$ and so $|f| \in \mathcal{L}(X, \mathcal{X}, \mu)$ by (a). Since $|f| = f^+ + f^-$, it follows by Proposition 3.5(b) and Definition 4.1 that

$$\left| \int f \, d\mu \right| = \left| \int f^+ d\mu - \int f^- d\mu \right| \leq \int f^+ d\mu + \int f^- d\mu = \int |f| \, d\mu. \qquad \square$$

As we pointed out in Section 3.2, the Monotone Convergence Theorem gave us the first evidences of linearity for the integral transformation in Proposition 3.5(a,b). Linearity is definitely accomplished in the next lemma.

Lemma 4.5. $\mathcal{L}$ *is a linear space and* $\int : \mathcal{L} \to \mathbb{R}$ *is a linear functional.*

Remarks: Before proving Lemma 4.5, note that what its statement says is twofold. First it says that the collection $\mathcal{L}(X, \mathcal{X}, \mu)$ is a(real) *linear space*. Indeed, Proposition 1.5 ensures that $\mathcal{M}(X, \mathcal{X})$ is a real linear space (when consisting to real-valued functions only — see the last paragraphs that close

Section 1.3). Since $\mathcal{L}(X, \mathcal{X}, \mu)$ is a subset of $\mathcal{M}(X, \mathcal{X})$, $\mathcal{L}(X, \mathcal{X}, \mu)$ is a linear space if and only if it is a linear manifold of $\mathcal{M}(X, \mathcal{X})$, which means that *if f and g are functions in $\mathcal{L}(X, \mathcal{X}, \mu)$ and γ is a any real number, then*

(a) $$\gamma f \text{ and } f+g \text{ lie in } \mathcal{L}(X, \mathcal{X}, \mu).$$

Consider the transformation $\int : \mathcal{L} \to \mathbb{R}$ that assigns to each function f in $\mathcal{L}(X, \mathcal{X}, \mu)$ the value of its integral $\int f\,d\mu$ in $\mathbb{R}$. A transformation of a (real or complex) linear space into $\mathbb{R}$ or $\mathbb{C}$ is called a (real or complex) *functional*. By Lemma 4.5, $\int : \mathcal{L} \to \mathbb{R}$ is a (real) *linear functional* (i.e., a *homogeneous* and *additive* functional), which means: (i) its domain $\mathcal{L}$ is a linear space, and (ii) if f and g are functions in $\mathcal{L}(X, \mathcal{X}, \mu)$ and γ is a real number, then

(b) $$\int \gamma f\,d\mu = \gamma \int f\,d\mu \quad \text{and} \quad \int (f+g)\,d\mu = \int f\,d\mu + \int g\,d\mu.$$

Thus the proof of Lemma 4.5 is reduced to proving (a) and (b).

Proof. First note that $\gamma f \in \mathcal{L}(X, \mathcal{X}, \mu)$ for every $\gamma \in \mathbb{R}$ if $f \in \mathcal{L}(X, \mathcal{X}, \mu)$, as is a particular case of Proposition 4.2(b). Since $-f = f^- - f^+$, we get

$$\gamma f = (\gamma f)^+ - (\gamma f)^- = \begin{cases} |\gamma| f^+ - |\gamma| f^-, & \gamma \geq 0, \\ |\gamma| f^- - |\gamma| f^+, & \gamma < 0, \end{cases}$$

where the functions $|\gamma| f^+$ and $|\gamma| f^-$ are in $\mathcal{M}(X, \mathcal{X})^+$, with $\int |\gamma| f^+ d\mu = |\gamma| \int f^+ d\mu$ and $\int |\gamma| f^- d\mu = |\gamma| \int f^- d\mu$ (Proposition 3.5(a)), and these integrals are finite since $f \in \mathcal{L}(X, \mathcal{X}, \mu)$. Hence, by Proposition 4.3,

$$\int (\pm |\gamma| f^+ - \pm |\gamma| f^-)\,d\mu = |\gamma| \left(\pm \int f^+ d\mu - \pm \int f^- d\mu \right),$$

and so,

$$\int \gamma f\,d\mu = \left\{ \begin{array}{ll} |\gamma| \left(\int f^+ d\mu - \int f^- d\mu \right), & \gamma \geq 0, \\ |\gamma| \left(\int f^- d\mu - \int f^+ d\mu \right), & \gamma < 0, \end{array} \right\} = \gamma \int f\,d\mu.$$

Thus homogeneity is proved. Now, to prove additivity, proceed as follows. Let f and g be in $\mathcal{L}(X, \mathcal{X}, \mu)$. Since $|f+g| \leq |f| + |g|$, and both $|f+g|$ and $|f| + |g|$ lie in $\mathcal{M}(X, \mathcal{X})^+$ (cf. Propositions 1.5 and 1.6), it follows by Problem 3.3(b) and Proposition 3.5(b) that $\int |f+g|\,d\mu \leq \int (|f| + |g|)\,d\mu = \int |f|\,d\mu + \int |g|\,d\mu$, and so $|f+g| \in \mathcal{L}(X, \mathcal{X}, \mu)$. In fact, since f and g are in $\mathcal{L}(X, \mathcal{X}, \mu)$, it follows that $|f|$ and $|g|$ are in $\mathcal{L}(X, \mathcal{X}, \mu)$ by Lemma 4.4, and so $\int |f+g|\,d\mu < \infty$. By Lemma 4.4 again, it follows that $f+g \in \mathcal{L}(X, \mathcal{X}, \mu)$ (since $f+g \in \mathcal{M}(X, \mathcal{X})$ according to Proposition 1.5). Now note that

$$f+g = (f^+ - f^-) + (g^+ - g^-) = (f^+ + g^+) - (f^- + g^-),$$

and also that $(f^+ + g^+)$ and $(f^- + g^-)$ lie in $\mathcal{M}(X, \mathcal{X})^+$ by Propositions 1.5 and 1.6, which have finite integrals by Proposition 3.5(b) — since f^+, f^-, g^+, and g^- have finite integrals because f and g lie in $\mathcal{L}(X, \mathcal{X}, \mu)$. Thus Propositions 4.3 and 3.5(b) and Definition 4.1 ensure that

$$\int (f + g)\, d\mu = \int (f^+ + g^+)\, d\mu - \int (f^- + g^-)\, d\mu$$
$$= \int f^+ d\mu - \int f^- d\mu + \int g^+ d\mu - \int g^- d\mu = \int f\, d\mu + \int g\, d\mu. \quad \square$$

In Proposition 3.5(c) we saw that given a measure, another measure is generated by a nonnegative measurable function. This has a natural extension for general real-valued (not necessarily nonnegative) functions, but this time a signed measure is generated instead. Indeed, the next proposition shows that given a measure, a signed measure is generated by a real-valued integrable function, called the *indefinite integral* of f with respect to μ.

Lemma 4.6. *Let $(X, \mathcal{X}, \mu)$ be a measure space. If $f \in \mathcal{L}(X, \mathcal{X}, \mu)$, then the real-valued set function $\nu: \mathcal{X} \to \mathbb{R}$ defined by*

$$\nu(E) = \int_E f\, d\mu \quad \text{for every} \quad E \in \mathcal{X}$$

is a signed measure.

Proof. If $f = f^+ - f^- \in \mathcal{L}(X, \mathcal{X}, \mu)$, then f^+ and f^- lie in $\mathcal{M}(X, \mathcal{X})^+$ and have finite integrals. Hence ν^+ and ν^- on $\mathcal{X}$ defined for each $E \in \mathcal{X}$ by

$$\nu^+(E) = \int_E f^+ d\mu \quad \text{and} \quad \nu^-(E) = \int_E f^- d\mu$$

are finite measures (cf. Proposition 3.5(c)). Therefore, by Definition 4.1, the set function ν on $\mathcal{X}$ defined for each $E \in \mathcal{X}$ by

$$\nu(E) = \int_E f\, d\mu = \int_E f^+ d\mu - \int_E f^- d\mu = \nu^+(E) - \nu^-(E)$$

is such that $\nu = \nu^+ - \nu^-$. Thus, as a linear combination of signed measures, $\nu: \mathcal{X} \to \mathbb{R}$ is itself a signed measure. $\square$

4.3 The Dominated Convergence Theorem

A very important convergence theorem for integrable functions is the forthcoming Dominated Convergence Theorem (also referred to as the Lebesgue Dominated Convergence Theorem). It goes along the line of the Monotone Convergence Theorem as in Corollary 3.10, now with no restriction to nonnegative functions, where nonnegativeness and convergence from below are replaced with integrability and dominated convergence. Chronologically, the original version of the Dominated Convergence Theorem was published by Lebesgue in 1904, prior to (and independently of) the original version of the Monotone Convergence Theorem, published by Beppo Levi in 1906 — these refer to Lebesgue measure space $(\mathbb{R}, \mathfrak{R}, \lambda)$. However, the Dominated Convergence Theorem can be easily proved — in a general abstract measure space $(X, \mathcal{X}, \mu)$ — as a consequence of the Monotone Convergence Theorem. In fact, we will prove it by using Fatou's Lemma, which in turn was proved in Lemma 3.6 as consequence of the Monotone Convergence Theorem.

Theorem 4.7. (Dominated Convergence Theorem). *Let $(X, \mathcal{X}, \mu)$ be a measure space. If $\{f_n\}$ is a sequence of real-valued functions in $\mathcal{M}(X, \mathcal{X})$ converging μ-almost everywhere to a real-valued function f in $\mathcal{M}(X, \mathcal{X})$, and if there exists a nonnegative function g in $\mathcal{L}(X, \mathcal{X}, \mu)$ such that $|f_n| \leq g$ for all n μ-almost everywhere, then each f_n and f lie in $\mathcal{L}(X, \mathcal{X}, \mu)$ and*

$$\int f \, d\mu = \lim_n \int f_n \, d\mu.$$

Proof. Suppose $|f_n| \leq g$ for all n μ-a.e. and $f_n \to f$ μ-a.e., so that $|f| \leq g$ μ-almost everywhere. Thus each f_n and f are integrable functions according to Problem 4.4(b) — which in fact is an immediate consequence of corollary of Lemma 4.4. Since $0 \leq g \pm f_n$ and $g \pm f_n \to g \pm f$ μ-a.e., use Fatou's Lemma (Lemma 3.9) to the sequences $\{g \pm f_n\}$, and apply Lemma 4.5 as follows:

$$\int g \, d\mu \pm \int f \, d\mu = \int (g \pm f) \, d\mu \leq \liminf_n \int (g \pm f_n) \, d\mu,$$

so that

$$\int g \, d\mu + \int f \, d\mu \leq \liminf_n \left(\int g \, d\mu + \int f_n \, d\mu \right) = \int g \, d\mu + \liminf_n \int f_n \, d\mu,$$

$$\int g \, d\mu - \int f \, d\mu \leq \liminf_n \left(\int g \, d\mu - \int f_n \, d\mu \right) = \int g \, d\mu - \limsup_n \int f_n \, d\mu,$$

and hence

$$\limsup_n \int f_n \, d\mu \leq \int f \, d\mu \leq \liminf_n \int f_n \, d\mu. \qquad \square$$

The Dominated Convergence Theorem plays a major part in integration theory. In particular, it is essential in the proof of completeness for the space L^p (Theorem 5.6), which is the main result of the next chapter. Applications of the Dominated Convergence Theorem will be frequent throughout the text from now on. As a special case, the next result, the Bounded Convergence Theorem, is an immediate consequence of Theorem 4.7.

Corollary 4.8. (Bounded Convergence Theorem). *Let* $(X, \mathcal{X}, \mu)$ *be a finite measure space* (i.e., a measure space equipped with a finite measure μ). *Suppose* $\{f_n\}$ *is a sequence of real-valued functions in* $\mathcal{M}(X, \mathcal{X})$ *converging* μ*-almost everywhere to a real-valued function* f *in* $\mathcal{M}(X, \mathcal{X})$*. If* $\{f_n\}$ *is bounded* μ*-almost everywhere* (i.e., if there exists a real number $\gamma > 0$ such that $|f_n| \leq \gamma$ for all n μ-a.e.), *then each* f_n *and* f *lie in* $\mathcal{L}(X, \mathcal{X}, \mu)$ *and*

$$\int f \, d\mu = \lim_n \int f_n \, d\mu.$$

Proof. If $(X, \mathcal{X}, \mu)$ is a finite measure space, and if there is a number $\gamma > 0$ such that $|f_n| \leq \gamma$ for all n μ-a.e. (i.e., if $\{f_n\}$ is bounded μ-almost everywhere), then the function $g\colon X \to \mathbb{R}$ such that $g(x) = \gamma$ for all $x \in X \backslash N$ for some $N \in \mathcal{X}$ with $\mu(N) = 0$ (i.e., the constant function $g = \gamma$ μ-a.e.) lies in $\mathcal{L}(X, \mathcal{X}, \mu)$ because $\int g \, d\mu = \gamma \mu(X) < \infty$. Now apply Theorem 4.7. □

4.4 Problems

Problem 4.1. *A real-valued bounded function defined on a closed and bounded interval of the real line is Riemann integrable if and only if the set of points at which it is not continuous has Lebesgue measure zero.* This is a well-known classical fundamental result (cf. Suggested Reading at the end of this chapter). Consider the Cantor set C and the Cantor-like set S of Problems 2.9 and 2.10. Recall that both sets C and S are totally disconnected. Let χ_C and χ_S from $[0, 1]$ to $\{0, 1\}$ be the characteristic functions of C and S, respectively. Which of these functions χ_C and χ_S is Riemann integrable? Are they Lebesgue integrable? What are their integrals?

Problem 4.2. Consider a sequence $\{f_n\}$ of functions in $\mathcal{L}(X, \mathcal{X}, \mu)$. If $\{f_n\}$ converges uniformly to $f \in \mathcal{L}(X, \mathcal{X}, \mu)$ and $\mu(X) < \infty$, then show that

$$\int f \, d\mu = \lim_n \int f_n \, d\mu,$$

and also show that this identity may fail without the assumption $\mu(X) < \infty$.

Hint: If $f - \varepsilon \leq f_n \leq f + \varepsilon$, then verify that $f^+ - \varepsilon \leq f_n^+ \leq f^+ + \varepsilon$ and $f^- - \varepsilon \leq f_n^- \leq f^- + \varepsilon$ for any $\varepsilon > 0$, and so verify that uniform convergence of $\{f_n\}$ to f implies uniform convergence of $\{f_n^+\}$ and $\{f_n^-\}$ to f^+ and f^-, respectively. Use Problem 3.6 and Definition 4.1 to prove the claimed identity, and use Problem 3.12(b) to verify that it requires that $\mu(X) < \infty$.

Problem 4.3. Let f be a function in $\mathcal{L}(X, \mathcal{X}, \mu)$. Show that the integral $\int f\, d\mu$ is unambiguously defined in terms of the integrals of any pair of functions f_1 and f_2 in $\mathcal{M}(X, \mathcal{X})^+ \cap \mathcal{L}(X, \mathcal{X}, \mu)$ — i.e., of any pair of nonnegative real-valued measurable functions with finite integrals — such that

$$f = f_1 - f_2.$$

Indeed, show that

$$\int f\, d\mu = \int f^+ d\mu - \int f^- d\mu = \int f_1\, d\mu - \int f_2\, d\mu.$$

Problem 4.4. Take a real-valued function f in $\mathcal{M}(X, \mathcal{X})$, and an arbitrary function $g \in \mathcal{L}(X, \mathcal{X}, \mu)$. Show that

(a) $0 \le f \le g$ μ-a.e. implies $f \in \mathcal{L}(X, \mathcal{X}, \mu)$ and $\int f\, d\mu \le \int g\, d\mu$,

(b) $|f| \le |g|$ μ-a.e. implies $f \in \mathcal{L}(X, \mathcal{X}, \mu)$ and $\int |f|\, d\mu \le \int |g|\, d\mu$,

(c) $f = g$ μ-a.e. implies $f \in \mathcal{L}(X, \mathcal{X}, \mu)$ and $\int f\, d\mu = \int g\, d\mu$.

Hints: (a) Use Problems 3.3(b) and 3.8(c) and Definition 4.1. (b) Apply item (a) and Lemma 4.4. (c) Use Proposition 4.2 and Lemma 4.5 to conclude that $\int (f-g)\, d\mu = 0$ and $f = (f-g) + g$ (since the functions are real-valued).

Problem 4.5. Take a pair of functions f and g in $\mathcal{L}(X, \mathcal{X}, \mu)$, a real number γ, an arbitrary measurable set $F \in \mathcal{X}$, and prove the following assertions.

(a) $\int_F \gamma f\, d\mu = \gamma \int_F f\, d\mu$ and $\int_F (f+g)\, d\mu = \int_F f\, d\mu + \int_F g\, d\mu$,

(b) $\int_E f\, d\mu \ge 0$ $(= 0)$ for all $E \in \mathcal{X}$ $\iff$ $f \ge 0$ $(= 0)$ μ-a.e.,

(c) $\int_E f\, d\mu = \int_E g\, d\mu$ for every $E \in \mathcal{X}$ $\iff$ $f = g$ μ-almost everywhere.

Hint: By Proposition 4.2(b), $f\chi_E \in \mathcal{L}(X, \mathcal{X}, \mu)$. Use Lemma 4.5 to prove (a) and Propositions 3.7(a) and 4.2(a) and Problem 4.4 to prove (b) and (c).

Problem 4.6. Prove that if f and g are functions in $\mathcal{L}(X, \mathcal{X}, \mu)$, then so are the functions $f \wedge g$ and $f \vee g$. (*Hint.* Problem 1.3, Lemmas 4.4 and 4.5.)

Problem 4.7. Let $(X, \mathcal{X}, \mu)$ be a measure space. A complex-valued function $f\colon X \to \mathbb{C}$ is *measurable* if its real and imaginary parts, f_1 and f_2, are real-valued measurable functions (Problem 1.7). A measurable complex-valued function $f = f_1 + i\, f_2$ is *integrable* if its real and imaginary parts are real-valued integrable functions ($f_1, f_2 \in \mathcal{L}(X, \mathcal{X}, \mu)$). If a complex-valued function is integrable, then the *integral* of f is defined as the complex number

$$\int f\,d\mu = \int f_1\,d\mu + i\int f_2\,d\mu.$$

Prove the complex version of Lemma 4.4. In other words, if f is measurable, then it is integrable if and only if $|f|$ is integrable and, in this case,

$$\left|\int f\,d\mu\right| \le \int |f|\,d\mu.$$

Conclude: f is integrable if and only if $|f| \in \mathcal{M}(X, \mathcal{X})$ and $\int |f|\,d\mu < \infty$ (i.e., $|f| \in \mathcal{L}(X, \mathcal{X}, \mu)$ — also see Section 10.1). Prove the complex version of Problem 4.4(b): if f and g are complex-valued functions, f measurable and g integrable, and $|f| \le |g|$ μ-a.e., then f is integrable and $\int |f|\,d\mu \le \int |g|\,d\mu$.

Hints: Note that $|f| \le |f_1| + |f_2|$, $|f_1| \le |f|$, and $|f_2| \le |f|$. Use Lemmas 4.4 and 4.5 to show that if f is integrable, then so is $|f_1| + |f_2|$, and hence $|f|$ is integrable by Problem 4.4(b) (since $|f| = (|f_1|^2 + |f_2|^2)^{\frac{1}{2}}$ is measurable). Conversely, if f is measurable and $|f|$ is integrable, then use Problem 4.4(b) and Lemma 4.4 to show that f_1 and f_2 are integrable. The properties f measurable, g integrable, and $|f| \le |g|$ μ-a.e. imply that f is integrable and $\int |f|\,d\mu \le \int |g|\,d\mu$. This also is a consequence of Problem 4.4(b). To prove that $\left|\int f\,d\mu\right| \le \int |f|\,d\mu$ proceed as follows. Write $\int f\,d\mu = \rho\, e^{i\theta}$ and consider the function $g = \mathrm{Re}\,(e^{-i\theta} f)\colon X \to \mathbb{R}$ so that $|g| \le |e^{-i\theta} f| = |f|$. Use Lemma 4.5 to verify that $\left|\int f\,d\mu\right| = \rho = e^{-i\theta}\int f\,d\mu = \int e^{-i\theta} f\,d\mu = \mathrm{Re}\int e^{-i\theta} f\,d\mu = \int \mathrm{Re}(e^{-i\theta} f)\,d\mu = \int g\,d\mu = \left|\int g\,d\mu\right| \le \int |g|\,d\mu \le \int |f|\,d\mu$.

Problem 4.8. Use Lemma 4.5 to prove its own complex version. That is, show that if f and g are complex-valued integrable functions (with respect to a measure μ) and γ is an arbitrary complex number, then γf and $f + g$ are again integrable complex-valued functions, and

$$\int \gamma f\,d\mu = \gamma\int f\,d\mu \quad\text{and}\quad \int (f+g)\,d\mu = \int f\,d\mu + \int g\,d\mu.$$

Problem 4.9. Prove the complex version of the Dominated Convergence Theorem (Theorem 4.7). If $\{f_n\}$ is a sequence of complex-valued measurable functions that converges pointwise to a complex-valued function f, and if g is a nonnegative integrable function (with respect to a measure μ) such that $|f_n| \le g$ for all n, then f is integrable and

$$\int f\,d\mu = \lim_n \int f_n\,d\mu.$$

Problem 4.10. Show that the indefinite integral of an integrable function, as defined in Lemma 4.6, is countably additive in the following sense. If f is in $\mathcal{L}(X, \mathcal{X}, \mu)$ and $\{E_n\}$ is a countable measurable partition of $E \in \mathcal{X}$, then

$$\int_E f\, d\mu = \sum\nolimits_n \int_{E_n} f\, d\mu.$$

Problem 4.11. Consider a measure space $(X, \mathcal{X}, \mu)$. Suppose a sequence $\{f_n\}$ of real-valued functions in $\mathcal{M}(X, \mathcal{X})$ is such that $\sum_{n=1}^m f_n \to f$ almost everywhere to a real-valued function f in $\mathcal{M}(X, \mathcal{X})$. Let g be a nonnegative function in $\mathcal{L}(X, \mathcal{X}, \mu)$. If $\left|\sum_{n=1}^m f_n\right| \leq g$ for all m almost everywhere, then f and each f_n are integrable (i.e., they lie in $\mathcal{L}(X, \mathcal{X}, \mu)$) and

$$\int f\, d\mu = \sum\nolimits_n \int f_n\, d\mu.$$

Problem 4.12. Take a sequence $\{f_n\}$ of functions in $\mathcal{L}(X, \mathcal{X}, \mu)$. If the series $\sum_{n=1}^\infty \int |f_n|\, d\mu$ of positive numbers converges (i.e., $\sum_{n=1}^\infty \int |f_n|\, d\mu < \infty$), then show that the sequence $\left\{\sum_{n=1}^m f_n\right\}$ of functions in $\mathcal{L}(X, \mathcal{X}, \mu)$ converges μ almost everywhere as $m \to \infty$ to a function f in $\mathcal{L}(X, \mathcal{X}, \mu)$, and

$$\int f\, d\mu = \sum\nolimits_n \int f_n\, d\mu.$$

Hint: Since the sequence $\left\{\sum_{n=1}^m \int |f_n|\, d\mu\right\}$ of positive numbers is bounded, it converges in $\mathbb{R}$. Apply the corollary of the Monotone Convergence Theorem stated in Problem 3.7 (i.e., the Beppo Levi Theorem) to verify that the sequence of positive numbers $\left\{\int \sum_{n=1}^m |f_n|\, d\mu\right\}$ converges in $\mathbb{R}$, and that the sequence $\left\{\sum_{n=1}^m |f_n|\right\}$ of functions in $\mathcal{M}(X, \mathcal{X})^+$ converges pointwise to a function h in $\mathcal{M}(X, \mathcal{X})^+$ such that $\int h\, d\mu = \sum_{n=1}^\infty \int |f_n|\, d\mu$. Set $N = \{x \in X \colon h(x) = +\infty\}$. Also set $g(x) = h(x)$ if $x \in X \backslash N$, and $g(x) = 0$ if $x \in N$. Apply Problems 3.8 and 3.9 to show that the nonnegative real-valued function g lies in $\mathcal{L}(X, \mathcal{X}, \mu)$, and the sequence $\left\{\sum_{n=1}^m |f_n|\right\}$ in $\mathcal{M}(X, \mathcal{X})^+$ converges almost everywhere to g, and so the sequence of real-valued functions $\left\{\sum_{n=1}^m f_n\right\}$ in $\mathcal{M}(X, \mathcal{X})$ converges almost everywhere to a function f in $\mathcal{M}(X, \mathcal{X})$ (because it converges absolutely almost everywhere). Show that $\left|\sum_{n=1}^m f_n\right| \leq g$ and apply the Dominated Convergence Theorem.

Problem 4.13. Let $f \in \mathcal{M}(X, \mathcal{X})$ be a real-valued function. For each positive integer n consider its n-truncation $f_n \in \mathcal{M}(X, \mathcal{X})$ as defined in Problem 1.5. Prove that (i) if $f \in \mathcal{L}(X, \mathcal{X}, \mu)$, then

$$\int f\, d\mu = \lim_n \int f_n\, d\mu.$$

Conversely, prove that (ii) if $\sup_n \int |f_n|\, d\mu < \infty$, then $f \in \mathcal{L}(X, \mathcal{X}, \mu)$.

Hint: Show that $\{f_n\}$ is a sequence of real-valued functions in $\mathcal{M}(X,\mathcal{X})$ that converges pointwise to the real-valued function f, and that $\{|f_n|\}$ is an increasing sequence of real-valued functions in $\mathcal{M}(X,\mathcal{X})^+$ that converges pointwise to $|f|$ (which implies that $|f_n| \le |f|$ for all n). To prove (i) verify that $|f|$ lies in $\mathcal{L}(X,\mathcal{X},\mu)$ and apply the Dominated Convergence Theorem. To prove the converse in (ii) apply the Monotone Convergence Theorem and conclude that f lies in $\mathcal{L}(X,\mathcal{X},\mu)$ since $\int |f|\,d\mu < \infty$.

Problem 4.14. If $\{f_n\}$ is a sequence of functions in $\mathcal{L}(X,\mathcal{X},\mu)$, and if f is a real-valued function in $\mathcal{M}(X,\mathcal{X})$, then prove that

$$\lim_n \int |f_n - f|\,d\mu = 0 \quad \text{implies} \quad f \in \mathcal{L}(X,\mathcal{X},\mu) \ \text{ and } \int |f|\,d\mu = \lim_n \int |f_n|\,d\mu.$$

(In the jargon of Chapter 5 this means: $f_n \to f$ in $L^1 \Longrightarrow \|f_n\|_1 \to \|f\|_1$.)

Hint: Recall that $\big||\alpha| - |\beta|\big| \le |\alpha - \beta|$ for all $\alpha, \beta \in \mathbb{R}$. Since f and f_n are real-valued functions, then apply Problem 3.3(b) and Proposition 3.5(b) to show that if $\lim_n \int |f_n - f|\,d\mu = 0$, then $\limsup_n \int |f_n|\,d\mu \le \int |f|\,d\mu \le \liminf_n \int |f_n|\,d\mu$. Also, since $\lim_n \int |f_n - f|\,d\mu = 0$ implies that $|f_n - f|$ lies in $\mathcal{L}(X,\mathcal{X},\mu)$, and since f_n lies in $\mathcal{L}(X,\mathcal{X},\mu)$, then use Lemmas 4.4 and 4.5 to conclude that the function f lies in $\mathcal{L}(X,\mathcal{X},\mu)$.

Problem 4.15. Consider the statement of the Dominated Convergence Theorem. Show that in addition to the results stated there we also have

$$\lim_n \int |f_n - f|\,d\mu = 0.$$

Hint: $|f_n - f| \to 0$ and $|f_n - f| \le 2g$ for all n almost everywhere.

Problem 4.16. Consider again the statement of the Dominated Convergence Theorem. Verify that the dominance assumption (viz., $|f_n| \le g$ for all n almost everywhere for some $g \in \mathcal{L}(X,\mathcal{X},\mu)$) cannot be dropped from the theorem statement (even under the assumption of uniform convergence — see Problem 4.2). In fact, take the Lebesgue measure space $(\mathbb{R},\mathfrak{R},\lambda)$, set $f = h = 0$, set $f_n = n\,\chi_{(0,\frac{1}{n}]}$, and set $h_n = \frac{1}{n}\chi_{[0,n]}$ for every integer $n \ge 1$ — all real-valued functions in $\mathcal{L}(\mathbb{R},\mathfrak{R},\lambda)$. Show that

(a) $\{f_n\}$ converges pointwise to f but $0 = \int f\,d\lambda \ne \lim_n \int f_n\,d\lambda = 1$,

(b) $\{h_n\}$ converges uniformly to g but $0 = \int h\,d\lambda \ne \lim_n \int h_n\,d\lambda = 1$.

Problem 4.17. Let $(X,\mathcal{X},\mu)$ be a measure space, and consider a function $f\colon X \times [0,1] \to \mathbb{R}$ such that, for each $s \in [0,1]$, the function $f(\cdot,s)\colon X \to \mathbb{R}$ is $\mathcal{X}$-measurable. If, for some $s_0 \in [0,1]$, $f(x,s_0) = \lim_{s \to s_0} f(x,s)$ for every

$x \in X$, and if $|f(x,s)| \leq g(x)$ for every $x \in X$ and all $s \in [0,1]$, for some g in $\mathcal{L}(X, \mathcal{X}, \mu)$, then use the Dominated Convergence Theorem to show that

$$\int f(x, s_0)\, d\mu = \lim_{s \to s_0} \int f(x,s)\, d\mu.$$

If, in addition, the function $f(x,\cdot)\colon [0,1] \to \mathbb{R}$ is continuous for each $x \in X$, then show that the function $F\colon [0,1] \to \mathbb{R}$, defined for each $s \in [0,1]$ by

$$F(s) = \int f(x,s)\, d\mu,$$

is continuous as well. (Continuity is with respect to the usual metric on $\mathbb{R}$).

Suggested Reading

Bartle [4], Berberian [7], Brown and Pearcy [8], Halmos [18], Kingman and Taylor [23], Royden [35], Rudin [36]. For the basic classic result stated in Problem 4.1, see [33, p. 23] (also see [1, p. 206], [41, p. 53]).

5

Banach Spaces L^p

5.1 Construction of L^1

A topology to equip the linear space $\mathcal{L}(X, \mathcal{X}, \mu)$ which will turn it into a Banach space is investigated in this chapter. Section 5.1 summarizes the basics on normed spaces that will be required in Chapter 5 (as well as in parts of Chapters and 6, 10, and 12). We assume the reader has been introduced to *linear spaces* (or *vector space*) before — see e.g., Lemma 4.5.

Definition 5.1. Let $\mathcal{L}$ be an arbitrary linear space over $\mathbb{F}$, where $\mathbb{F}$ stands either for the real field $\mathbb{R}$ or the complex field $\mathbb{C}$. A real-valued function

$$\|\;\|\colon \mathcal{L} \to \mathbb{R}$$

is a *norm* on $\mathcal{L}$ if the following conditions (referred to as the *norm axioms*) are satisfied for all vectors f and g in $\mathcal{L}$ and all scalars γ in $\mathbb{F}$.

(i) $\|f\| \geq 0$ (*nonnegativeness*),

(ii) $\|f\| > 0$ if $f \neq 0$ (*positiveness*),

(iii) $\|\gamma f\| = |\gamma|\,\|f\|$ (*absolute homogeneity*),

(iv) $\|f + g\| \leq \|f\| + \|g\|$ (*subadditivity — triangle inequality*).

C.S. Kubrusly, *Essentials of Measure Theory*,
DOI 10.1007/978-3-319-22506-7_5

Any linear space $\mathcal{L}$ equipped with a norm $\| \ \|$ on it is a *normed space* (or a *normed linear space*, or a *normed vector space*). A linear space is a *real* or a *complex linear space* if $\mathbb{F} = \mathbb{R}$ or $\mathbb{F} = \mathbb{C}$ and, when equipped with a norm, it is called a *real* or *complex normed space*.

Elements of any linear space $\mathcal{L}$ are called *vectors*, and elements of a field $\mathbb{F}$ are called *scalars*. If $\mathcal{L}$ is a linear space equipped with a norm $\| \ \|$, then the resulting normed space is denoted by $(\mathcal{L}, \| \ \|)$ or simply by $\mathcal{L}$ if the norm is clear in the context. Observe from axioms (i), (ii), and (iii) that if a function $\| \ \| \colon \mathcal{L} \to \mathbb{R}$ is a norm, then

$$\|f\| = 0 \quad \text{if and only if} \quad f = 0.$$

If a function $\| \ \| \colon \mathcal{L} \to \mathbb{R}$ satisfies the three axioms (i), (iii), and (iv) but not necessarily axiom (ii), then it is called a *seminorm* (or a *pseudonorm*). In other words, a seminorm does vanish at the origin (as a norm does) but a seminorm may also vanish at a nonzero vector (as a norm never does). A linear space $\mathcal{L}$ equipped with a seminorm is called a *seminormed space*.

Definition 5.2. A sequence of vectors $\{f_n\}$ in a normed space $(\mathcal{L}, \| \ \|)$ *converges* to a vector f in $\mathcal{L}$ if for each real number $\varepsilon > 0$ there exists a positive integer n_ε such that

$$n \geq n_\varepsilon \quad \text{implies} \quad \|f_n - f\| < \varepsilon.$$

If a sequence of vectors $\{f_n\}$ converges to a vector $f \in \mathcal{L}$, then it is said to be a *convergent sequence* and f is said to be the *limit* of $\{f_n\}$. In order to distinguish it among other convergence modes, we refer to the preceding concept as *norm convergence* (or *convergence in the norm topology*). A sequence $\{f_n\}$ is a *Cauchy sequence* in $(\mathcal{L}, \| \ \|)$ (or satisfies the *Cauchy criterion*) if for each real number $\varepsilon > 0$ there is a positive integer n_ε such that

$$n, m \geq n_\varepsilon \quad \text{implies} \quad \|f_m - f_n\| < \varepsilon.$$

A common notation for the Cauchy criterion is $\lim_{m,n} \|f_m - f_n\| = 0$. A sequence $\{f_n\}$ is a *bounded sequence* in $(\mathcal{L}, \| \ \|)$ if $\sup_n \|f_n\| < \infty$.

Proposition 5.3. *Consider an arbitrary normed space* $(\mathcal{L}, \| \ \|)$.

(a) *Every convergent sequence in* $(\mathcal{L}, \| \ \|)$ *is a Cauchy sequence.*

(b) *Every Cauchy sequence in* $(\mathcal{L}, \| \ \|)$ *is bounded.*

(c) *If a Cauchy sequence in* $(\mathcal{L}, \| \ \|)$ *has a subsequence that converges in* $(\mathcal{L}, \| \ \|)$, *then the sequence itself converges in* $(\mathcal{L}, \| \ \|)$ *and its limit coincides with the limit of that convergent subsequence.*

Proof. Take a normed space $(\mathcal{L}, \|\ \|)$ and a sequence $\{f_n\}$ of vectors in $\mathcal{L}$.

(a) Take any $\varepsilon > 0$. If $\{f_n\}$ converges to $f \in L$, then there exists $n_\varepsilon \geq 1$ such that $\|f_n - f\| < \frac{\varepsilon}{2}$ for all $n \geq n_\varepsilon$. Since $\|f_m - f_n\| \leq \|f_m - f\| + \|f - f_n\|$ by the triangle inequality, it follows that $\|f_m - f_n\| < \varepsilon$ whenever $m, n \geq n_\varepsilon$.

(b) Suppose $\{f_n\}$ is Cauchy. Then there is an $n_1 > 1$ such that $\|f_m - f_n\| < 1$ for all $m, n \geq n_1$. Let $\beta \in \mathbb{R}$ be the maximum of $\{\|f_m - f_n\| \in \mathbb{R} \colon m, n < n_1\}$ (a finite set). Thus $\|f_m - f_n\| \leq \|f_m - f_{n_1}\| + \|f_{n_1} - f_n\| \leq 2\max\{1, \beta\}$ for all m, n by the triangle inequality. But $\|f_m\| \leq \|f_m - f_1\| + \|f_1\|$ for all m.

(c) Consider a subsequence $\{f_{n_k}\}$ of a Cauchy sequence $\{f_n\}$ that converges to $f \in \mathcal{L}$ (i.e., $\|f_{n_k} - f\| \to 0$ as $k \to \infty$). Take any $\varepsilon > 0$. Since $\{f_n\}$ is a Cauchy sequence, there is a positive integer n_ε such that $\|f_m - f_n\| < \frac{\varepsilon}{2}$ for all $m, n \geq n_\varepsilon$. Since $\{f_{n_k}\}$ converges to f, there is a positive integer k_ε such that $\|f_{n_k} - f\| < \frac{\varepsilon}{2}$ for all $k \geq k_\varepsilon$. Therefore, if j is any integer for which $j \geq k_\varepsilon$ and $n_j \geq n_\varepsilon$ (for instance, if $j = \max\{n_\varepsilon, k_\varepsilon\}$), then $\|f_n - f\| \leq \|f_n - f_{n_j}\| + \|f_{n_j} - f\| < \varepsilon$ for every $n \geq n_\varepsilon$ by the triangle inequality, which implies that the sequence $\{f_n\}$ converges to f. □

By Proposition 5.3(a), every convergent sequence is a Cauchy sequence. However, the converse may fail (see Problems 5.13 and 5.15). But there are normed spaces with the crucial property that every Cauchy sequence converges. Normed spaces possessing this special property are called *complete*: a normed space $\mathcal{L}$ is complete if every Cauchy sequence in $\mathcal{L}$ is a convergent sequence in $\mathcal{L}$. A *Banach space* is precisely a complete normed space.

Proposition 5.4. *Set* $\mathcal{L} = \mathcal{L}(X, \mathcal{X}, \mu)$. *The function* $\|\ \| \colon \mathcal{L} \to \mathbb{R}$ *defined by*

$$\|f\| = \int |f|\, d\mu \quad \text{for every} \quad f \in \mathcal{L}(X, \mathcal{X}, \mu)$$

is a seminorm on the linear space $\mathcal{L}(X, \mathcal{X}, \mu)$, *which is such that*

$$\|f\| = 0 \quad \text{if and only if} \quad f = 0 \ \mu\text{-almost everywhere.}$$

Proof. Recall from Lemma 4.5 that $\mathcal{L}(X, \mathcal{X}, \mu)$ is a real linear space. Set $\mathcal{L} = \mathcal{L}(X, \mathcal{X}, \mu)$. Observe that the function $\|\ \| \colon \mathcal{L} \to \mathbb{R}$ is well defined since the integral $\int |f|\, d\mu$ exists in $\mathbb{R}$ for every $f \in \mathcal{L}(X, \mathcal{X}, \mu)$ by Lemma 4.4. Let f and g be arbitrary functions in $\mathcal{L}(X, \mathcal{X}, \mu)$, and take any scalar γ in $\mathbb{R}$. Note that $|\gamma f(x)| = |\gamma||f(x)|$ for every $x \in X$, which means $|\gamma f| = |\gamma||f|$. Since the triangle inequality holds in $\mathbb{R}$, that is, since $|\alpha + \beta| \leq |\alpha| + |\beta|$ for every pair $\{\alpha, \beta\}$ of real numbers, it follows that $|f(x) + g(x)| \leq |f(x)| + |g(x)|$ for every $x \in X$, which means $|f + g| \leq |f| + |g|$. Now verify the norm axioms.

Axiom (i) in Definition 5.1, $\|f\| \geq 0$, is immediate (Problem 3.3). For axioms (iii) and (iv) recall that $\int : \mathcal{L} \to \mathbb{R}$ is a linear functional (Lemma 4.5). Thus,

$$\|\gamma f\| = \int |\gamma f|\, d\mu = \int |\gamma||f|\, d\mu = |\gamma| \int |f|\, d\mu = |\gamma|\|f\|,$$

$$\|f+g\| = \int |f+g|\, d\mu \leq \int (|f|+|g|)\, d\mu = \int |f|\, d\mu + \int |g|\, d\mu = \|f\| + \|g\|$$

(cf. Problem 3.3). Therefore $\|\ \|$ is a seminorm on $\mathcal{L}$. Moreover, Proposition 3.7(a) ensures that $\|f\| = 0$ if and only if $f = 0$ μ-almost everywhere. □

Nevertheless, this seminorm $\|\ \|$ is not a norm on $\mathcal{L}(X, \mathcal{X}, \mu)$. In fact, it does not satisfy axiom (ii) in Definition 5.1: there may be a function f in $\mathcal{L}(X, \mathcal{X}, \mu)$ such that $f \neq 0$ and $\|f\| = 0$ (e.g., take a Lebesgue integrable function f in $\mathcal{L}(\mathbb{R}, \Re, \lambda)$ such that $f(x) = 0$ for all $x \in \mathbb{R}$ except at the origin, where $f(0) = 1$). In order to make this seminorm on $\mathcal{L}(X, \mathcal{X}, \mu)$ into a norm we need to redefine the concept of equality between functions in $\mathcal{L}(X, \mathcal{X}, \mu)$ (other than the usual pointwise definition) so that axiom (ii) is satisfied.

Take a measure space $(X, \mathcal{X}, \mu)$, and let f and g be arbitrary real-valued functions in $\mathcal{M}(X, \mathcal{X})$. We say that f and g are *equivalent* (or *μ-equivalent*), denoted by $f \sim g$, if $f = g$ almost everywhere (i.e., μ-a.e.). This $\sim$ is in fact an equivalence relation on $\mathcal{M}(X, \mathcal{X})$. For every *real-valued* function f in $\mathcal{M}(X, \mathcal{X})$, let $[f]$ be the *equivalence class* of f (with respect to μ),

$$[f] = \{ f' \in \mathcal{M}(X, \mathcal{X}) \colon f' \sim f \}.$$

This $[f]$ is the subset of $\mathcal{M}(X, \mathcal{X})$ consisting of all functions in $\mathcal{M}(X, \mathcal{X})$ that are μ-equivalent to f. The following necessary and sufficient conditions for equality between equivalence classes are readily verified. Indeed,

$$[f] = [g] \iff f \sim g \iff f = g \ \ \mu\text{-almost everywhere.}$$

Problem 4.4(c) says that if f is a function in $\mathcal{L}(X, \mathcal{X}, \mu)$, then so is every f' in $[f]$ and $\int f' d\mu = \int f\, d\mu$. Therefore, if f is in $\mathcal{L}(X, \mathcal{X}, \mu)$, then so is every g' in $[g]$ whenever $[f] = [g]$ and $\int g' d\mu = \int f\, d\mu$. Thus set

$$L^1 = L^1(\mu) = L^1(X, \mathcal{X}, \mu) = \{ [f] \subseteq \mathcal{M}(X, \mathcal{X}) \colon f \in \mathcal{L}(X, \mathcal{X}, \mu) \},$$

which is the collection of all equivalence classes of functions in $\mathcal{L}(X, \mathcal{X}, \mu)$. This collection $L^1 = L^1(X, \mathcal{X}, \mu)$ is also referred to as the *quotient space* of $\mathcal{L}(X, \mathcal{X}, \mu)$ modulo $\sim$ and is also denoted by $\mathcal{L}(X, \mathcal{X}, \mu)/\!\sim$. Since $\mathcal{L}(X, \mathcal{X}, \mu)$ is a linear space, it can be shown that $L^1(X, \mathcal{X}, \mu)$ is made into a linear space when scalar multiplication and vector addition are defined by

$$\gamma[f] = [\gamma f] \quad \text{and} \quad [f] + [g] = [f + g]$$

for every $[f]$ and $[g]$ in $L^1(X,\mathcal{X},\mu)$ and every scalar γ. Observe that the origin $[0]$ of the linear space $L^1(X,\mathcal{X},\mu)$ is a linear manifold,

$$[0] = \{f \in \mathcal{L}(X,\mathcal{X},\mu)\colon\ f = 0\ \mu\text{-a.e.}\},$$

of the linear space $\mathcal{L}(X,\mathcal{X},\mu)$, and $L^1(X,\mathcal{X},\mu)$ can still be viewed as the *quotient space* of $\mathcal{L}(X,\mathcal{X},\mu)$ modulo $[0]$, also denoted by $\mathcal{L}(X,\mathcal{X},\mu)/[0]$. The seminorm on $\mathcal{L}(X,\mathcal{X},\mu)$ given in Proposition 5.4 induces a norm in $L^1(X,\mathcal{X},\mu)$. In fact, consider the function $\|\ \|_1\colon L^1 \to \mathbb{R}$ defined by

$$\|[f]\|_1 = \int |f|\,d\mu \quad \text{for every} \quad [f] \in L^1(X,\mathcal{X},\mu),$$

where $f \in \mathcal{L}(X,\mathcal{X},\mu)$ is any representative of the equivalence class $[f]$.

Proposition 5.5. *The function $\|\ \|_1$ is a norm on the linear space L^1.*

Proof. Consider the seminorm $\|\ \|$ on $\mathcal{L}(X,\mathcal{X},\mu)$ of Proposition 5.4. Thus the function $\|\ \|_1\colon L^1 \to \mathbb{R}$ is well defined. Actually, for any $[f]$ in $L^1(X,\mathcal{X},\mu)$,

$$\|[f]\|_1 = \|f\|,$$

whose value does not depend on the representative f of the equivalence class $[f]$. Indeed, if f and f' are functions in $\mathcal{L}(X,\mathcal{X},\mu)$ such that $f = f'$ μ-a.e., then $|f'| = |f|$ μ-a.e. (since $\big||f| - |f'|\big| \le |f - f'|$), and so $\int |f'|\,d\mu = \int |f|\,d\mu$ (Problem 4.4(c)). Note that $\|\ \|_1$ satisfies all the axioms (i), (ii), (iii), and (iv) of Definition 5.1. In fact, take arbitrary classes $[f]$ and $[g]$ in $L^1(X,\mathcal{X},\mu)$, and an arbitrary scalar $\gamma \in \mathbb{R}$ so that, by Proposition 5.4,

$$\|[f]\|_1 = \|f\| \ge 0,$$

$$\|[f]\|_1 = 0 \iff \|f\| = 0 \iff f = 0\ \mu\text{-a.e.} \iff [f] = [0],$$

$$\|\gamma[f]\|_1 = \|[\gamma f]\|_1 = \|\gamma f\| = |\gamma|\|f\| = |\gamma|\|[f]\|_1,$$

$$\|[f] + [g]\|_1 = \|[f + g]\|_1 = \|f + g\| \le \|f\| + \|g\| = \|[f]\|_1 + \|[g]\|_1. \qquad \square$$

5.2 Spaces L^p and L^∞

Consider a measure space $(X,\mathcal{X},\mu)$. Extending the construction of L^1 in the previous section, we now define the linear spaces L^p for each $p \ge 1$ and L^∞, equip them with norms, and show that they are Banach spaces.

We begin with the spaces L^p. Take an arbitrary real number $p \ge 1$. A *real-valued* function f in $\mathcal{M}(X,\mathcal{X})$ is *p-integrable* if $f^p \in \mathcal{L}(X,\mathcal{X},\mu)$ or, equivalently, if $\int |f|^p d\mu < \infty$ (cf. Lemma 4.4). Along the same line used to construct L^1 (which is the particular case of L^p for $p = 1$), set

$$L^p = L^p(\mu) = L^p(X, \mathcal{X}, \mu) = \{[f] \subseteq \mathcal{M}(X, \mathcal{X}) \colon f^p \in \mathcal{L}(X, \mathcal{X}, \mu)\},$$

the collection of all equivalence classes of p-integrable functions. In other words, $L^p(X, \mathcal{X}, \mu)$ is the collection of all equivalence classes of real-valued functions f in $\mathcal{M}(X, \mathcal{X})$ for which $\int |f|^p d\mu < \infty$ for every representative f of $[f]$. Thus consider the function $\| \, \|_p \colon L^p \to \mathbb{R}$ defined by

$$\|[f]\|_p = \left(\int |f|^p \, d\mu\right)^{\frac{1}{p}} \quad \text{for every} \quad [f] \in L^p(X, \mathcal{X}, \mu),$$

where f is any representative of the equivalence class $[f]$ in $L^p(X, \mathcal{X}, \mu)$.

Now we consider the space L^∞. An extended real-valued function f in $\mathcal{M}(X, \mathcal{X})$ is *essentially bounded* if it is bounded almost everywhere. Roughly speaking, if $\sup_{x \in X} |f(x)| < \infty$ μ-a.e. on X. Precisely this means that there is a real number $\beta > 0$ such that $|f| \le \beta$ μ-almost everywhere. If an extended real-valued function f in $\mathcal{M}(X, \mathcal{X})$ is essentially bounded, then set

$$\operatorname{ess\,sup}|f| = \inf\{\beta \ge 0 \colon |f| \le \beta \ \mu\text{-a.e.}\} = \inf_{N \in \mathcal{X}} \ \sup_{x \in X \setminus N} |f(x)|$$

in $\mathbb{R}$, where $\inf_{N \in \mathcal{X}}$ is taken over all $N \in \mathcal{X}$ such that $\mu(N) = 0$. Next set

$$L^\infty = L^\infty(\mu) = L^\infty(X, \mathcal{X}, \mu) = \{[f] \subseteq \mathcal{M}(X, \mathcal{X}) \colon \sup_{x \in X} |f(x)| < \infty \ \mu\text{-a.e.}\},$$

the collection of all equivalence classes of essentially bounded extended real-valued functions. In other words, $L^\infty(X, \mathcal{X}, \mu)$ is the collection of all equivalence classes of extended real-valued functions f in $\mathcal{M}(X, \mathcal{X})$ for which $\operatorname{ess\,sup}|f| < \infty$ for every representative f of $[f]$. Thus consider the function $\| \, \|_\infty \colon L^\infty \to \mathbb{R}$ defined by

$$\|[f]\|_\infty = \operatorname{ess\,sup}|f| \quad \text{for every} \quad [f] \in L^\infty(X, \mathcal{X}, \mu),$$

where f is any representative of the equivalence class $[f]$ in $L^\infty(X, \mathcal{X}, \mu)$.

We have seen in the previous section that since $\mathcal{L}(X, \mathcal{X}, \mu)$ is a linear space, then $L^1(X, \mathcal{X}, \mu)$ was made into a linear space, with scalar multiplication and vector addition of equivalence classes defined as before. This extends immediately to $L^p(X, \mathcal{X}, \mu)$, so that $L^p(X, \mathcal{X}, \mu)$ is made into a linear space as well. Similarly, $L^\infty(X, \mathcal{X}, \mu)$ is also made into a linear space under the same definition of scalar multiplication and vector addition.

Note that the elements of $L^p(X, \mathcal{X}, \mu)$ and $L^\infty(X, \mathcal{X}, \mu)$ are equivalence classes of functions (and not functions themselves). If f is any function of an equivalence class $[f]$, then it is usual and convenient to write f for $[f]$. Since the common usage is simpler, we follow it, and refer to a "function f" in $L^p(X,\mathcal{X},\mu)$ or in $L^\infty(X,\mathcal{X},\mu)$ instead of "an equivalence class $[f]$ that contains f". Thus we write $\|f\|_p$ and $\|f\|_\infty$ instead of $\|[f]\|_p$ and $\|[f]\|_\infty$.

Take any real number $p>1$. Let $q=\frac{p}{p-1}>1$ be the unique solution to the equation $\frac{1}{p}+\frac{1}{q}=1$ (or, equivalently, the unique solution to the equation $p+q=pq$). In this case p and q are *Hölder conjugates* of each other. We will show that $\|\;\|_p$ and $\|\;\|_\infty$ are norms on the linear spaces L^p and L^∞, respectively, but first we need the following fundamental inequalities, which the reader is asked to prove following the hints to Problems 5.1 and 5.3.

Proposition 5.6. (Hölder inequality). *If* $p,q>1$ *are Hölder conjugates, and if* $f\in L^p$ *and* $g\in L^q$, *then* $fg\in L^1$ *and*

$$\|fg\|_1\le\|f\|_p\|g\|_q.$$

If $f\in L^1$ *and* $g\in L^\infty$, *then* $fg\in L^1$ *and*

$$\|fg\|_1\le\|f\|_1\|g\|_\infty.$$

Remark: A very special case. For $p=q=2$, the Hölder inequality leads to the *Schwarz* (or *Cauchy–Schwarz*) *inequality*. An *inner product* on the *real* linear space L^2 is a bilinear functional $\langle\;,\;\rangle\colon L^2\times L^2\to\mathbb{R}$ given by $\langle f;g\rangle=\int fg\,d\mu$ for every $f,g\in L^2$. Indeed, If f and g lie in L^2, then $fg\in L^1$ and

$$|\langle f;g\rangle| = \left|\int fg\,d\mu\right| \le \int|fg|\,d\mu = \|fg\|_1 \le \|f\|_2\|g\|_2,$$

where $\langle f;g\rangle=\int fg\,d\mu$ is the *inner product* of f and g in L^2. If $\mu(X)=1$, then $|\langle f;\chi_X\rangle|\le\|f\|_1\le\|f\|_2$, and so $(\int f\,d\mu)^2\le(\int|f|\,d\mu)^2\le\int|f|^2\,d\mu$.

Proposition 5.7. (Minkowski inequality). *Take any real number* $p\ge1$. *If* $f,g\in L^p$, *then* $f+g\in L^p$ *and*

$$\|f+g\|_p\le\|f\|_p+\|g\|_p.$$

If $f,g\in L^\infty$, *then* $f+g\in L^\infty$ *and*

$$\|f+g\|_\infty\le\|f\|_\infty+\|g\|_\infty.$$

Lemma 5.8. *The functions* $\|\;\|_p$ *and* $\|\;\|_\infty$ *are norms on* L^p *and* L^∞.

Proof. Consider the real linear spaces L^p for each $p\ge1$ and L^∞. Proposition 5.5 ensures that $\|\;\|_1$ is a norm on L^1. Exactly the same argument shows that $\| \|_p$ is a norm on L^p for each $p > 1$, where the triangle inequality is precisely the Minkowski inequality of Proposition 5.7. To verify that $\| \|_\infty$ is a norm on L^∞, note that properties (i) and (iii) of Definition 5.1 are trivially verified, (iv) is again the Minkowski inequality of Proposition 5.7, and (ii) follows since $\operatorname{ess\,sup}|f| = 0$ means $|f| = 0$ μ-almost everywhere. □

5.3 The Riesz–Fischer Completeness Theorem

A central result in integration theory is the *Riesz Theorem*, also referred to as the *Riesz–Fischer Theorem*, or the *Completeness Theorem* (Theorem 5.9). It says that the linear spaces L^p and L^∞ equipped with the norms $\| \|_p$ and $\| \|_\infty$ are *complete* normed spaces (where Cauchy sequences converge).

Theorem 5.9. *$(L^p, \| \|_p)$ for each $p \geq 1$ and $(L^\infty, \| \|_\infty)$ are Banach spaces.*

Proof. Take any real number $p \geq 1$. According to Lemma 5.8, consider the normed spaces $(L^p(X, \mathcal{X}, \mu), \| \|_p)$ and $(L^\infty(X, \mathcal{X}, \mu), \| \|_\infty)$. The proof is split into two parts. Part (a) shows that $(L^p, \| \|_p)$ is complete, and part (b) shows that $(L^\infty, \| \|_\infty)$ is also complete.

(a) Let $\{f_n\}$ be an arbitrary Cauchy sequence in L^p (Definition 5.2). Thus for any integer $k \geq 1$ there is another integer $n_k \geq 1$ for which

$$\|f_m - f_n\|_p < \left(\tfrac{1}{2}\right)^k \quad \text{whenever} \quad m, n \geq n_k.$$

Then there exists a subsequence $\{f_{n_k}\}$ of $\{f_n\}$ such that

$$\|f_{n_{k+1}} - f_{n_k}\|_p < \left(\tfrac{1}{2}\right)^k$$

for each $k \geq 1$. Let $g\colon X \to \overline{\mathbb{R}}$ be a function defined for each $x \in X$ by

$$g(x) = |f_{n_1}(x)| + \sum_{k=1}^{\infty} \left|f_{n_{k+1}}(x) - f_{n_k}(x)\right|.$$

First note that g is a well-defined extended real-valued nonnegative $\mathcal{X}$-measurable function on X (i.e., a function in $\mathcal{M}(X, \mathcal{X})^+$ — cf. Proposition 1.8). Thus the Monotone Convergence Theorem as in Corollary 3.10 ensures that

$$\int g^p \, d\mu = \lim_n \int \Big(|f_{n_1}| + \sum_{k=1}^{n} |f_{n_{k+1}} - f_{n_k}| \Big)^p d\mu.$$

Recall that each $|f_{n_k}|$ is in L^p, and so $|f_{n_1}| + \sum_{k=1}^{n} |f_{n_{k+1}} - f_{n_k}|$ is in the linear space L^p for each integer $n \geq 1$. The Minkowski inequality of Proposition 5.7 (i.e., the triangle inequality) plus a trivial induction ensure that $\|\sum_{k=1}^{n} g_k\|_p \le \sum_{k=1}^{n} \|g_k\|_p$ for every $n \ge 1$ if $\{g_k\}$ is a sequence in L^p. Then

$$\bigg(\int\Big(|f_{n_1}|+\sum_{k=1}^{n}\big|f_{n_{k+1}}-f_{n_k}\big|\Big)^p d\mu\bigg)^{\frac{1}{p}} = \bigg\||f_{n_1}|+\sum_{k=1}^{n}\big|f_{n_{k+1}}-f_{n_k}\big|\bigg\|_p$$
$$< \|f_{n_1}\|_p+\sum_{k=1}^{n}\big(\tfrac{1}{2}\big)^k = \|f_{n_1}\|_p+1$$

for every integer $n \ge 1$. The preceding two expressions ensure that $\int g^p\,d\mu < (\|f_{n_1}\|_p+1)^p < \infty$, and so the set $E=\{x\in X\colon g(x)<\infty\}$ lies in $\mathcal{X}$ with $\mu(X\backslash E)=0$ by Problem 3.9(b). This implies that $\sum_{k=1}^{\infty}(f_{n_{k+1}}(x)-f_{n_k}(x))$ converges for every $x\in E$ (since it converges absolutely). Thus set

$$f(x)=\begin{cases} f_{n_1}(x)+\sum_{k=1}^{\infty}\big(f_{n_{k+1}}(x)-f_{n_k}(x)\big)=\lim_k f_{n_{k+1}}(x), & x\in E,\\ 0, & x\notin E,\end{cases}$$

defining a real-valued function $f\in\mathcal{M}(X,\mathcal{X})$ in L^p. Indeed, $|f|\le g$, and so

$$\int|f|^p\,d\mu \le \int g^p\,d\mu < \big(\|f_{n_1}\|_p+1\big)^p < \infty$$

(cf. Problem 3.3(b)). Hence $f\in L^p$. Observe that
(i) $\{f_{n_k}\}$ converges almost everywhere to f (i.e., $|f_{n_k}-f|\to 0$ μ-a.e.),
(ii) $|f_{n_k}|^p \le g^p$ for all k, where g^p is a nonnegative function in $\mathcal{L}(X,\mathcal{X},\mu)$.
In fact, $f_{n_k}(x)\to f(x)$ for every $x\in E$ and, for all k,

$$|f_{n_k}| \le \bigg|\sum_{j=k}^{\infty}\big(|f_{n_j}|-|f_{n_{j+1}}|\big)\bigg| \le \sum_{j=k}^{\infty}|f_{n_{j+1}}-f_{n_j}| \le \sum_{j=1}^{\infty}|f_{n_{j+1}}-f_{n_j}| \le g.$$

Also, since $|f_{n_k}-f|^p \le (|f_{n_k}|+|f|)^p \le (g+|f|)^p$ by (ii), and $|f|\le g$,

(iii) $|f_{n_k}-f|^p \le (2g)^p$ for all k, where $(2g)^p$ lies in $\mathcal{L}(X,\mathcal{X},\mu)$.

Therefore, since $|f_{n_k}-f|^p\to 0$ μ-a.e. by (i), and according to (iii), it follows by the Dominated Convergence Theorem (Theorem 4.7) that

$$\|f_{n_k}-f\|_p^p = \int|f_{n_k}-f|^p\,d\mu \to 0 \quad \text{as} \quad k\to\infty,$$

and so the subsequence $\{f_{n_k}\}$ of $\{f_n\}$ converges in $(L^p,\|\,\|_p)$ to $f\in L^p$. Use Proposition 5.3(c) to infer that the arbitrary Cauchy sequence $\{f_n\}$ converges in $(L^p,\|\,\|_p)$, and so the normed space $(L^p,\|\,\|_p)$ is complete.

(b) Let $\{f_n\}$ be any sequence of functions in L^∞. Recall that a countable collection of sets of measure zero is again a set of measure zero. Thus, there exists a set $N\in\mathcal{X}$ with $\mu(N)=0$ such that

$$|f_n(x)| \leq \|f_n\|_\infty \quad \text{and} \quad |f_m(x) - f_n(x)| \leq \|f_m - f_n\|_\infty$$

for all $x \in X \backslash N$, for every $m, n \geq 1$. If $\{f_n\}$ is a Cauchy sequence, then for each $\varepsilon > 0$ there exists a positive integer n_ε such that

$$\|f_m - f_n\|_\infty < \varepsilon \quad \text{whenever} \quad n, m \geq n_\varepsilon,$$

and hence

$$\sup_{\substack{x \in X \backslash N \\ m,n \geq n_\varepsilon}} \big|f_m(x) - f_n(x)\big| < \varepsilon.$$

Therefore, the scalar sequence $\{f_n(x)\}$ is a real-valued Cauchy sequence for every x in $X \backslash N$. Since $\mathbb{R}$ (with its usual norm $|\ |$) is a complete normed space, it follows that $\{f_n(x)\}$ converges in $\mathbb{R}$ for every $x \in X \backslash N$. Thus set

$$f(x) = \begin{cases} \lim_n f_n(x), & x \in X \backslash N, \\ 0, & x \in N. \end{cases}$$

This defines a real-valued function f in $\mathcal{M}(X, \mathcal{X})$ which, in fact, lies in L^∞. Actually, take x, y arbitrary in $X \backslash N$, and an arbitrary $n \geq n_\varepsilon$. Note that

$$|f(x) - f(y)| \leq |f(x) - f_n(x)| + |f_n(x) - f_n(y)| + |f_n(y) - f(y)|.$$

Since the function $|\ |: \mathbb{R} \to \mathbb{R}$ is continuous, it follows that

$$|f(x) - f_n(x)| = \big|\lim_m f_m(x) - f_n(x)\big| = \lim_m |f_m(x) - f_n(x)| \leq \varepsilon,$$

and since each f_n lies in L^∞, it also follows that

$$|f_n(x) - f_n(y)| \leq |f_n(x)| + |f_n(y)| \leq 2\|f_n\|_\infty.$$

Thus

$$|f(x)| \leq |f(y)| + |f(x) - f(y)| \leq |f(y)| + 2(\varepsilon + \|f_{n_\varepsilon}\|_\infty),$$

which implies that $f \in L^\infty$. Moreover, for all $n \geq n_\varepsilon$,

$$\|f - f_n\|_\infty = \sup_{x \in X \backslash N} |f(x) - f_n(x)| \leq \varepsilon.$$

Outcome: the arbitrary Cauchy sequence $\{f_n\}$ converges in $(L^\infty, \|\ \|_\infty)$ to $f \in L^\infty$, and so the normed space $(L^\infty, \|\ \|_\infty)$ is complete. □

5.4 Problems

Problem 5.1. Prove the *Hölder inequality*. Let $(X,\mathcal{X},\mu)$ be a measure space and take a pair of real-valued measurable functions f and g in $\mathcal{M}(X,\mathcal{X})$. If $f^p\in\mathcal{L}(X,\mathcal{X},\mu)$ and $g^q\in\mathcal{L}(X,\mathcal{X},\mu)$, where $p>1$ and $q>1$ are Hölder conjugates, then show that $fg\in\mathcal{L}(X,\mathcal{X},\mu)$ and

$$\int |fg|\,d\mu \leq \Big(\int |f|^p\,d\mu\Big)^{\frac{1}{p}}\Big(\int |g|^q\,d\mu\Big)^{\frac{1}{q}}.$$

Hint: First prove the *Young inequality* which says that $\alpha\beta\leq\frac{\alpha^p}{p}+\frac{\beta^q}{q}$ for every positive real numbers α and β whenever p and q are Hölder conjugates.

If $f\in\mathcal{L}(X,\mathcal{X},\mu)$ and g is essentially bounded, then $fg\in\mathcal{L}(X,\mathcal{X},\mu)$ and

$$\int |fg|\,d\mu \leq \text{ess sup}\,|g|\int |f|\,d\mu.$$

Hint: Proposition 4.2(b) and Problem 4.4 (or Problem 3.8(c)).

Problem 5.2. Consider the second inequality of Problem 5.1. If f^p lies in $\mathcal{L}(X,\mathcal{X},\mu)$ for some $p\geq 1$ (or if f is essentially bounded) and g is essentially bounded, then $(fg)^p$ lies in $\mathcal{L}(X,\mathcal{X},\mu)$ (or fg is essentially bounded) and

$$\int |fg|^p\,d\mu \leq \text{ess sup}\,|g|^p\int |f|^p\,d\mu\ \ \big(\text{or}\ \ \text{ess sup}\,|fg|\leq \text{ess sup}\,|f|\ \text{ess sup}\,|g|\,\big).$$

Problem 5.3. Prove the *Minkowski inequality*. Let $(X,\mathcal{X},\mu)$ be a measure space and take p-integrable functions f and g (i.e., real-valued functions in $\mathcal{M}(X,\mathcal{X})$ such that $\int|f|^p\,d\mu<\infty$ and $\int|g|^p\,d\mu<\infty$) for an arbitrary real number $p\geq 1$. Show that $f+g$ is p-integrable (i.e., $\int|f+g|^p\,d\mu<\infty$) and

$$\Big(\int |f+g|^p\,d\mu\Big)^{\frac{1}{p}} \leq \Big(\int |f|^p\,d\mu\Big)^{\frac{1}{p}}+\Big(\int |g|^p\,d\mu\Big)^{\frac{1}{p}}.$$

Hint: The special case of $p=1$ was proved in Proposition 5.4. To prove the case of $p>1$ proceed as follows. Take any α and β in $\mathbb{R}$. Since $|\alpha+\beta|^p\leq 2^p(|\alpha|^p+|\beta|^p)$, show that $\int|f+g|^p\,d\mu<\infty$. Since $|\alpha+\beta|^p\leq(|\alpha|+|\beta|)^p=(|\alpha|+|\beta|)^{p-1}|\alpha|+(|\alpha|+|\beta|)^{p-1}|\beta|$, and recalling that $(p-1)q=p$ if q is the Hölder conjugate of p, use Problem 5.1 to prove the claimed inequality.

Moreover, show that if f and g are essentially bounded, then

$$\text{ess sup}\,|f+g| \leq \text{ess sup}\,|f| + \text{ess sup}\,|g|.$$

Problem 5.4. The *Littlewood second principle* says that "every" function is nearly simple. That is, if $f \in L^p(X, \mathcal{X}, \mu)$ for some $p \geq 1$, then for every $\varepsilon > 0$ there is a measurable simple function φ_ε such that $\|f - \varphi_\varepsilon\|_p < \varepsilon$. Prove it.

Hint: If $0 \leq |f| \in L^p$, then Problem 1.6 ensures the existence of an increasing sequence $\{\varphi_n\}$ of measurable simple functions converging pointwise to $|f|$. So $(|f| - \varphi_n)^p \to 0$ pointwise and $0 \leq (|f| - \varphi_n)^p \leq |f|^p$. Use the Dominated Convergence Theorem (Theorem 4.7) to infer that $\|f - \varphi_n\|_p \to 0$.

Problem 5.5. Let ℓ^p be the set of all scalar-valued (real or complex) sequences $x = \{\xi_k\}$ such that $\sum_{k=1}^\infty |\xi_k|^p < \infty$ (i.e., the set of all scalar-valued *p-summable sequences*), for each real number $p \geq 1$. Let ℓ^∞ be the set of all scalar-valued sequences $x = \{\xi_k\}$ such that $\sup_{k\geq 1} |\xi_k| < \infty$ (i.e., the set of all scalar-valued *bounded sequences*). These are (real or complex) linear spaces. Consider the measure space $(\mathbb{N}, \wp(\mathbb{N}), \mu)$, where μ is the counting measure of Example 2B. Use Problem 3.4 and show that we may identify

$$\ell^p = L^p(\mathbb{N}, \wp(\mathbb{N}), \mu) \quad \text{and} \quad \ell^\infty = L^\infty(\mathbb{N}, \wp(\mathbb{N}), \mu),$$

where each equivalence class in L^p and L^∞ contains just one element.

Problem 5.6. Use the previous problem and Lemma 5.8 to show that

$$\|x\|_p = \Big(\sum_{k=1}^\infty |\xi_k|^p\Big)^{\frac{1}{p}} \quad \text{for every sequence } x = \{\xi_k\} \in \ell^p,$$

$$\|x\|_\infty = \sup_{k\geq 1} |\xi_k| \quad \text{for every sequence } x = \{\xi_k\} \in \ell^\infty,$$

define norms $\|\ \|_p$ and $\|\ \|_\infty$ on ℓ^p and on ℓ^∞, and then apply Theorem 3.9 to verify that $(\ell^p, \|\ \|_p)$ for every $p \geq 1$ and $(\ell^\infty, \|\ \|_\infty)$ are Banach spaces.

Problem 5.7. In particular, the Hölder and Minkowski inequalities (Propositions 5.6 and 5.7) hold for sequences in ℓ^p and ℓ^∞ equipped with the norms of the previous problem. Now prove the *Jensen inequality* for sequences, which says that if p and q are real numbers such that $0 < p < q$, then

$$\Big(\sum_{k=1}^\infty |\xi_k|^q\Big)^{\frac{1}{q}} \leq \Big(\sum_{k=1}^\infty |\xi_k|^p\Big)^{\frac{1}{p}}$$

for every scalar-valued sequence $x = \{\xi_k\}$ such that $\sum_{k=1}^\infty |\xi_k|^p < \infty$.

Hint: Prove that $\sum_{k=1}^\infty \alpha_k^r \leq \big(\sum_{k=1}^\infty \alpha_k\big)^r$ for each real $r \geq 1$ if the sequence of *nonnegative* real numbers $\{\alpha_k\}$ is such that $\sum_{k=1}^\infty \alpha_k < \infty$ (i.e., whenever the series $\sum_{k=1}^\infty \alpha_k$ converges).

Moreover, for real numbers q and p show that

$$1 < p < q \quad \text{implies} \quad \ell^1 \subset \ell^p \subset \ell^q \subset \ell^\infty.$$

Hint: Take $\{\frac{1}{k}\} \in \ell^p \backslash \ell^1$ for $p > 1$ to verify that these are proper inclusions.

Problem 5.8. Let p, q, r be real numbers such that $1 \leq r < \min\{p, q\}$ and

$$\frac{1}{p} + \frac{1}{q} = \frac{1}{r}.$$

Prove the *generalized Hölder inequality*, which reads as follows. If $f \in L^p$ and $g \in L^q$, then $fg \in L^r$, and

$$\|fg\|_r \leq \|f\|_p \|g\|_q.$$

Hint: Show that $\frac{p}{r}$ and $\frac{q}{r}$ are Hölder conjugates.

Also, if $(X, \mathcal{X}, \mu)$ is a *finite* measure space (i.e., $\mu(X) < \infty$), then prove that

$$1 < r < p \quad \text{implies} \quad L^\infty \subseteq L^p \subseteq L^r \subseteq L^1.$$

Hint: $\|f\|_r \leq \|f\|_p \, \mu(X)^{\frac{p-r}{pr}}$ by the generalized Hölder inequality.

Problem 5.9. Consider an arbitrary measure space $(X, \mathcal{X}, \mu)$. Prove that the following assertions are pairwise equivalent.

(a) $\mu(X) < \infty$.

(b) $L^\infty \subseteq L^p$ for every $p \geq 1$.

(c) $L^\infty \subseteq L^p$ for some $p \geq 1$.

Problem 5.10. Prove that if $(X, \mathcal{X}, \mu)$ is a *finite* measure space, then

$$\lim_{p \to \infty} \|f\|_p = \|f\|_\infty \quad \text{for every} \quad f \in L^\infty.$$

Hint: Problem 5.9. $|f|^p \leq |f|^{p-1}|f|$ implies $\int |f|^p \, d\mu \leq \|f\|_\infty^{p-1} \int |f| \, d\mu$, and $\|f\|_p \leq \|f\|_\infty^{1-(1/p)} \|f\|_1^{1/p} = \|f\|_\infty (\|f\|_1 / \|f\|_\infty)^{1/p}$. So $\limsup \|f\|_p \leq \|f\|_\infty$. For γ in $(0, \|f\|_\infty)$ the set $E = \{x \in X \colon \gamma < |f(x)|\}$ is such that $\mu(E)^{1/p} \gamma \leq (\int_E |f|^p \, d\mu)^{1/p} \leq \|f\|_p$. So $\gamma \leq \liminf \|f\|_p$ and $\|f\|_\infty = \sup \gamma \leq \liminf \|f\|_p$.

Problem 5.11. Let $(X, \mathcal{X}, \mu)$ be an arbitrary measure space. Prove that

$$L^r \cap L^q \subseteq L^p \quad \text{whenever} \quad 1 \leq r \leq p \leq q.$$

Hint: Take $E = \{x \in X\colon |f(x)| \le 1\}$ and $F = \{x \in X\colon 1 < |f(x)|\}$ in $\mathcal{X}$ (cf. Proposition 1.6). Show that $\int_E |f|^p\,d\mu + \int_F |f|^p\,d\mu \le \int_E |f|^r\,d\mu + \int_F |f|^q\,d\mu$.

Problem 5.12. Let $(\mathbb{R}, \mathfrak{R}, \lambda)$ be the Lebesgue measure space. Take an arbitrary $\mathfrak{R}$-measurable set E and consider the restriction $\lambda|_E$ of the Lebesgue measure λ to the σ-algebra $\mathcal{E} = \wp(E) \cap \mathfrak{R}$ of all Borel subsets of E as in Problem 2.11. Take any real number $p \ge 1$ and consider the *Lebesgue spaces*:

$$L^p(E) = L^p(E, \mathcal{E}, \lambda|_E) \quad \text{and} \quad L^\infty(E) = L^\infty(E, \mathcal{E}, \lambda|_E).$$

(a) For each $p \ge 1$ (or $p = \infty$), a function $f \in \mathcal{M}(\mathbb{R}, \mathfrak{R})$ is *locally* L^p if $f \in L^p(E)$ for every bounded set $E \in \mathfrak{R}$. Verify that, if $f \in L^p(\mathbb{R}, \mathfrak{R}, \lambda)$ for some p, then f is locally L^p, and show that the converse fails.

(b) Set $E = [0,1]$. Show that the inclusions of Problem 5.8 are proper:

$$L^\infty([0,1]) \subset L^p([0,1]) \subset L^r([0,1]) \subset L^1([0,1])$$

whenever $1 < r < p$, by exhibiting functions in

$$L^2([0,1]) \backslash L^\infty([0,1]) \quad \text{and} \quad L^1([0,1]) \backslash L^2([0,1]).$$

(c) On the other hand, for an unbounded $\mathfrak{R}$-measurable set E we get the opposite. Exhibit functions in

$$L^\infty([1,\infty)) \backslash L^2([1,\infty)) \quad \text{and} \quad L^2([1,\infty)) \backslash L^1([1,\infty)).$$

Hints: $\int \frac{dx}{\sqrt{x}} = 2\sqrt{x}$, $\int \frac{dx}{x} = \log(x)$, and $\int \frac{dx}{x^2} = -\frac{1}{x}$.

Problem 5.13. Consider the set $C[0,1]$ consisting of all real-valued *continuous* functions $f\colon [0,1] \to \mathbb{R}$ defined on the closed and bounded interval $[0,1]$. First verify that $C[0,1]$ is a linear space when vector addition and scalar multiplication are pointwise defined. Now show that the functions $\|\ \|_p\colon C[0,1] \to \mathbb{R}$ for each real $p \ge 1$ and $\|\ \|_\infty\colon C[0,1] \to \mathbb{R}$, given by

$$\|f\|_p = \left(\int_{[0,1]} |f(x)|^p\,dx \right)^{\frac{1}{p}} \quad \text{and} \quad \|f\|_\infty = \max_{x \in [0,1]} |f(x)|,$$

are well defined for every $f \in C[0,1]$, and are norms on $C[0,1]$ (cf. Minkowski inequality). Also show that it makes no difference whether the preceding integral is Riemann or Lebesgue. Next prove the following assertions.

(a) The normed space $(C[0,1], \|\ \|_p)$ is not complete for any $p \ge 1$.

Hint: Take the sequence $\{f_n\}$ of functions in $C[0,1]$ given by

$$f_n(x) = \begin{cases} 1, & x \in [0, \frac{1}{2}], \\ n+1-2nx, & x \in [\frac{1}{2}, \frac{n+1}{2n}], \\ 0, & x \in [\frac{n+1}{2n}, 1]. \end{cases}$$

Show that $\{f_n\}$ is a Cauchy sequence in $(C[0,1], \| \|_p)$ but does not converge in $(C[0,1], \| \|_p)$ to any (continuous) function in $C[0,1]$.

(b) However, $(C[0,1], \| \|_\infty)$ is a Banach space. (*Hint:* Proof of Theorem 5.9.)

Problem 5.14. Take an arbitrary $p \geq 1$. Now consider the Lebesgue Space $L^p([0,1])$ of all p-integrable functions on $[0,1]$ (defined in Problem 5.12). Let $R^p[0,1]$ denote the subset of $L^p[0,1]$ consisting of all (equivalence classes of) Riemann integrable functions f for which $|f|^p$ has a finite Riemann integral. $R^p[0,1]$ is a linear manifold of $L^p[0,1]$, and so it is a linear space. Equip it with the norm $\| \|_p$ and consider the normed spaces $(R^p[0,1], \| \|_p)$. Let $\{f_n\}$ be a sequence of real-valued functions on $[0,1]$ defined by

$$f_n(x) = \begin{cases} 1, & x = \frac{k}{n!} \in [0,1] \text{ for some integer } k \geq 0, \\ 0, & \text{otherwise.} \end{cases}$$

Verify that each f_n lies in $R^p[0,1]$. (*Hint:* Problem 4.1.) Apply the same argument to show that the Dirichlet function f on $[0,1]$,

$$f(x) = \chi_{[0,1]\cap\mathbb{Q}}(x) = \begin{cases} 1, & x \in [0,1] \cap \mathbb{Q}, \\ 0, & x \in [0,1] \backslash \mathbb{Q}, \end{cases}$$

lies in $L^p[0,1] \backslash R^p[0,1]$. (*Hint:* $[0,1] \backslash \mathbb{Q}$ is totally disconnected and of measure 1 — cf. Problems 2.7(b) and 4.1.) Then show that

$$f_n \to f \quad \text{pointwise.}$$

Hint: If $f_n(x) = 1$, then $x = \frac{k}{n!}$, so $x = \frac{k(n+1)}{(n+1)!}$ and hence $f_{n+1}(x) = 1$.

Can we infer from this problem that $(R^p[0,1], \| \|_p)$ is not complete?

Problem 5.15. Consider the setup of Problem 5.14, where $(R^p[0,1], \| \|_p)$ is a linear manifold of the Banach space $(L^p[0,1], \| \|_p)$. We show that the normed space $(R^p[0,1], \| \|_p)$ is not complete for any $p \geq 1$. To begin with, take the decreasing collection $\{S_n\}$ of closed subsets of $S_0 = [0,1]$ used to build up the Cantor-like set $S = \bigcap_n S_n$ of Problem 2.10. Observe that

$$\lambda(S_n) = 1 - \sum_{i=0}^{n-1} \frac{2^i}{4^{i+1}} = \frac{1}{2} + \frac{1}{2^{n+1}}$$

is the length of S_n (and so the Lebesgue measure of S_n) for each $n \geq 1$. Take the sequence $\{f_n\}$ of characteristic functions of S_n for every $n \geq 1$,

$$f_n(x) = \chi_{S_n}(x) = \begin{cases} 1, & x \in S_n, \\ 0, & x \in S_0 \backslash S_n, \end{cases}$$

and verify that each f_n belongs to $R^p[0,1]$ for every $p \geq 1$ (see Problem 4.1). Let f be the characteristic function of S; that is,

$$f(x) = \chi_S(x) = \begin{cases} 1, & x \in S, \\ 0, & x \in S_0 \backslash S. \end{cases}$$

(a) Show that $\{f_n\}$ is a Cauchy sequence in $(R^p[0,1], \| \ \|_p)$.

Hint: Verify that $\|f_m - f_n\|_p^p \leq \frac{1}{2^m+1}$ whenever $m \leq n$.

(b) Show that $f \in L^p[0,1]$ and $\{f_n\}$ converges in $(L^p[0,1], \| \ \|_p)$ to f.

Hint: Verify that $\|f_n - f\|_p^p = \lambda(S_n \backslash S) = \frac{1}{2^{n+1}}$ since $f_n - f = \chi_{S_n \backslash S}$.

(c) Show that $f \notin R^p[0,1]$.   (*Hint:* Problems 2.10 and 4.1.)

Use (a), (b), (c) to infer that there is a Cauchy sequence $\{f_n\}$ of functions in $(R^p[0,1], \| \ \|_p)$ that does not converge in $(R^p[0,1], \| \ \|_p)$. Thus conclude that for any $p \geq 1$ the normed space $(R^p[0,1], \| \ \|_p)$ is not a Banach space:

$(R^p[0,1], \| \ \|_p)$ is an incomplete normed space.

Remark: We have promised in Chapter 3 to show that the integral

$$\int (\cdot) \, d\mu \colon L^1 \to \mathbb{R}$$

is a *continuous linear functional*. In fact, linearity follows from Lemma 4.5 (since linearity of $\mathcal{L}$ was extended to L^1 in Proposition 5.5, and so linearity of the integral functional on L^1 follows by using the same argument of Lemma 4.5). We will now verify continuity for the *integral functional* (cf. Problem 5.16(a)) — this will be extended in Proposition 10.G). Recall that *a map between metric spaces is continuous if and only if it preserves convergence.* In particular, the integral $\int (\cdot) \, d\mu \colon L^1 \to \mathbb{R}$ is a *continuous functional* if and only if, whenever a sequence $\{f_n\}$ of functions in L^1 converges in L^1 to $f \in L^1$, then the real sequence $\{\int f_n \, d\mu\}$ converges in $\mathbb{R}$ to $\int f_n \, d\mu \in \mathbb{R}$.

Problem 5.16. Let $(X, \mathcal{X}, \mu)$ be a measure space and take any $p \geq 1$.

(a) If a sequence $\{f_n\}$ of functions in L^1 converges in L^1 to $f \in L^1$, then

$$\int f \, d\mu = \lim_n \int f_n \, d\mu.$$

Hint: $\left|\int f_n \, d\mu - \int f \, d\mu\right| \leq \int |f_n - f| \, d\mu = \|f_n - f\|_1$.

(b) If a sequence $\{f_n\}$ of functions in L^p converges in L^p to $f \in L^p$, then

$$\int |f|^p \, d\mu = \lim_n \int |f_n|^p \, d\mu.$$

Hint: Verify that, by the triangle inequality, $\left|\|f_n\|_p - \|f\|_p\right| \leq \|f_n - f\|_p$.

If, in addition, $\mu(X) < \infty$, then

$$\int f \, d\mu = \lim_n \int f_n \, d\mu.$$

Hint: Hölder inequality with a constant function g, and the hint to (a).

Problem 5.17. Prove the *Riesz Theorem*, which reads as follows. Consider a measure space $(X, \mathcal{X}, \mu)$, take an arbitrary $p \geq 1$, and suppose a sequence $\{f_n\}$ of functions in L^p converges μ-almost everywhere to $f \in L^p$. Then

$$\int |f|^p \, d\mu = \lim_n \int |f_n|^p \, d\mu$$

if and only if $\{f_n\}$ converges in L^p to f. In other words,

$$f_n \in L^p \to f \in L^p \quad \mu\text{-a.e.} \quad \Longrightarrow \quad \left\{ \|f_n\|_p \to \|f\|_p \Longleftrightarrow \|f_n - f\|_p \right\}.$$

Hint: $0 \leq h_n = 2^p(|f_n|^p + |f|^p) - |f_n - f|^p \in L^p \to 2^{p+1}|f|^p \in L^p$ a.e. (hint to Problem 5.3 ensures $0 \leq h_n$, $f_n \to f$ a.e. implies a.e. convergence). If $\|f_n\|_p \to \|f\|_p$ then, by Lemma 3.9, $2^{p+1}\int |f|^p \, d\mu \leq \liminf_n \int h_n \, d\mu = 2^{p+1}\int |f|^p \, d\mu - \limsup_n \int |f_n - f|^p \, d\mu$. This implies $\lim_n \int |f_n - f|^p \, d\mu = 0$.

See Problem 4.14. Now give another solution to Problem 4.15: under the assumption of the Dominated Convergence Theorem, $f_n \to f$ in L^1.

Suggested Reading

Bartle [4], Bauer [6], Brown and Pearcy [8], Halmos [18], Royden [35], Rudin [36]. See also [7], [13], [16], [17], [41], [42] and, for an introduction to Banach spaces, e.g., [26, Chapter 4].

6

Convergence of Functions

6.1 Four Basic Convergence Notions

Major convergence concepts for sequences of *real-valued* functions will be considered in this chapter. We have already met four convergence concepts so far (viz., pointwise, uniform, almost everywhere, and convergence in L^p). These are reviewed and compared in this section. Further concepts, namely, convergence in measure, uniform almost everywhere, and almost uniform convergence, will be discussed and compared in subsequent sections.

Definition 6.1. Take a sequence $\{f_n\}$ of real-valued functions $f_n\colon X \to \mathbb{R}$ on a set X. The sequence $\{f_n\}$ *converges pointwise* to a real-valued function $f\colon X \to \mathbb{R}$ on X if the real-valued sequence $\{f_n(x)\}$ converges in $\mathbb{R}$ to the real number $f(x)$ for every $x \in X$. That is, if $|f_n(x) - f(x)| \to 0$ as $n \to \infty$ for every $x \in X$. In other words, $\{f_n\}$ converges pointwise to f if for every $\varepsilon > 0$ and each $x \in X$ there is a positive integer $n_{\varepsilon,x}$ such that

$$n \geq n_{\varepsilon,x} \quad \text{implies} \quad |f_n(x) - f(x)| < \varepsilon.$$

In this case we write $f_n \to f$ pointwise. If the integer $n_{\varepsilon,x}$ does not depend on x, then $\{f_n\}$ *converges uniformly* to f, and we write $f_n \to f$ uniformly. Thus a sequence of functions $\{f_n\}$ converges uniformly to a real-valued function $f\colon X \to \mathbb{R}$ if $\sup_{x\in X} |f_n(x) - f(x)| \to 0$ as $n \to \infty$ or, equivalently, if for every $\varepsilon > 0$ there is a positive integer n_ε such that

$$n \geq n_\varepsilon \quad \text{implies} \quad \sup_{x\in X} |f_n(x) - f(x)| < \varepsilon.$$

C.S. Kubrusly, *Essentials of Measure Theory*,
DOI 10.1007/978-3-319-22506-7_6

Observe that pointwise convergence is convergence in the normed space $(\mathbb{R}, |\ |)$. Since $(\mathbb{R}, |\ |)$ is a Banach space (cf. Section 5.1), a real-valued sequence $\{f_n(x)\}$ converges to $f(x)$ in $\mathbb{R}$ for every $x \in X$ if and only if $\{f_n(x)\}$ is a Cauchy sequence in $(\mathbb{R}, |\ |)$ for every $x \in X$. This means that for every $\varepsilon > 0$ and each $x \in X$ there is a positive integer $n_{\varepsilon,x}$ such that

$$m, n \geq n_{\varepsilon,x} \quad \text{implies} \quad |f_m(x) - f_n(x)| < \varepsilon,$$

which is denoted as $\lim_{m,n} |f_m(x) - f_n(x)| = 0$ for every $x \in X$. If such an $n_{\varepsilon,x}$ does not depend on x, then $\{f_n\}$ is a *uniform Cauchy sequence*. This means that for every $\varepsilon > 0$ there is a positive integer n_ε such that

$$m, n \geq n_\varepsilon \quad \text{implies} \quad \sup_{x \in X} |f_m(x) - f_n(x)| < \varepsilon,$$

which is denoted as $\lim_{m,n} \sup_{x \in X} |f_m(x) - f_n(x)| = 0$. It is clear that if $f_n \to f$ uniformly, then $\{f_n\}$ is a uniform Cauchy sequence. Actually, by the triangle inequality, for every m and n,

$$\sup_{x \in X} |f_m(x) - f_n(x)| \leq \sup_{x \in X} |f_m(x) - f(x)| + \sup_{x \in X} |f(x) - f_n(x)|.$$

Conversely, if $\{f_n\}$ is a uniform Cauchy sequence, then $\{f_n(x)\}$ is a real-valued Cauchy sequence, and so it converges in $\mathbb{R}$ to a real number, say $f(x)$, for every $x \in X$. This defines a function $f\colon X \to \mathbb{R}$ such that $f_n \to f$. Since $\{f_n\}$ is a uniform Cauchy sequence, it follows that $f_n \to f$ uniformly. In fact, for arbitrary $x \in X$ and $\varepsilon > 0$ there is a positive integer n_ε for which

$$m \geq n_\varepsilon \quad \text{implies} \quad |f_m(x) - f(x)| = \lim_n |f_m(x) - f_n(x)| < \varepsilon.$$

Examples throughout the chapter will compare the several convergence notions. Problems 6.6 and 6.7 summarize all possible implications among some convergence notions that are based on measure-theoretical concepts.

Example 6A. $f_n \to f$ uniformly $\underset{\not\Leftarrow}{\Rightarrow}$ $f_n \to f$ pointwise.

It is clear that uniform convergence implies pointwise convergence (to the *same* and unique limit). It is readily verified that the converse fails. For instance, for each integer $n \geq 1$ let $f_n\colon [0,1] \to \mathbb{R}$ be given by $f_n(x) = x^n$ if $x \in [0,1)$ and $f_n(x) = 0$ if $x = 1$. The sequence $\{f_n\}$ converges pointwise to the null function 0 (i.e., $0\colon [0,1] \to \mathbb{R}$ such that $0(x) = 0$ for all $x \in [0,1]$) but it does not converge to 0 uniformly (since $\sup_{x \in [0,1]} |f_n(x)| = 1$ for all $n \geq 1$). Hence it does not converge uniformly to any function.

The previous convergence notions are all measure free. The next notions require a measure space. So, from now on, take a measure space $(X, \mathcal{X}, \mu)$, and by a measurable function we mean an $\mathcal{X}$-measurable function on X.

Definition 6.2. A sequence $\{f_n\}$ of real-valued functions $f_n\colon X \to \mathbb{R}$ *converges almost everywhere* with respect to the measure $\mu\colon \mathcal{X} \to \overline{\mathbb{R}}$ (or μ-*almost everywhere*) to a real-valued function $f\colon X \to \mathbb{R}$ if the real-valued sequence $\{f_n(x)\}$ converges in $\mathbb{R}$ to the real number $f(x)$ for every x except in a set of measure zero. In this case we write $f_n \to f$ μ-a.e. (or $f_n \to f$ a.e. if the measure μ is clear or is not relevant in the context). That is, $\{f_n\}$ converges almost everywhere to f if there exists a set $N \in \mathcal{X}$ with $\mu(N) = 0$ such that for every $\varepsilon > 0$ and each $x \in X \backslash N$ there is a positive integer $n_{\varepsilon,x}$ such that

$$n \geq n_{\varepsilon,x} \quad \text{implies} \quad |f_n(x) - f(x)| < \varepsilon.$$

A sequence $\{f_n\}$ of real-valued functions on a set X *converges in* $L^p(X, \mathcal{X}, \mu)$ for some $p \geq 1$ if it converges in the Banach space $(L^p, \| \, \|_p)$. In this case we write $f_n \to f$ in L^p. In other words, $\{f_n\}$ converges in L^p if $\|f_n - f\|_p \to 0$ as $n \to \infty$ for some real-valued function $f \in L^p$. That is, for each $\varepsilon > 0$ there is a positive integer n_ε such that

$$n \geq n_\varepsilon \quad \text{implies} \quad \|f_n - f\|_p < \varepsilon.$$

Example 6B. $\quad \{f_n\}$ converges pointwise $\underset{\not\Leftarrow}{\Longrightarrow}$ $\{f_n\}$ converges a.e.,

where, for any measure space $(X, \mathcal{X}, \mu)$, the implication

$$f_n \to f \text{ pointwise} \quad \Longrightarrow \quad f_n \to f \text{ a.e.}$$

holds trivially, since $\mu(\varnothing) = 0$ for every measure μ on any σ-algebra $\mathcal{X}$. We will now see *how* the converse fails. Recall that the limit of a sequence that converges pointwise or uniformly is unique, and so is the limit of a sequence that converges in L^p, where in this case uniqueness is understood almost everywhere (the class of equivalence $[f]$ containing f is unique). But unlike pointwise, uniform, and convergence in L^p, the notion of almost everywhere convergence does not imply uniqueness of the almost everywhere limit. For instance, take the function $f\colon \mathbb{R} \to \mathbb{R}$ defined by $f(x) = 0$ for all $x \in \mathbb{R}\backslash\{0\}$ and $f(0) = 1$. Let $\{f_n\}$ be a constant sequence with $f_n = f$ for all $n \geq 1$. Consider the Lebesgue measure space $(\mathbb{R}, \Re, \lambda)$. The sequence $\{f_n\}$ trivially converges pointwise to f (which is its unique pointwise limit of $\{f_n\}$), and $\{f_n\}$ converges a.e. to any function $f'\colon \mathbb{R} \to \mathbb{R}$ such that $f' = f$ a.e. (i.e., $f'(x) = f(x)$ for every $x \in \mathbb{R}\backslash N$ for some $N \in \Re$ such that $\lambda(N) = 0$). In particular, it converges almost everywhere to the null function $0\colon \mathbb{R} \to \mathbb{R}$ (since $\lambda(\{0\}) = 0$ and $f(x) = 0$ for all $x \in \mathbb{R}\backslash\{0\}$), and also to f itself (since $\lambda(\varnothing) = 0$). Thus, $f_n \to 0$ λ-a.e. and $f_n \to f \neq 0$ pointwise (and so λ-a.e.). Now note that we might argue that such an example would be meaningless had we decided to work with equivalence classes of functions,

$$[f] = \{f'\colon \mathbb{R} \to \mathbb{R}\colon\ f' = f \ \lambda\text{-a.e.}\},$$

rather than with single functions. In fact, this would yield uniqueness for the almost everywhere limit in this particular example, but would not be enough to avoid the failure of the converse in the preceding implication; the converse in the preceding implication fails anyway. Indeed, if $g_n = (-1)^n f$, then $g_n \to 0$ λ-a.e. and $\{g_n\}$ does not converge pointwise (to any function).

In general, the notions of uniform convergence and convergence in L^p are not related (see Problem 6.1):

$$\{f_n\} \text{ converges uniformly } \;\substack{\not\Rightarrow \\ \not\Leftarrow}\; \{f_n\} \text{ converges in } L^p.$$

However, if $\{f_n\}$ converges both uniformly and in L^p, then the limits coincide almost everywhere. Moreover, in a *finite* measure space uniform convergence implies convergence in L^p to the same (a.e.) limit.

Proposition 6.3. *Take any $p \geq 1$. Consider a sequence $\{f_n\}$ of functions in L^p, and let f, f', f'' be real-valued measurable functions.*

(a) *If $f_n \to f'$ uniformly and $f_n \to f''$ in L^p, then $f' = f''$ almost everywhere.*

(b) *If $\mu(X) < \infty$ and $f_n \to f$ uniformly, then $f \in L^p$ and $f_n \to f$ in L^p.*

Proof. Take real-valued $\mathcal{X}$-measurable functions f_n, f, f', and f'' on X.

(a) $|f' - f''| \leq |f' - f_n| + |f_n - f''| \leq \sup_{x \in X} |f'(x) - f_n(x)| + |f_n - f''|$ for each n. Since $f_n \to f'$ uniformly, we get $|f' - f''| \leq \limsup_n |f_n - f''|$. Since $f_n \to f''$ in L^p, it follows (cf. Problem 3.3(b) and Theorem 4.7) that

$$0 \leq \int |f' - f''|^p \, d\mu \leq \limsup_n \int |f_n - f''|^p \, d\mu = \lim_n \|f_n - f''\|_p^p = 0,$$

and so $f' = f''$ a.e. (cf. Propositions 1.5, 1.6, and 3.7(a)).

(b) If each f_n lies in L^p for an arbitrary $p \geq 1$, then

$$\|f_m - f_n\|_p^p = \int |f_m - f_n|^p \, d\mu \leq \sup_{x \in X} |f_m(x) - f_n(x)|^p \, \mu(X)$$

for every pair of positive integers m and n. Since $\{f_n\}$ is uniformly Cauchy (because $f_n \to f$ uniformly) and $\mu(X) < \infty$, it follows that $\{f_n\}$ is a Cauchy sequence in the Banach space $(L^p, \| \, \|_p)$, and so it converges in L^p; and the L^p limit coincides a.e. with the uniform limit f (i.e., it is in $[f]$) by (a). □

According to Problem 6.2(a), observe that even if $\mu(X) < \infty$,

$$\{f_n\} \text{ converges pointwise } \;\not\Rightarrow\; \{f_n\} \text{ converges in } L^p.$$

But under the assumption of *dominated convergence*, just convergence almost everywhere is enough to ensure convergence in L^p.

Proposition 6.4. *Take any* $p \geq 1$. *Consider a sequence* $\{f_n\}$ *of functions in* L^p, *let* f *be a real-valued measurable function, and take* g *in* L^p.

$$|f_n| \leq g \text{ for all } n \text{ and } f_n \to f \text{ a.e.} \implies f \in L^p \text{ and } f_n \to f \text{ in } L^p.$$

Proof. The dominance assumption, namely, $|f_n| \leq g$ for all n, is equivalent to almost everywhere dominance; that is, $|f_n| \leq g$ for all n a.e., since the functions f_n and g are in L^p, where inequalities (and equalities) are understood in the sense of equivalence classes (and so they are always interpreted almost everywhere). Since $|f_n|^p \leq g^p$ for all n a.e. and $f_n \to f$ a.e., it follows that $|f|^p \leq g^p$ almost everywhere. Thus, if $g \in L^p$, then f^p is integrable (cf. Problem 4.4(b)); that is, $f \in L^p$. Moreover,

$$|f_n - f|^p \leq \big(|f_n| + |f|\big)^p \leq (2g)^p \in L^1 \quad \text{and} \quad |f_n - f|^p \to 0 \ \mu\text{-a.e.},$$

so $f_n \to f$ in L^p by the Dominated Convergence Theorem (Theorem 4.7):

$$\lim_n \|f_n - f\|_p^p = \lim_n \int |f_n - f|^p \, d\mu = 0. \qquad \square$$

Remark: All constant functions lie in every L^p if the measure is *finite*. This yields the following *uniformly bounded* version of the previous proposition.

$$\sup_n |f_n| < \infty, \ \mu(X) < \infty \text{ and } f_n \to f \ \mu\text{-a.e.} \implies f_n \to f \in L^p \text{ in } L^p.$$

Convergence in L^p is of crucial importance since it is convergence in the norm topology of the Banach space $(L^p, \| \, \|_p)$. Generally, it does not imply uniform convergence, nor is it implied by uniform convergence (cf. Problem 6.1), although in a finite measure space it is weaker than uniform convergence (Proposition 6.3). Even in a finite measure space it is not implied by almost everywhere convergence, and not even by pointwise convergence (cf. Problem 6.2(a)). But, under the dominance hypothesis it becomes weaker than almost everywhere convergence (Proposition 6.4). However,

$$\{f_n\} \text{ converges in } L^p \quad \not\Longrightarrow \quad \{f_n\} \text{ converges a.e.},$$

even under finite measure and dominance condition (cf. Problem 6.2(b)). On the other hand, convergence in measure (defined in the next section) is weaker than L^p-convergence.

6.2 Convergence in Measure

Throughout this section, let $(X, \mathcal{X}, \mu)$ be a measure space and let $\{f_n\}$ be a sequence of real-valued $\mathcal{X}$-measurable functions (which means that each function $f_n\colon X \to \mathbb{R}$ lies in $\mathcal{M}(X, \mathcal{X})$).

Definition 6.5. A sequence $\{f_n\}$ of real-valued functions in $\mathcal{M}(X, \mathcal{X})$ *converges in measure* to a real-valued function f in $\mathcal{M}(X, \mathcal{X})$ if

$$\lim_n \mu(\{x \in X\colon |f_n(x) - f(x)| \geq \alpha\}) = 0$$

for every $\alpha > 0$. In this case we write $f_n \to f$ in measure. The sequence $\{f_n\}$ is *Cauchy in measure* if for every $\alpha > 0$,

$$\lim_{m,n} \mu(\{x \in X\colon |f_m(x) - f_n(x)| \geq \alpha\}) = 0.$$

Since the functions f_n and f lie in $\mathcal{M}(X, \mathcal{X})$, the set

$$F_n(\alpha) = \{x \in X\colon |f_n(x) - f(x)| \geq \alpha\}$$

lies in $\mathcal{X}$ for every integer $n \geq 1$ and each real $\alpha > 0$. Thus convergence in measure means $\lim_n \mu(F_n(\alpha)) = 0$, which implies that the sequence $\{\mu(F_n(\alpha))\}$ is eventually real-valued. So $f_n \to f$ in measure if and only if for every $\varepsilon > 0$ and every $\alpha > 0$ there is a positive integer $n_{\varepsilon,\alpha}$ such that

$$n \geq n_{\varepsilon,\alpha} \quad \text{implies} \quad \mu(\{x \in X\colon |f_n(x) - f(x)| \geq \alpha\}) < \varepsilon.$$

Similarly, the sequence $\{f_n\}$ is Cauchy in measure if and only if for every $\varepsilon > 0$ and every $\alpha > 0$ there exists a positive integer $n_{\varepsilon,\alpha}$ such that

$$m, n \geq n_{\varepsilon,\alpha} \quad \text{implies} \quad \mu(\{x \in X\colon |f_m(x) - f_n(x)| \geq \alpha\}) < \varepsilon.$$

Example 6C. $f_n \to f$ in L^p $\;\substack{\Longrightarrow \\ \not\Longleftarrow}\;$ $f_n \to f$ in measure.

Convergence in L^p implies convergence in measure to the same limit. In fact, for every $\alpha > 0$ and every integer $n \geq 1$,

$$\alpha^p \mu(F_n(\alpha)) = \int_{F_n(\alpha)} \alpha^p \, d\mu \leq \int_{F_n(\alpha)} |f_n - f|^p \, d\mu \leq \int |f_n - f|^p \, d\mu = \|f_n - f\|_p^p.$$

The converse, however, fails even under the assumption that $\mu(X) < \infty$. Actually, the sequence $\{f_n\}$ of Problem 6.2(a) does not converge in L^p but it is readily verified that it converges in measure to the null function.

Example 6D. $f_n \to f$ uniformly $\underset{\not\Leftarrow}{\Rightarrow}$ $f_n \to f$ in measure.

Indeed, if $f_n \to f$ uniformly, then for any $\varepsilon > 0$ there is an n_ε such that

$$n \geq n_\varepsilon \implies \sup_{x \in X} |f_n(x) - f(x)| < \varepsilon \implies F_n(\varepsilon) = \varnothing \implies \mu(F_n(\varepsilon)) = 0.$$

This means that the sequence $\{\mu(F_n(\alpha))\}$ not only converges to zero but is eventually null for every $\alpha > 0$. Again, the converse fails. For example, the sequence $\{f_n\}$ of Problem 6.1(b) converges in L^p, and so it converges in measure, but it does not converge uniformly. Observe that the sequence $\{f_n\}$ of Problem 6.1(a) also yields another example of a sequence that converges in measure (since it converges uniformly) but not in L^p.

Proposition 6.6. *Let $\{f_n\}$ be a sequence of real-valued measurable functions.*

(a) *If $\{f_n\}$ converges in measure, then it is Cauchy in measure.*

(b) *If $\{f_n\}$ converges in measure, then the limit is unique almost everywhere.*

(c) *If $\{f_n\}$ is Cauchy in measure and has a subsequence that converges in measure, then it converges in measure itself and its limit coincides with the limit of that subsequence.*

Proof. Take real-valued measurable functions f_n, f, and f'. Set $F_n(\alpha) = \{x \in X\colon |f_n(x) - f(x)| \geq \alpha\}$ and $F_n'(\alpha) = \{x \in X\colon |f_n(x) - f'(x)| \geq \alpha\}$.

(a) Since $|f_m(x) - f_n(x)| \leq |f_m(x) - f(x)| + |f_n(x) - f(x)|$, by the triangle inequality, we get $\{x \in X\colon |f_m(x) - f_n(x)| \geq \alpha\} \subseteq F_m(\frac{\alpha}{2}) \cup F_n(\frac{\alpha}{2})$, and so

$$\mu(\{x \in X\colon |f_m(x) - f_n(x)| \geq \alpha\}) \leq \mu(F_m(\tfrac{\alpha}{2})) + \mu(F_n(\tfrac{\alpha}{2})).$$

Suppose $f_n \to f$ in measure. Then, by definition, $\lim_n \mu(F_n(\frac{\alpha}{2})) = 0$, and hence $\lim_{m,n} \mu(\{x \in X\colon |f_m(x) - f_n(x)| \geq \alpha\}) = 0$, for every $\alpha > 0$.

(b) Since $|f(x) - f'(x)| \leq |f_n(x) - f(x)| + |f_n(x) - f'(x)|$ (triangle inequality again), we get $\{x \in X\colon |f(x) - f'(x)| \geq \alpha\} \subseteq F_n(\frac{\alpha}{2}) \cup F_n'(\frac{\alpha}{2})$, and so

$$\mu(\{x \in X\colon |f(x) - f'(x)| \geq \alpha\}) \leq \mu(F_n(\tfrac{\alpha}{2})) + \mu(F_n'(\tfrac{\alpha}{2})).$$

Suppose $\{f_n\}$ converges in measure to both f and f', then $\lim_n \mu(F_n(\frac{\alpha}{2})) = \lim_n \mu(F_n'(\frac{\alpha}{2})) = 0$. This implies that $\mu(\{x \in X\colon |f(x) - f'(x)| \geq \alpha\}) = 0$ for every $\alpha > 0$. Thus, $f' = f$ a.e. (which is a consequence of the inclusion $\{x \in X\colon |f(x) - f'(x)| > 0\} \subseteq \bigcup_{k=1}^\infty \{x \in X\colon |f(x) - f'(x)| \geq \frac{1}{k}\}$).

(c) Take a subsequence $\{f_{n_k}\}$ of a sequence $\{f_n\}$. Let $\alpha > 0$ and $\varepsilon > 0$ be arbitrary positive numbers. Suppose $\{f_n\}$ is Cauchy in measure. Thus there is an integer $n_{\varepsilon,\alpha}$ for which $\mu\{x \in X\colon |f_m(x) - f_n(x)| \geq \frac{\alpha}{2}\} < \frac{\varepsilon}{2}$ whenever $m, n \geq n_{\varepsilon,\alpha}$. If $\{f_{n_k}\}$ converges in measure to f, then there is an integer $k_{\varepsilon,\alpha}$ such that $\mu\{x \in X\colon |f_{n_k}(x) - f(x)| \geq \frac{\alpha}{2}\} < \frac{\varepsilon}{2}$ for every $k \geq k_{\varepsilon,\alpha}$. Hence, if j is an integer such that $j \geq k_{\varepsilon,\alpha}$ and $n_j \geq n_{\varepsilon,\alpha}$, then

$$\mu\{x \in X\colon |f_n(x) - f_{n_j}(x)| \geq \tfrac{\alpha}{2}\} < \tfrac{\varepsilon}{2},$$
$$\mu\{x \in X\colon |f_{n_j}(x) - f(x)| \geq \tfrac{\alpha}{2}\} < \tfrac{\varepsilon}{2},$$

for $n \geq n_{\varepsilon,\alpha}$. Since $|f_n(x) - f(x)| \leq |f_n(x) - f_{n_j}(x)| + |f_{n_j}(x) - f(x)|$,

$$\begin{aligned}&\{x \in X\colon |f_n(x) - f(x)| \geq \alpha\}\\ &\subseteq \{x \in X\colon |f_n(x) - f_{n_j}(x)| \geq \tfrac{\alpha}{2}\} \cup \{x \in X\colon |f_{n_j}(x) - f(x)| \geq \tfrac{\alpha}{2}\},\end{aligned}$$

so that

$$\begin{aligned}&\mu(\{x \in X\colon |f_n(x) - f(x)| \geq \alpha\})\\ &\leq \mu(\{x \in X\colon |f_n(x) - f_{n_j}(x)| \geq \tfrac{\alpha}{2}\}) + \mu(\{x \in X\colon |f_{n_j}(x) - f(x)| \geq \tfrac{\alpha}{2}\}),\end{aligned}$$

and hence $\mu(\{x \in X\colon |f_n(x) - f(x)| \geq \alpha\}) < \varepsilon$ for all $n \geq n_{\varepsilon,\alpha}$. Then $\lim_n \mu(F_n(\alpha)) = 0$, and so $\{f_n\}$ converges in measure to f. □

Convergences in measure and almost everywhere are not related.

$$\{f_n\} \text{ converges in measure} \quad \not\Longrightarrow \quad \{f_n\} \text{ converges a.e.},$$

even in the case of $\mu(X) < \infty$. Indeed, the sequence $\{f_n\}$ of Problem 6.2(b) acts on a finite measure space, converges in measure (since it converges in L^p) to the null function, but $\{f_n(x)\}$ fails to converge for every x, and so $\{f_n\}$ does not converge a.e. (thus it does not converge pointwise). Even though uniform convergence implies convergence in measure (Example 6D),

$$\{f_n\} \text{ converges pointwise} \quad \not\Longrightarrow \quad \{f_n\} \text{ converges in measure}$$

(and, consequently, convergence a.e. does not imply convergence in measure). In fact, the sequence $\{f_n\}$ of Problem 6.3(a,b) converges pointwise (i.e., everywhere in $\mathbb{R}$, and so a.e.) to the null function, but it is not Cauchy in measure and so it does not converge in measure by Proposition 6.6(a). Observe that if a sequence does not converge in measure, then it does not converge both in L^p and uniformly (see Examples 6C and 6D). But convergence almost everywhere implies convergence in measure whenever $\mu(X) < \infty$, as it will be verified in the forthcoming Propositions 6.12 and 6.13.

Proposition 6.7. *If a sequence of real-valued measurable functions is Cauchy in measure, then it has a subsequence that converges both in measure and almost everywhere.*

Proof. Let $(X, \mathcal{X}, \mu)$ be a measure space, and take a sequence $\{f_n\}$ of real-valued functions in $\mathcal{M}(X, \mathcal{X})$. Suppose $\{f_n\}$ is Cauchy in measure. Take an arbitrary integer $k \geq 1$. Thus there is another integer $n_k \geq 1$ such that

$$\mu(\{x \in X \colon |f_m(x) - f_n(x)| \geq (\tfrac{1}{2})^k\}) < (\tfrac{1}{2})^k \quad \text{whenever} \quad m, n \geq n_k.$$

This implies that there is a subsequence $\{f_{n_k}\}$ of $\{f_n\}$ for which

$$\mu(\{x \in X \colon |f_{n_{k+1}}(x) - f_{n_k}(x)| \geq (\tfrac{1}{2})^k\}) < (\tfrac{1}{2})^k.$$

Now, for each integer $k \geq 1$, consider the set $E_k \in \mathcal{X}$ given by

$$E_k = \bigcup_{j=k}^{\infty} \{x \in X \colon |f_{n_{j+1}}(x) - f_{n_j}(x)| \geq (\tfrac{1}{2})^j\},$$

so that $\mu(E_k) < \sum_{j=k}^{\infty} (\frac{1}{2})^j = (\frac{1}{2})^{k-1}$. Set $N = \bigcap_{k=1}^{\infty} E_k \in \mathcal{X}$, which is such that $\mu(N) = 0$ (because $N \subseteq E_k$ and so $\mu(N) \leq \mu(E_k)$, for all $k \geq 1$). Since

$$f_{n_i}(x) - f_{n_j}(x) = \sum_{\ell=1}^{i-j} f_{n_{i-\ell+1}}(x) - f_{n_{i-\ell}}(x) = \sum_{\ell=j}^{i-1} f_{n_{\ell+1}}(x) - f_{n_\ell}(x)$$

for every $1 \leq j < i$, it follows that if $x \in X \backslash E_k$ and $k \leq j < i$, then

$$|f_{n_i}(x) - f_{n_j}(x)| \leq \sum_{\ell=j}^{i-1} |f_{n_{\ell+1}}(x) - f_{n_\ell}(x)| < \sum_{\ell=j}^{i-1} (\tfrac{1}{2})^\ell < (\tfrac{1}{2})^{j-1}.$$

Thus, if x lies in $X \backslash N = X \backslash \bigcap_{k=1}^{\infty} E_k = \bigcup_{k=1}^{\infty} (X \backslash E_k)$ or, equivalently, if x lies $X \backslash E_k$ for some $k \geq 1$, then the above inequality holds for every pair of distinct integers $i, j \geq k$. This leads to the following two results: The one in (i) ensures that $\{f_{n_k}\}$ converges almost everywhere, and the one in (ii) ensures that $\{f_{n_k}\}$ converges in measure (to the same limit f).

(i) By the above inequality $\{f_{n_k}(x)\}$ is a Cauchy sequence in $\mathbb{R}$, and so it converges in $\mathbb{R}$ for every x in $X \backslash N$. Since $\mu(N) = 0$, we get

$$f_{n_k} \to f \;\; \mu\text{-a.e.}, \quad \text{where} \quad f(x) = \begin{cases} \lim_k f_{n_k}(x), & x \in X \backslash N, \\ 0, & x \in N, \end{cases}$$

defining the real-valued measurable function f on X.

(ii) Take any $k \geq 1$. The above convergence ensures that $f_{n_j}(x) \to f(x)$ for every x in $X \backslash E_k$, and this implies that for every $j \geq k$

$$|f_{n_j}(x) - f(x)| = \lim_i |f_{n_i}(x) - f_{n_j}(x)| \leq (\tfrac{1}{2})^{j-1} \leq (\tfrac{1}{2})^{k-1}.$$

Moreover, since $\mu(E_k) < (\frac{1}{2})^{k-1}$ for each $k \geq 1$, it follows that for every $\varepsilon > 0$ and $\alpha > 0$ there exists an integer $k' = k_{\varepsilon,\alpha} \geq 1$ such that

$$\mu(E_{k'}) < \left(\tfrac{1}{2}\right)^{k'-1} < \min\{\varepsilon, \alpha\}.$$

However (since $|f_{n_j}(x) - f(x)| \le \left(\frac{1}{2}\right)^{k'-1}$ for every $j \ge k'$), if $j \ge k'$, then

$$\{x \in X : |f_{n_j}(x) - f(x)| \ge \alpha\} \subseteq \{x \in X : |f_{n_j}(x) - f(x)| \ge \left(\tfrac{1}{2}\right)^{k'}\} \subseteq E_{k'}.$$

Therefore,

$$\mu(\{x \in X : |f_{n_j}(x) - f(x)| \ge \alpha\}) \le \mu(E_{k'}) < \varepsilon$$

for all $j \ge k' = k_{\varepsilon,\alpha}$, which means that $f_{n_j} \to f$ in measure. □

The following theorem is an important consequence of Proposition 6.7, which is referred to as the *Riesz–Weyl Theorem*. It says that *a sequence converges in measure if and only if it is Cauchy in measure.*

Theorem 6.8. *If a sequence of real-valued measurable functions is Cauchy in measure, then it converges in measure.*

Proof. Propositions 6.6(c) and 6.7. □

We have already seen that convergence in L^p implies convergence in measure, but the converse fails even in a finite measure space (see Example 6C). However, dominated convergence in measure implies convergence in L^p, as a consequence of another application of Proposition 6.7.

Proposition 6.9. *Take any $p \ge 1$. If $\{f_n\}$ is a sequence of functions in L^p, if f is a real-valued measurable function, and if g lies in L^p, then*

$$|f_n| \le g \text{ for all } n \text{ and } f_n \to f \text{ in measure} \implies f \in L^p \text{ and } f_n \to f \text{ in } L^p.$$

Proof. Consider the dominance assumption, and note that in this context plain dominance (i.e., $|f_n| \le g$ for all n) is equivalent to almost everywhere dominance (i.e., $|f_n| \le g$ for all n a.e.) — cf. proof of Proposition 6.4. We carry on a proof by contradiction. If $\{f_n\}$ does not converge in L^p to f, then there is a subsequence $\{h_k\}$ of $\{f_n\}$ and a real $\varepsilon > 0$ such that

$$\|h_k - f\|_p \ge \varepsilon \quad \text{for every} \quad k \ge 1. \qquad (*)$$

Suppose $\{f_n\}$ converges in measure to f. Thus every subsequence of $\{f_n\}$ converges in measure to f. In particular, $\{h_k\}$ converges in measure to f. Propositions 6.6(a) and 6.7 ensure that $\{h_k\}$ has a subsequence $\{h_{k_j}\}$ that converges both in measure and almost everywhere. Since $\{h_{k_j}\}$ converges in measure, Proposition 6.6 ensures that it must converge to f. Since $\{h_{k_j}\}$ also converges almost everywhere to f, and since $|h_{k_j}| \le g \in L^p$ for all j, it follows by Proposition 6.4 that f lies in L^p and $\{h_{k_j}\}$ converges in L^p to f, which contradicts the assertion in $(*)$. Then $\{f_n\}$ converges in L^p to f. □

6.3 Almost Uniform Convergence

Consider a measure space $(X, \mathcal{X}, \mu)$. Let $\{f_n\}$ be a sequence of real-valued functions on X, and let f be a real-valued function on X. We say that $\{f_n\}$ *converges uniformly almost everywhere* f $\{f_n\}$ converges uniformly to f on $X \backslash N$ (i.e., $\lim_n \sup_{x \in X \backslash N} |f_n(x) - f(x)| = 0$) for some set N in $\mathcal{X}$ with $\mu(N) = 0$. Equivalently, if there exists a set N in $\mathcal{X}$ with $\mu(N) = 0$ such that for every $\varepsilon > 0$ there is a positive integer n_ε for which

$$n \geq n_\varepsilon \quad \text{implies} \quad \sup_{x \in X \backslash N} |f_n(x) - f(x)| < \varepsilon.$$

In other words, if the sequence converges uniformly on the complement of a set of measure zero. However, we will be dealing in this section with a weaker notion of convergence, which requires uniform convergence on the complement of sets that have arbitrarily small measure. Actually, we have already met this notion in the proof of Proposition 6.7, part (ii).

Definition 6.10. A sequence $\{f_n\}$ of real-valued functions f_n on X *converges almost uniformly* (with respect to μ) to a real-valued function f on X if for each $\delta > 0$ there is a set E_δ in $\mathcal{X}$ with $\mu(E_\delta) < \delta$ such that $\{f_n\}$ converges uniformly to f on $X \backslash E_\delta$ (i.e., $\lim_n \sup_{x \in X \backslash E_\delta} |f_n(x) - f(x)| = 0$). Equivalently, if for every $\delta > 0$ and every $\varepsilon > 0$ there is a set E_δ in $\mathcal{X}$ with $\mu(E_\delta) < \delta$ and a positive integer $n_{\varepsilon,\delta}$ such that

$$n \geq n_{\varepsilon,\delta} \quad \text{implies} \quad \sup_{x \in X \backslash E_\delta} |f_n(x) - f(x)| < \varepsilon.$$

In this case write $f_n \to f$ a.u. (or $f_n \to f$ μ-a.u.). A sequence $\{f_n\}$ is *almost uniformly Cauchy* if for each $\delta > 0$ there is a set E_δ in $\mathcal{X}$ with $\mu(E_\delta) < \delta$ such that $\{f_n\}$ is a uniform Cauchy sequence on $X \backslash E_\delta$ (which means that $\lim_{m,n} \sup_{x \in X \backslash E_\delta} |f_m(x) - f_n(x)| = 0$). Equivalently, if for every $\delta > 0$ and every $\varepsilon > 0$ there is an $E_\delta \in \mathcal{X}$ with $\mu(E_\delta) < \delta$ and an $n_{\varepsilon,\delta} \geq 1$ such that

$$m, n \geq n_{\varepsilon,\delta} \quad \text{implies} \quad \sup_{x \in X \backslash E_\delta} |f_m(x) - f_n(x)| < \varepsilon.$$

Example 6E. Consider the string of implications,

$$f_n \to f \text{ uniformly} \quad \begin{smallmatrix}\Longrightarrow \\ \not\Longleftarrow\end{smallmatrix} \quad f_n \to f \text{ uniformly a.e.} \quad \begin{smallmatrix}\Longrightarrow \\ \not\Longleftarrow\end{smallmatrix} \quad f_n \to f \text{ a.u.},$$

where uniform convergence trivially implies uniform almost everywhere convergence (set $N = \varnothing$), which in turn trivially implies almost uniform convergence (set $E_\delta = N$). The converses, however, fail even if $\mu(X) < \infty$. Actually, take the *finite* Lebesgue measure space $\big([0,1], \wp([0,1]) \cap \Re, \lambda\big)$ and let $\{f_n\}$

be a sequence of real-valued functions on $[0,1]$ such that $f_n(x)=0$ for all $x\neq 0$ and $f_n(0)=(-1)^n$ for every $n\geq 1$. It is clear that $\{f_n\}$ does not converge pointwise, and so it does not converge uniformly, but it converges uniformly almost everywhere. It is also clear that the sequence $\{f_n\}$ of Example 6A converges almost uniformly to the null function 0, but it does not converge to 0 uniformly almost everywhere since $\sup_{[0,1]\backslash N}|f_n(x)|=1$ for any Borel set $N\subseteq[0,1]$ with $\lambda(N)=0$, for all n.

Note that if $f_n\to f$ almost uniformly, then $\{f_n\}$ is an almost uniform Cauchy sequence. In fact, for each m and n,

$$\sup_{x\in X\backslash E_\delta}|f_m(x)-f_n(x)|\leq \sup_{x\in X\backslash E_\delta}|f_m(x)-f(x)|+\sup_{x\in X\backslash E_\delta}|f(x)-f_n(x)|.$$

The next result ensures the converse, and therefore *a sequence converges almost uniformly if and only if it is an almost uniform Cauchy sequence.*

Proposition 6.11. *Let $\{f_n\}$ be a sequence of real-valued functions.*

(a) *If $\{f_n\}$ is almost uniformly Cauchy, then it converges almost uniformly, and it also converges almost everywhere to the same real-valued limit f.*

(b) *If each function f_n is measurable, then so is the limit function.*

Proof.

(a) Take a measure space $(X,\mathcal{X},\mu)$. Suppose a sequence $\{f_n\}$ of real-valued functions on X is almost uniform Cauchy (with respect to μ). Then for each integer $k\geq 1$ there is a set E_k in $\mathcal{X}$ with $\mu(E_k)<\frac{1}{k}$ such that

$$\lim_{m,n}\sup_{x\in X\backslash E_k}|f_m(x)-f_n(x)|=0.$$

Set $N=\bigcap_{k=1}^\infty E_k$ in $\mathcal{X}$ so that $\mu(N)\leq\mu(E_k)<\frac{1}{k}$ for every $k\geq 1$, and so $\mu(N)=0$. If $x\in X\backslash N=X\backslash\bigcap_{k=1}^\infty E_k=\bigcup_{k=1}^\infty(X\backslash E_k)$, then the real-valued sequence $\{f_n(x)\}$ is Cauchy in $\mathbb{R}$, and hence it converges in $\mathbb{R}$. Thus,

$$f_n\to f \text{ a.e.,}\quad \text{where}\quad f(x)=\begin{cases}\lim_n f_n(x), & x\in X\backslash N,\\ 0, & x\in N.\end{cases}$$

This convergence defines a real-valued function $f\colon X\to\mathbb{R}$ on X,

$$f=\lim_n\big(f_n\chi_{X\backslash N}\big).$$

Since $\{f_n\}$ is almost uniformly Cauchy, and since $f_n(x)\to f(x)$ for every x on each $X\backslash E_k\subseteq X\backslash N$, it follows that $f_n(x)\to f(x)$ uniformly on $X\backslash E_k$. In fact, for $k\geq 1$ and $\varepsilon>0$ arbitrary there exists $n_{\varepsilon,k}\geq 1$ for which

$$m, n \geq n_{\varepsilon,k} \quad \text{implies} \quad \sup_{x \in X \backslash E_k} |f_m(x) - f_n(x)| < \varepsilon,$$

and so, for every $x \in X \backslash E_k$,

$$m \geq n_{\varepsilon,k} \quad \text{implies} \quad |f_m(x) - f(x)| = \lim_n |f_m(x) - f_n(x)| < \varepsilon,$$

which implies $\sup_{x \in X \backslash E_k} |f_m(x) - f(x)| < \varepsilon$. This ensures that

$$f_n \to f \text{ a.u.},$$

since, for each $\delta > 0$, take k large enough so that $\frac{1}{k} \leq \delta$ and set $E_\delta = E_k$ in $\mathcal{X}$ so that $\mu(E_\delta) < \delta$ and $\{f_n\}$ converges uniformly on $X \backslash E_\delta$.

(b) Since $f = \lim_n f_n$ μ-a.e., it follows by Propositions 1.8 and 1.9 that f is $\mathcal{X}$-measurable whenever each f_n is $\mathcal{X}$-measurable. □

So almost uniform convergence implies almost everywhere convergence,

$$f_n \to f \text{ a.u.} \quad \underset{\not\Leftarrow}{\Longrightarrow} \quad f_n \to f \text{ a.e.},$$

but the converse fails in general. Indeed, even pointwise convergence does not imply almost uniform convergence (cf. Problem 6.3). However, the converse holds in a finite measure space, as it will be seen in Proposition 6.13.

Proposition 6.12. *Consider a sequence $\{f_n\}$ of measurable functions. If $f_n \to f$ almost uniformly, then $f_n \to f$ in measure.*

Proof. Let $(X, \mathcal{X}, \mu)$ be a measure space. For each $\alpha > 0$, consider the set $F_n(\alpha) = \{x \in X \colon |f_n(x) - f(x)| \geq \alpha\}$ in $\mathcal{X}$. If $f_n \to f$ a.u., then for each $\delta > 0$ there is a set $E_\delta \in \mathcal{X}$ with $\mu(E_\delta) < \delta$ and an integer $n_{\alpha,\delta} \geq 1$ such that

$$n \geq n_{\alpha,\delta} \quad \text{implies} \quad \sup_{x \in X \backslash E_\delta} |f_n(x) - f(x)| < \alpha.$$

Thus $F_n(\alpha) \subseteq E_\delta$, and hence $\mu(F_n(\alpha)) < \mu(E_\delta) < \delta$ for every $n \geq n_{\alpha,\delta}$. Therefore $\lim_n \mu(F_n(\alpha)) = 0$, which means that $f_n \to f$ in measure. □

So almost uniform convergence also implies convergence in measure,

$$f_n \to f \text{ a.u.} \quad \underset{\not\Leftarrow}{\Longrightarrow} \quad f_n \to f \text{ in measure},$$

but the converse fails even if $\mu(X) < \infty$. In fact, the sequence of Problem 6.2(b), which acts on a finite measure space, converges in measure (since it converges in L^p), but does not converge a.u. (since it does not converge a.e.). Furthermore, convergences almost uniform and in L^p are not related,

$$\{f_n\} \text{ converges a.u.} \quad \underset{\not\Leftarrow}{\not\Rightarrow} \quad \{f_n\} \text{ converges in } L^p,$$

even if $\mu(X) < \infty$. Take the finite measure space of Problem 6.2(a). If $f_n = (n+1)\chi_{[1/(n+1),\, 2/(n+1)]}$ for each $n \geq 1$, then $\{f_n\}$ converges almost uniformly to the null function but it does not converge in L^p (Problem 6.2(a)). Conversely, we saw above that convergence in L^p (in a finite measure space) does not imply almost uniform convergence. However, almost uniform dominated convergence implies convergence in L^p by Propositions 6.9 and 6.12.

By Proposition 6.11, almost uniform convergence implies almost everywhere convergence. We close this chapter by showing that the converse holds in a finite measure space. This is referred to as the *Egoroff Theorem* (Proposition 6.13). The finite measure assumption in the Egoroff Theorem can be replaced with dominated convergence (Corollary 6.14).

Proposition 6.13. *Consider a sequence $\{f_n\}$ of measurable functions. If $\mu(X) < \infty$ and $f_n \to f$ almost everywhere, then $f_n \to f$ almost uniformly.*

Proof. Let $(X, \mathcal{X}, \mu)$ be a measure space. Suppose $\{f_n\}$ converges almost everywhere to f. This means that $f_n(x) \to f(x)$ for every x in $X \backslash N \in \mathcal{X}$ for some $N \in \mathcal{X}$ with $\mu(N) = 0$. That is, $f_n \to f$ pointwise on the complement $X' = X \backslash N$ of a set of measure zero, which implies that $f'_n \to f'$ pointwise on X with $f'_n = f_n \chi_{X'}$ for each n and $f' = f\chi_{X'}$. That is, $f'_n(x) \to f'(x)$ for every $x \in X$. Take an arbitrary positive integer m and set, for each $n \geq 1$,

$$F'_n(m) = \big\{x \in X' \colon |f_n(x) - f(x)| \geq \tfrac{1}{m}\big\} = \big\{x \in X \colon |f'_n(x) - f'(x)| \geq \tfrac{1}{m}\big\}.$$

Recall that f' and f'_n are measurable (since f and f_n are). Thus $|f'_n - f'|$ is a measurable function. Then each $F'_n(m)$ is a measurable set, and so

$$E'_n(m) = \bigcup_{k=n}^{\infty} F'_k(m)$$

is a measurable set for each $n \geq 1$. Hence $\{E'_n(m)\}$ is a decreasing sequence of sets in $\mathcal{X}$ (i.e., $E'_{n+1}(m) \subseteq E'_n(m) \in \mathcal{X}$). Take an arbitrary x in X. Since $f'_n(x) \to f'(x)$, it follows that there exists an integer $n_{m,x} \geq 1$ such that

$$k \geq n_{m,x} \implies |f'_k(x) - f'(x)| < \tfrac{1}{m} \implies x \notin F'_k(m) \implies x \notin E'_{n_{m,x}}(m),$$

and so $x \notin \bigcap_{n=1}^{\infty} E'_n(m)$. Thus

$$\bigcap_{n=1}^{\infty} E'_n(m) = \varnothing, \quad \text{which implies} \quad \mu\Big(\bigcap_{n=1}^{\infty} E'_n(m)\Big) = 0.$$

From now on suppose μ is a finite measure. Proposition 2.2(d) ensures that

$$\mu(X) < \infty \implies \mu(E'_1(m)) < \infty \implies \lim_n \mu(E'_n(m)) = 0.$$

Then for every $\varepsilon > 0$ there is an integer $n_{\varepsilon,m} \geq 1$ such that $\mu(E'_n(m)) < \frac{\varepsilon}{2^m}$ whenever $n \geq n_{\varepsilon,m}$. Since this happens for an arbitrary integer $m \geq 1$, set

$$E'_\varepsilon = \bigcup_{m=1}^{\infty} E'_{n_{\varepsilon,m}}(m)$$

in $\mathcal{X}$ so that

$$\mu(E'_\varepsilon) = \mu\Big(\bigcup_{m=1}^{\infty} E'_{n_{\varepsilon,m}}(m)\Big) \leq \sum_{m=1}^{\infty} \mu(E'_{n_{\varepsilon,m}}(m)) < \sum_{m=1}^{\infty} \tfrac{\varepsilon}{2^m} = \varepsilon.$$

Suppose $x \in X\backslash E'_\varepsilon$, so that $x \notin E'_{n_{\varepsilon,m}}(m)$. Hence

$$n \geq n_{\varepsilon,m} \quad\Longrightarrow\quad x \notin F'_n(m) \quad\Longrightarrow\quad |f'_n(x) - f'(x)| < \tfrac{1}{m}.$$

Thus $\{f'_n\}$ converges uniformly to f' on $X\backslash E'_\varepsilon$. Take E_ε in $\mathcal{X}$ given by

$$E_\varepsilon = E'_\varepsilon \cup N.$$

Since $X\backslash E_\varepsilon = X\backslash(E'_\varepsilon \cup N) = (X\backslash E'_\varepsilon) \cap (X\backslash N) = (X\backslash E'_\varepsilon) \cap X'$, we get

$$\sup_{x \in X\backslash E_\varepsilon} |f_n(x) - f(x)| = \sup_{x \in X\backslash E'_\varepsilon} |f'_n(x) - f'(x)|.$$

Then $\{f_n\}$ converges uniformly to f on $X\backslash E_\varepsilon$ (because $\{f'_n\}$ converges uniformly to f' on $X\backslash E'_\varepsilon$). Since $\mu(E_\varepsilon) \leq \mu(E'_\varepsilon) + \mu(N) = \mu(E'_\varepsilon) < \varepsilon$, it follows that $\{f_n\}$ converges almost uniformly. □

Corollary 6.14. *Take a sequence $\{f_n\}$ of measurable functions. Suppose $|f_n| \leq g \in L^p$ and $f_n \to f$ almost everywhere. Then $f_n \to f$ almost uniformly.*

Proof. Consider the proof of Proposition 6.13. The assumption $\mu(X) < \infty$ was used there only to ensure that $\mu(E'_1(m)) < \infty$. We now show that $\mu(E'_1(m)) < \infty$ still holds if we assume dominated convergence instead, and so we are reduced to the previous proof. If $|f_n| \leq g \in L^p$ and $f_n \to f$ a.e., then $|f| \leq g$ (so that $f_n, f \in L^p$) and $|f_n - f| \leq |f_n| + |f| \leq 2g$ (see Proposition 6.4). By setting $G'(m) = \{x \in X' \colon 2g(x) \geq \frac{1}{m}\}$ in $\mathcal{X}$, we get

$$F'_n(m) = \{x \in X' \colon \tfrac{1}{m} \leq |f_n(x) - f(x)|\} \subseteq \{x \in X' \colon \tfrac{1}{m} \leq 2g(x)\} = G'(m)$$

for all n, and hence

$$E'_1(m) = \bigcup_{n=1}^{\infty} F'_n(m) \subseteq G'(m).$$

Since $\int g^p\,d\mu < \infty$, it follows that $\mu(\{x \in X \colon g^p(x) \geq \varepsilon\}) < \infty$ for every $\varepsilon > 0$ (Problem 3.9), and so $\mu(G'(m)) < \infty$. Thus $\mu(E'_1(m)) < \infty$. □

6.4 Problems

Problem 6.1. Let $(\mathbb{R}, \Re, \lambda)$ be the Lebesgue measure space.

(a) If $f_n = n^{-1/p}\chi_{[0,n]}$ for each $n \geq 1$, then $\{f_n\}$ converges uniformly to the null function but does not converge in L^p for any $p \geq 1$.

Hint: If $n = 2m$, then show that $\|f_n - f_m\|_p^p \geq \frac{n-m}{n} = \frac{1}{2}$.

(b) If $f_n = n^{1/p}\chi_{[n, n+(1/n^2)]}$ for each $n \geq 1$, then $\{f_n\}$ converges in L^p to the null function 0 for every $p \geq 1$ but does not converge uniformly.

Hint: Show that $\{f_n\}$ converges pointwise to 0 but $\sup_{x\in\mathbb{R}}|f_n(x)| = n^{1/p}$.

Problem 6.2. Take the *finite* measure space $\big([0,1], \wp([0,1]) \cap \Re, \lambda\big)$, where λ is the restriction of Lebesgue measure on the σ-algebra $\wp([0,1]) \cap \Re$ (cf. Problem 2.11). In other words, let λ be the Lebesgue measure acting on the Borel subsets of $[0,1]$, which is a probability measure.

(a) If $f_n = (n+1)\chi_{[1/(n+1),\, 2/(n+1)]}$ for $n \geq 1$, then $\{f_n\}$ converges pointwise to the null function 0 but does not converge in L^p for any $p \geq 1$.

Hint: If $n \geq 2m+1$, then $\|f_n - f_m\|_p^p \geq (2^{p-1}+1)(m+1)^{p-1}$.

Observe that the sequence $\{f_n\}$ does not converge uniformly (since it converges pointwise to 0 but $\sup_{x\in[0,1]}|f_n(x)| = (n+1)$), which is a consequence of Proposition 6.3(b) since $\{f_n\}$ does not converge in L^p.

(b) Consider the intervals $E_{k,j} = \big[\frac{j-1}{k}, \frac{j}{k}\big]$ for each pair of integers j and k such that $1 \leq j \leq k$. For each $k \geq 1$ take the finite sequence $\{E_{k,j}\}_{1\leq j\leq k}$. Stack these finite sequences to get the infinite sequence of intervals

$$\begin{aligned}\{E_n\}_{n\geq 1} &= \big\{\{E_{k,j}\}_{1\leq j\leq k}\big\}_{k\geq 1}\\ &= \big\{\{E_{1,1}\}, \{E_{2,1}, E_{2,2}\}, \{E_{3,1}, E_{3,2}, E_{3,3}\}, \{E_{4,1}, E_{4,2}, E_{4,3}, E_{4,4}\}, \ldots\big\}\\ &= \big\{[0,1], [0,\tfrac{1}{2}], [\tfrac{1}{2},1], [0,\tfrac{1}{3}], [\tfrac{1}{3},\tfrac{2}{3}], [\tfrac{2}{3},1], [0,\tfrac{1}{4}], [\tfrac{1}{4},\tfrac{2}{4}], [\tfrac{2}{4},\tfrac{3}{4}], [\tfrac{3}{4},1], \ldots\big\}.\end{aligned}$$

Show that if $f_n = \chi_{E_n}$ for each $n \geq 1$, then $\{f_n\}$ converges in L^p to the null function for every $p \geq 1$, but the real-valued sequence $\{f_n(x)\}$ does not converge for every x in $[0,1]$ (i.e., $\{f_n\}$ does not converge pointwise everywhere, and so it does not converge almost everywhere).

Hint: First note that the real-valued sequence $\{\lambda(E_n)\}$ is bounded and decreasing, thus convergent. For every $m \geq 1$ there is an n_m such that $\lambda(E_{n_m}) \leq \frac{1}{m}$. Hence $\|f_n\|_p \to 0$. Next take an arbitrary x in $[0,1]$. The real-valued sequence $\{f_n(x)\}$ has a subsequence constantly equal to 1 and another constantly equal to 0. Thus $\{f_n(x)\}$ does not converge.

Problem 6.3. Consider the Lebesgue measure space $(\mathbb{R}, \mathfrak{R}, \lambda)$ and, for each $n \geq 1$, take the characteristic function $f_n = \chi_{[n, n+1]}$. Prove the assertions.

(a) $\{f_n\}$ converges pointwise (and so a.e.) to the null function.

(b) $\{f_n\}$ does not converge in measure (so not uniformly and not in L^p).

Hint: Verify that $\lambda(\{x \in X : |f_m(x) - f_n(x)| \geq \frac{1}{2}\}) = 2$ for every $m \neq n$. Thus conclude that $\{f_n\}$ is not Cauchy in measure (Proposition 6.6(a)).

(c) $\{f_n\}$ does not converge almost uniformly (and so not uniformly a.e.).

Problem 6.4. Let $g : \mathbb{R} \to \mathbb{R}$ and $f_n : \mathbb{R} \to \mathbb{R}$ for each positive integer n be real-valued functions on $\mathbb{R}$ given by

$$g(x) = \begin{cases} 0, & x \leq 0, \\ \frac{1}{\sqrt{x}}, & x \in (0, 1], \\ \frac{1}{x^2}, & x \in [1, \infty), \end{cases} \qquad f_n(x) = f_{g,n}(x) = \begin{cases} n, & g(x) \geq n, \\ 0, & \text{otherwise.} \end{cases}$$

Consider the Lebesgue measure space $(\mathbb{R}, \mathfrak{R}, \lambda)$. Show that $g \in L^1$, and

(a) $\{f_n\}$ is dominated by g and converges pointwise to the null function 0,

(b) $\{f_n\}$ converges to 0 almost uniformly but does not converge uniformly almost everywhere.

Hint: $\sup_{x \in X \setminus N} |f_n(x)| \to \infty$ for every set N of Lebesgue measure zero.

Problem 6.5. The *symmetric difference* of two sets A and B is the set

$$A \nabla B = (A \backslash B) \cup (B \backslash A) = (A \cup B) \backslash (A \cap B).$$

Consider a measure space $(X, \mathcal{X}, \mu)$ and let E and F be arbitrary sets in $\mathcal{X}$. We say that the sets E and F are *equivalent* (or *μ-equivalent*), denoted by $E \sim F$, if $\mu(E \nabla F) = 0$. The relation $\sim$ is in fact an equivalence relation on $\mathcal{X}$. Define the function $d : \mathcal{X} \times \mathcal{X} \to \mathbb{R}$ by $d(E, F) = \mu(E \nabla F)$ for every E, F in $\mathcal{X}$. Verify that $d(E, F) \geq 0$, $d(E, F) = d(F, E)$, and (triangle inequality) $d(E, F) \leq d(E, G) + d(G, F)$ for every E, F, G in $\mathcal{X}$. (Observe that d is a *pseudometric* on $\mathcal{X}$ — cf. Section 11.1 — and so it induces a *metric* on the quotient space $\mathcal{X}/\!\sim$.) Take a sequence $\{E_n\}$ of sets in $\mathcal{X}$ and, for each $n \geq 1$, set $f_n = \chi_{E_n} : X \to \mathbb{R}$, the characteristic function of E_n. Show that $\{\chi_{E_n}\}$ is Cauchy in measure if and only if $\lim_{m,n} \mu(E_m \nabla E_n) = 0$. That is,

$$\{f_n\} \text{ is Cauchy in measure if and only if } \lim_{m,n} d(E_m, E_n) = 0.$$

Problem 6.6. The following diagrams show the relationship among almost everywhere convergence, almost uniform convergence, convergence in L^p, and convergence in measure. These will be denoted by (a.e.), (a.u.), (L^p), and (μ), respectively. The first diagram considers the general case (with no additional assumption), the second one considers the case of finite measure (i.e., if $\mu(X) < \infty$), and the third diagram considers the case of dominated convergence (i.e., when $|f_n| \leq g \in L^p$).

$$
\begin{array}{ccc}
\text{(a.e.)} & \Longleftarrow & \text{(a.u.)} \\
 & & \Downarrow \\
(L^p) & \Longrightarrow & (\mu)
\end{array}
\qquad
\begin{array}{ccc}
\text{(a.e.)} & \Longleftrightarrow & \text{(a.u.)} \\
 & & \Downarrow \\
(L^p) & \Longrightarrow & (\mu)
\end{array}
\qquad
\begin{array}{ccc}
\text{(a.e.)} & \Longleftrightarrow & \text{(a.u.)} \\
 & & \Downarrow \\
(L^p) & \Longleftrightarrow & (\mu)
\end{array}
$$

General case ***Finite measure*** ***Dominated convergence***

As usual, the arrows mean implication. Show that the above diagrams are correct and complete in the sense that all arrows are true and no arrow can be added except for the trivial ones (i.e., up to *modus ponens* — for example, it is obvious from those diagrams that (a.e.) implies (μ) in a finite measure space and also that (a.e.) implies (L^p) under the dominance assumption).

Problem 6.7. Now let (u.a.e.) denote uniform convergence almost everywhere. Show that the implications

$$\text{(u.a.e.)} \Longrightarrow \text{(a.e.)} \qquad \text{and} \qquad \text{(u.a.e.)} \Longrightarrow \text{(a.u.)}$$

hold true, and their converses fail even under finite measure and dominance assumptions. Also, (u.a.e.) implies (L^p) under finite measure or dominance but not in general, and the converse fails even under both assumptions.

Problem 6.8. Show that the Dominated Convergence Theorem holds if almost everywhere convergence is replaced with convergence in measure.

Hint: Consider the Dominated Convergence Theorem (Theorem 4.7). Use Problem 6.6 (specifically, Proposition 6.9) and Problem 5.16 for $p = 1$.

Problem 6.9. Prove the *Vitali Convergence Theorem*, which reads as follows. Take an arbitrary $p \geq 1$. If $\{f_n\}$ is a sequence of real-valued function in $L^p(X, \mathcal{X}, \mu)$, then $f_n \to f$ in L^p for some $f \in L^p$ if and only if the following three assumptions hold true.

(a) $f_n \to f$ in measure.

(b) For each $\varepsilon > 0$ there is an $E_\varepsilon \in \mathcal{X}$ with $\mu(E_\varepsilon) < \infty$ such that for all $n \geq 1$

$$\int_F |f_n|^p \, d\mu < \varepsilon^p \quad \text{for every} \quad F \in \mathcal{X} \text{ for which } F \cap E_\varepsilon = \varnothing.$$

(c) For each $\varepsilon > 0$ there is a $\delta_\varepsilon > 0$ such that for all $n \geq 1$

$$\int_E |f_n|^p \, d\mu < \varepsilon^p \quad \text{for every} \quad E \in \mathcal{X} \text{ with } \mu(E) < \delta_\varepsilon.$$

Hint: To prove that assumptions (a), (b), (c) imply $f_n \to f$ in L^p proceed as follows. Use the Minkowski inequality to show that assumption (b) implies $\|f_m - f_n\|_p = \big(\int_{E_\varepsilon} |f_n - f_m|^p\big)^{\frac{1}{p}} + 2\varepsilon$ for every $m, n \geq 1$. By the Minkowski inequality also show that (a) and (c) imply $\big(\int_{E_\varepsilon} |f_n - f_m|^p\big)^{\frac{1}{p}} \leq 3\varepsilon$ for every $m, n \geq n_\varepsilon$ for some $n_\varepsilon \geq 1$. Thus conclude that $\{f_n\}$ is Cauchy in L^p. Apply Proposition 6.6 for uniqueness of the limit f.

Problem 6.10. Take any $p \geq 1$. If a sequence $\{f_n\}$ of functions in L^p is such that $f_n \to f \in L^p$ a.e. and $\|f_n\|_p \to \|f\|_p$, then $f_n \to f$ in L^p. Prove.

Hint: Use Proposition 6.4. (Compare with Problems 4.14 and 5.17.)

Suggested Reading

Bartle [4], Berberian [7], Halmos [18], Kingman and Taylor [23], Munroe [30].

7

Decomposition of Measures

7.1 The Jordan Decomposition Theorem

Take a signed measure $\nu\colon \mathcal{X} \to \mathbb{R}$ on a σ-algebra $\mathcal{X}$ of subsets of a set X. According to Definition 2.3, signed measures are *real-valued* set functions. We saw in Section 2.2 that if μ and λ are finite measures, then $\nu = \mu - \lambda$ is a signed measure. In this section we show that all signed measures ν admit a decomposition into a difference of two finite measures.

Definition 7.1. Consider a signed measure ν on a σ-algebra $\mathcal{X}$. A set A^+ in $\mathcal{X}$ is *positive* with respect to ν if $\nu(A^+ \cap E) \geq 0$ for all E in $\mathcal{X}$. A set A^- in $\mathcal{X}$ is *negative* with respect to ν if $\nu(A^- \cap E) \leq 0$ for all E in $\mathcal{X}$. A set N in $\mathcal{X}$ is *null* with respect to ν if $\nu(N \cap E) = 0$ for all E.

In other words, let ν be an arbitrary signed measure on an arbitrary σ-algebra $\mathcal{X}$ of subsets of an arbitrary set X. A measurable set is *positive, negative,* or *null* if each measurable subset of it has nonnegative, nonpositive, or null measure, respectively. The set X always has a measurable partition consisting of a positive and a negative set with respect to ν.

Theorem 7.2. (Hahn Decomposition Theorem). Let $\mathcal{X}$ be a σ-algebra of subsets of a set X. *If ν is a signed measure on $\mathcal{X}$, then there exists a measurable partition $\{A^+, A^-\}$ of X such that A^+ is positive and A^- is negative with respect to ν.*

Proof. Consider a signed measure ν on $\mathcal{X}$. We show that there exists a pair of sets A^+ and A^- in $\mathcal{X}$ such that $A^+ \cup A^- = X$, $A^+ \cap A^- = \varnothing$, A^+ is

C.S. Kubrusly, *Essentials of Measure Theory*,
DOI 10.1007/978-3-319-22506-7_7

positive, and A^- is negative. Let $\mathcal{A}^+ \subseteq \mathcal{X}$ be the collection of all positive sets with respect to ν, which is not empty (since it contains the empty set). Set $\alpha = \sup_{A\in\mathcal{A}^+}\nu(A)$ and take a sequence $\{A_n\}$ of sets in $\mathcal{A}^+$ such that $\sup_n \nu(A_n) = \alpha$. Thus $A^+ = \bigcup_n A_n$ is a positive set (Problem 7.2(b)) with $0 \le \nu(A^+) = \alpha < \infty$ (since $\nu(A_n) \le \nu(A^+) \le \alpha$ for all n — Problem 7.3). Take its complement $A^- = X\backslash A^+$. If A^- is a negative set, then we are done.

Claim. $A^- = X\backslash A^+$ is a negative set.

Proof. Suppose A^- is not negative. Then it has a measurable subset E_0 such that $\nu(E_0) > 0$. If E_0 is a positive set, then $\nu(A^+ \cup E_0) > \alpha$ (because $A^+ \cap E_0 = \varnothing$), which is a contradiction ($\alpha = \sup_{A\in\mathcal{A}}\nu(A)$). Thus E_0 is not positive, so it has measurable subsets of negative measure. Let n_0 be the smallest positive integer such that E_0 has a measurable subset of measure not greater than $-\frac{1}{n_0}$, say E_1 with $\nu(E_1) \le -\frac{1}{n_0}$. Observe that

$$\nu(E_0\backslash E_1) = \nu(E_0) - \nu(E_1) > \nu(E_0) > 0$$

(cf. Problem 7.1(a)). If $E_0\backslash E_1$ is a positive set, then $\nu(A^+ \cup (E_0\backslash E_1)) > \alpha$ (because $A^+ \cap (E_0\backslash E_1) = \varnothing$), which is again a contradiction. Thus $E_0\backslash E_1$ is not positive, so it has measurable subsets of negative measure. Let n_1 be the smallest positive integer such that $E_0\backslash E_1$ has a measurable subset of measure not greater than $-\frac{1}{n_1}$, say E_2 with $\nu(E_2) \le -\frac{1}{n_1}$. Again, note that

$$\nu(E_0\backslash(E_1\cup E_2)) = \nu(E_0) - \nu(E_1\cup E_2) = \nu(E_0) - (\nu(E_1)+\nu(E_2)) > \nu(E_0) > 0$$

(because $E_1 \cap E_2 = \varnothing$). As before, $E_0\backslash(E_1 \cup E_2)$ is not positive, so it has measurable subsets of negative measure. Let n_2 be the smallest positive integer such that $E_0\backslash(E_1 \cup E_2)$ has a measurable subset of measure not greater than $-\frac{1}{n_2}$, say E_3 with $\nu(E_3) \le -\frac{1}{n_2}$. This leads to the inductive construction of a sequence $\{E_k\}_{k=1}^\infty$ of pairwise disjoint measurable sets and a sequence $\{n_k\}_{k=1}^\infty$ of integers with each n_k being the smallest positive integer for which $E_0\backslash\bigcup_{i=1}^k E_i$ has a measurable subset of measure not greater than $-\frac{1}{n_k}$. Moreover, $\nu(E_{k+1}) \le -\frac{1}{n_k}$ for every $k \ge 0$, and so $\sum_{k=0}^\infty \frac{1}{n_k} < \infty$. In fact, by setting $E = \bigcup_{k=1}^\infty E_k$ in $\mathcal{X}$ we get

$$-\infty < \nu(E) = \sum_{k=1}^\infty \nu(E_k) \le -\sum_{k=0}^\infty \tfrac{1}{n_k} < 0,$$

since $\{E_k\}_{k=1}^\infty$ consists of disjoint sets. Thus $\frac{1}{n_k} \to 0$ as $k \to \infty$. Note that

$$\nu(E_0\backslash E) = \nu(E_0) - \nu(E) > \nu(E_0) > 0.$$

The set $E_0\backslash E$ is indeed positive. In fact, suppose $E_0\backslash E$ has a measurable subset of negative measure, say F with $\nu(F) < 0$. Since $n_k \to \infty$ as $k \to \infty$, take k large enough so that $\frac{1}{n_k-1} < -\nu(F)$; that is, $\nu(F) < -\frac{1}{n_k-1}$. But $F \subseteq E_0\backslash E \subseteq E_0\backslash \bigcup_{i=1}^{k} E_i$, so $E_0\backslash \bigcup_{i=1}^{k} E_i$ has a measurable subset of measure less than $-\frac{1}{n_k-1}$, which contradicts the fact that n_k is the smallest positive integer for which $E_0\backslash \bigcup_{i=1}^{k} E_i$ has a measurable subset of measure not greater than $-\frac{1}{n_k}$. Thus every measurable subset of $E_0\backslash E$ has a nonnegative measure, and so $E_0\backslash E$ is a positive set. Therefore, since $A^+\cap (E_0\backslash E) = \varnothing$ (because $E_0 \subseteq A^-$) and $\nu(E_0\backslash E) > 0$, it follows that $\nu(A^+\cup (E_0\backslash E)) > \alpha$, which is again a contradiction. Outcome: A^- must be a negative set. □

Let $(X, \mathcal{X})$ be a measurable space. A measurable partition $\{A^+, A^-\}$ of X, where $A^+ \in \mathcal{X}$ is positive and $A^- \in \mathcal{X}$ is negative with respect to a signed measure ν on $\mathcal{X}$, is called a *Hahn decomposition* of X with respect to ν. Given a signed measure ν on $\mathcal{X}$, a Hahn decomposition of X is not unique (if there exists a nonempty null set with respect to ν). In fact, if $\{A^+, A^-\}$ is a Hahn decomposition of X and N is a null set, then $\{A^+\cup N, A^-\backslash N\}$ and $\{A^+\backslash N, A^-\cup N\}$ are also Hahn decompositions of X (all with respect to ν). However, this lack of uniqueness is indistinguishable for the signed measure ν, and so it is not a disadvantage of the Hahn decomposition.

Proposition 7.3. *Suppose $\{A_1^+, A_1^-\}$ and $\{A_2^+, A_2^-\}$ are Hahn decompositions of X with respect to a signed measure ν on $\mathcal{X}$. Then, for every $E \in \mathcal{X}$,*

$$\nu(A_1^+ \cap E) = \nu(A_2^+ \cap E) \quad \text{and} \quad \nu(A_1^- \cap E) = \nu(A_2^- \cap E).$$

Proof. If A, B, C are arbitrary sets, then $\{A\cap(B\backslash C), A\cap B\cap C\}$ is a partition of $A \cap B$. In $\mathcal{X}$, if $\nu(A\cap (B\backslash C)) = 0$, then $\nu(A\cap B) = \nu(A\cap B\cap C)$. Since $E\cap (A_1^+\backslash A_2^+) \subseteq A_1^+ \cap A_2^-$ and $E\cap (A_2^+\backslash A_1^+) \subseteq A_2^+ \cap A_1^-$, it follows that

$$\nu(E\cap (A_1^+\backslash A_2^+)) = 0 \quad \text{and} \quad \nu(E\cap (A_2^+\backslash A_1^+)) = 0.$$

Thus

$$\nu(E\cap A_1^+) = \nu(E\cap A_1^+\cap A_2^+) \quad \text{and} \quad \nu(E\cap A_2^+) = \nu(E\cap A_2^+\cap A_1^+),$$

and so

$$\nu(E\cap A_1^+) = \nu(E\cap A_2^+).$$

Analogously, replacing A_1^+ with A_1^-, and A_2^+ with A_2^-, we get

$$\nu(E\cap A_1^-) = \nu(E\cap A_2^-).$$

□

Consider a Hahn decomposition $\{A^+, A^-\}$ of X with respect to a signed measure ν on $\mathcal{X}$. Problem 2.11 ensures that the set functions $\nu^+ : \mathcal{X} \to \mathbb{R}$ and $\nu^- : \mathcal{X} \to \mathbb{R}$ defined for every set E in $\mathcal{X}$ by

$$\nu^+(E) = \nu(A^+\cap E) \quad \text{and} \quad \nu^-(E) = -\nu(A^-\cap E)$$

are finite measures on $\mathcal{X}$. The measures ν^+ and ν^- are called *positive variation* and *negative variation* of ν, respectively. Note: (1) ν^+ and ν^- are unambiguously defined (their definitions do not depend on the Hahn decomposition $\{A^+, A^-\}$ by Proposition 7.3), and (2) $\nu^+(A^-) = \nu^-(A^+) = 0$, signifying that ν^+ and ν^- are *singular* (as it will be defined in Definition 7.9).

Theorem 7.4. (Jordan Decomposition Theorem). *Let $(X, \mathcal{X})$ be a measurable space. Suppose ν is a signed measure on $\mathcal{X}$. Then*

$$\nu = \nu^+ - \nu^-,$$

where ν^+ and ν^- are the positive and negative variations of ν. If

$$\nu = \lambda - \mu,$$

where λ and μ are finite measures on $\mathcal{X}$, then

$$\nu^+ \leq \lambda \quad \text{and} \quad \nu^- \leq \mu.$$

Proof. Take an arbitrary Hahn decomposition $\{A^+, A^-\}$ of X with respect to a signed measure ν on $\mathcal{X}$. Since $\{A^+, A^-\}$ is a partition of X, it follows that $(A^+ \cap E) \cup (A^- \cap E) = E$ and $(A^+ \cap E) \cap (A^- \cap E) = \varnothing$, and hence

$$\nu(E) = \nu(A^+ \cap E) + \nu(A^- \cap E) = \nu^+(E) - \nu^-(E),$$

for every $E \in \mathcal{X}$. If λ and μ are finite measures on $\mathcal{X}$ (so that $0 \leq \lambda(E) < \infty$ and $0 \leq \mu(E) < \infty$ for every $E \in \mathcal{X}$) such that $\nu = \lambda - \mu$, then

$$\nu^+(E) = \nu(A^+ \cap E) = \lambda(A^+ \cap E) - \mu(A^+ \cap E) \leq \lambda(A^+ \cap E) \leq \lambda(E),$$
$$\nu^-(E) = -\nu(A^- \cap E) = -\lambda(A^- \cap E) + \mu(A^+ \cap E) \leq \mu(A^+ \cap E) \leq \mu(E),$$

for every $E \in \mathcal{X}$ (Proposition 2.2(a)). Therefore, $\nu^+ \leq \lambda$ and $\nu^- \leq \mu$. $\square$

The sum of finite measures is again a finite measure. The *total variation* of a signed measure $\nu\colon \mathcal{X} \to \mathbb{R}$ is the finite measure $|\nu|\colon \mathcal{X} \to \mathbb{R}$ defined by

$$|\nu| = \nu^+ + \nu^-.$$

Example 7A. Let ν be a signed measure on $\mathcal{X}$. The total variation $|\nu|$ coincides with the (ordinary) variation μ discussed in Example 2I. In fact, as in Example 2I, the (ordinary) variation μ is the measure defined by

$$\mu(E) = \sup_{\{E^+, E^-\} \in \boldsymbol{E}(2)} \big(\nu(E^+) - \nu(E^-)\big) \quad \text{for every} \quad E \in \mathcal{X},$$

where the supremum is taken over all measurable partitions $\{E^+, E^-\}$ of E consisting of two sets such that $\nu(E^+) \geq 0$ and $\nu(E^-) \leq 0$. If $\{A^+, A^-\}$ is any Hahn decomposition of X with respect to ν and $\mathcal{E} = \wp(E) \cap \mathcal{X}$, then

$$\nu^+(E) + \nu^-(E) = \nu(A^+ \cap E) - \nu(A^- \cap E) \leq \mu(E)$$
$$\leq \sup_{F \in \mathcal{E}} \nu(F) - \inf_{F \in \mathcal{E}} \nu(F) = \nu^+(E) + \nu^-(E),$$

where the last identity follows from Theorem 7.4 (via Problem 7.5), and so

$$\mu(E) = |\nu|(E) = \nu^+(E) + \nu^-(E) \quad \text{for every} \quad E \in \mathcal{X}.$$

Proposition 7.5. *Consider a measure space $(X, \mathcal{X}, \mu)$, and take a function $f \in \mathcal{L}(X, \mathcal{X}, \mu)$. If $\nu\colon \mathcal{X} \to \mathbb{R}$ is the signed measure defined in* Lemma 4.6,

$$\nu(E) = \int_E f\, d\mu \quad \text{for every} \quad E \in \mathcal{X},$$

then the measures ν^+, ν^-, and $|\nu|$ are given for each $E \in \mathcal{X}$ by

$$\nu^+(E) = \int_E f^+ d\mu, \qquad \nu^-(E) = \int_E f^- d\mu, \qquad \text{and} \quad |\nu|(E) = \int_E |f|\, d\mu.$$

Proof. Take the sets $F_+ = \{x \in X\colon f(x) > 0\}$, $F_- = \{x \in X\colon f(x) < 0\}$, and $F_0 = \{x \in X\colon f(x) = 0\}$, and set

$$F^+ = F_+ \cup F_0 = \{x \in X\colon f \geq 0\} \quad \text{and} \quad F^- = F_- \cup F_0 = \{x \in X\colon f \leq 0\}.$$

Thus $\{F^+, F_-\}$ is a measurable partition of X such that $\nu(F^+ \cap E) \geq 0$ and $\nu(F_- \cap E) \leq 0$ for each $E \in \mathcal{X}$, and so $\{F^+, F_-\}$ is a Hahn decomposition of X with respect to ν. Since $\int_{F_- \cap E} f\, d\mu = \int_{F^- \cap E} f\, d\mu$ (cf. Proposition 3.7(a), Problem 3.8(a), and Definition 4.1), it follows for every $E \in \mathcal{X}$ that

$$\nu^+(E) = \nu(F^+ \cap E) = \int_{F^+ \cap E} f\, d\mu = \int_E f\chi_{F^+}\, d\mu = \int_E f^+ d\mu,$$

$$\nu^-(E) = -\nu(F_- \cap E) = -\int_{F_- \cap E} f\, d\mu = \int_E -f\chi_{F^-}\, d\mu = \int_E f^- d\mu,$$

$$|\nu|(E) = \nu^+(E) + \nu^-(E) = \int_E (f^+ + f^-)\, d\mu = \int_E |f|\, d\mu. \qquad \square$$

7.2 The Radon–Nikodým Theorem

Definition 7.6. Take a measurable space $(X, \mathcal{X})$ and let $E \in \mathcal{X}$ be an arbitrary measurable set. A measure λ on $\mathcal{X}$ is *absolutely continuous* with respect to a measure μ on $\mathcal{X}$ if

$$\mu(E) = 0 \quad \text{implies} \quad \lambda(E) = 0$$

(i.e., $\lambda(E) = 0$ for every $E \in \mathcal{X}$ such that $\mu(E) = 0$). Notation: $\lambda \ll \mu$.

Let λ and μ be measures on $\mathcal{X}$, let E be an arbitrary measurable set in $\mathcal{X}$, and consider the following statements.

(a) For every $\varepsilon > 0$ there is a $\delta_\varepsilon > 0$ (which does not depend on E) such that

$$\mu(E) < \delta_\varepsilon \quad \text{implies} \quad \lambda(E) < \varepsilon.$$

(b) The measure λ is absolutely continuous with respect to μ (i.e., $\lambda \ll \mu$).

These are equivalent if λ is a finite measure, as it will be shown in Proposition 7.7 below, thus justifying the terminology "absolute continuity".

Proposition 7.7. *Consider the above assertions.*

Claim: (a) *implies* (b), *and* (b) *implies* (a) *if λ is finite.*

Proof. Suppose (a) holds. If $\mu(E) = 0$ for some $E \in \mathcal{X}$, then $\lambda(E) < \varepsilon$ for all $\varepsilon > 0$, which means that $\lambda(E) = 0$. Therefore (a) implies (b). Conversely, Suppose (a) fails. Thus there exists an $\varepsilon > 0$ such that for every $\delta > 0$ there exists an $E_\delta \in \mathcal{X}$ for which $\mu(E_\delta) < \delta$ and $\lambda(E_\delta) \geq \varepsilon$. In particular, for every $n \geq 1$ there exists an $E_n \in \mathcal{X}$ such that $\mu(E_n) < \frac{1}{2^n}$ and $\lambda(E_n) \geq \varepsilon$. Set $F_n = \bigcup_{k=n}^{\infty} E_k$ in $\mathcal{X}$ so that $\mu(F_n) \leq \sum_{k=n}^{\infty} \mu(E_k) < \sum_{k=n}^{\infty} \frac{1}{2^k} = \frac{1}{2^{n-1}}$ and $\lambda(F_n) \geq \lambda(E_k) \geq \varepsilon$ for every $n \geq 1$. Set $F = \bigcap_{n=1}^{\infty} F_n$ in $\mathcal{X}$. Since $\{F_n\}$ is a decreasing sequence of sets in $\mathcal{X}$, and since $\mu(F_1) \leq 1$ and $\lambda(F_1) < \infty$ if λ is a finite measure, it then follows by Proposition 2.2(d) that

$$\mu(F) = \lim_n \mu(F_n) = 0 \quad \text{and} \quad \lambda(F) = \lim_n \lambda(F_n) \geq \varepsilon,$$

and so (b) fails. Equivalently, (b) implies (a) if λ is a finite measure. □

Propositions 3.5(c) and 3.7(b) ensure that *if μ is a measure on $\mathcal{X}$ and f is a function in $\mathcal{M}(X, \mathcal{X})^+$* (a nonnegative extended real-valued measurable function), *then the set function λ on $\mathcal{X}$ defined by*

$$\lambda(E) = \int_E f \, d\mu \quad \textit{for every} \quad E \in \mathcal{X}$$

is a measure which is absolutely continuous with respect to μ. What comes as a nice and perhaps unexpected result is that the converse holds if the measures λ and μ are σ-finite. That is, in this case, there exists a function f in $\mathcal{M}(X, \mathcal{X})^+$ such that λ is expressed as an integral of f with respect to μ. Also, the function f is unique μ-a.e. (which means that if g in $\mathcal{M}(X, \mathcal{X})^+$ is such that $\lambda(E) = \int_E g \, d\mu$ for every $E \in \mathcal{X}$, then $g = f$ μ-almost everywhere). This converse is a fundamental result in measure theory, which we see next.

Theorem 7.8. (Radon–Nikodým Theorem). *Take any measurable space $(X, \mathcal{X})$. If λ and μ are σ-finite measures on $\mathcal{X}$, and if λ is absolutely continuous with respect to μ, then there exists a unique (μ-almost everywhere unique) real-valued function f in $\mathcal{M}(X, \mathcal{X})^+$ such that*

$$\lambda(E) = \int_E f\,d\mu \quad \text{for every} \quad E \in \mathcal{X}.$$

Proof. Suppose λ and μ are measures on $\mathcal{X}$ such that $\lambda \ll \mu$. We split the proof into two parts. Part (a) proves the theorem for finite measures. Part (b) extends the proof for σ-finite measures.

(a) Take an arbitrary real number $\alpha > 0$. Suppose λ and μ are finite measures. Thus $\nu_\alpha = \lambda - \alpha\mu$ is a (real-valued) signed measure. Let $\{A_\alpha^+, A_\alpha^-\}$ be a Hahn decomposition for X with respect to the signed measure ν_α. Consider a sequence $\{E_k\}_{k\geq 1}$ of sets in $\mathcal{X}$ recursively defined by

$$E_{k+1} = A^-_{(k+1)\alpha} \setminus \bigcup_{j=1}^{k} E_j \quad \text{with} \quad E_1 = A^-_\alpha.$$

It is immediately verified by induction that

(i) $$\{E_k\}_{k\geq 1} \text{ is a sequence of disjoint sets,}$$

(ii) $$\bigcup_{j=1}^{k} E_j = \bigcup_{j=1}^{k} A^-_{j\alpha} \quad \text{for every} \quad k \geq 1,$$

and hence

$$E_k = A^-_{k\alpha} \setminus \bigcup_{j=1}^{k-1} A^-_{j\alpha} = A^-_{k\alpha} \cap \bigcap_{j=1}^{k-1} A^+_{j\alpha} \quad \text{for every} \quad k \geq 2.$$

Then $E_k \in (A^-_{k\alpha} \cap A^+_{(k-1)\alpha})$, which implies that for each set $E \subseteq (\mathcal{X} \cap E_k)$, $\lambda(E) - k\alpha\mu(E) \leq 0$ and $\lambda(E) - (k-1)\alpha\mu(E) \geq 0$, and so

(iii) $$(k-1)\alpha\mu(E) \leq \lambda(E) \leq k\alpha\mu(E),$$

for every $E \subseteq (\mathcal{X} \cap E_k)$ and every $k \geq 2$. Set (cf. property (ii))

$$F = X \setminus \bigcup_{j=1}^{\infty} E_j = X \setminus \bigcup_{j=1}^{\infty} A^-_{j\alpha} = \bigcap_{j=1}^{\infty} A^+_{j\alpha} \subseteq A^+_{k\alpha} \quad \text{for all} \quad k \geq 1$$

so that $\lambda(F) - k\alpha\mu(F) \geq 0$, equivalently, $0 \leq k\alpha\mu(F) \leq \lambda(F)$, for all $k \geq 1$. Since λ is a finite measure, it follows that $\mu(F) = 0$. Since $\lambda \ll \mu$, we get

(iv) $$\lambda(F) = 0.$$

Take the nonnegative real-valued function f_α in $\mathcal{M}^+ = \mathcal{M}(X, \mathcal{X})^+$ given by

$$f_\alpha(x) = \begin{cases} (k-1)\alpha, & x \in E_k \text{ for some } k \geq 1, \\ 0, & x \in F = X \setminus \bigcup_{k=1}^{\infty} E_k. \end{cases}$$

Observe that $\{F, E_k;\, k\geq 1\}$ is a measurable partition of X by property (i). Take an arbitrary $E\in\mathcal{X}$ so that $\{E\cap F, E\cap E_k;\, k\geq 1\}$ is a measurable partition of E. Thus (cf. Problems 3.3(a) and 4.10)

$$\begin{aligned}\int_E f_\alpha\,d\mu &= \int_{\bigcup_{k=1}^\infty (E\cap E_k)} f_\alpha\,d\mu = \sum_{k=1}^\infty \int_{E\cap E_k} f_\alpha\,d\mu\\ &= \sum_{k=1}^\infty (k-1)\alpha\int_{E\cap E_k} d\mu = \sum_{k=2}^\infty (k-1)\alpha\,\mu(E\cap E_k)\\ &\leq \sum_{k=2}^\infty \lambda(E\cap E_k) = \lambda\big(\textstyle\bigcup_{k=2}^\infty (E\cap E_n)\big) \leq \lambda(E)\end{aligned}$$

by properties (iii) and (i). Similarly, by properties (i), (iii), and (iv) we get

$$\begin{aligned}\lambda(E) &= \int_E d\lambda = \int_{\bigcup_{k=1}^\infty (E\cap E_k)} d\lambda = \lambda\big(\textstyle\bigcup_{k=1}^\infty (E\cap E_k)\big) = \displaystyle\sum_{k=1}^\infty \lambda(E\cap E_k)\\ &\leq \sum_{k=1}^\infty k\,\alpha\,\mu(E\cap E_k) = \sum_{k=1}^\infty \int_{E\cap E_k} k\,\alpha\,d\mu = \sum_{k=1}^\infty \int_{E\cap E_k} (f_\alpha+\alpha)\,d\mu\\ &= \int_{\bigcup_{k=1}^\infty (E\cap E_k)} (f_\alpha+\alpha)\,d\mu \leq \int_E (f_\alpha+\alpha)\,d\mu = \int_E f_\alpha\,d\mu + \alpha\,\mu(E).\end{aligned}$$

Take an arbitrary integer $n\geq 1$, set $\alpha = \big(\frac{1}{2}\big)^n$ and $f_n = f_{(1/2)^n}$. The previously displayed inequalities ensure that

$$\text{(v)}\qquad \int_E f_n\,d\mu \leq \lambda(E) \leq \int_E f_n\,d\mu + \big(\tfrac{1}{2}\big)^n \mu(X)$$

for all $n\geq 1$. Hence, for an arbitrary pair of positive integers m and n,

$$\begin{aligned}\int_E f_n\,d\mu &\leq \lambda(E) \leq \int_E f_m\,d\mu + \big(\tfrac{1}{2}\big)^m \mu(X),\\ \int_E f_m\,d\mu &\leq \lambda(E) \leq \int_E f_n\,d\mu + \big(\tfrac{1}{2}\big)^n \mu(X),\end{aligned}$$

so that

$$\begin{aligned}\int_E f_n\,d\mu - \int_E f_m\,d\mu &\leq \lambda(E) - \int_E f_m\,d\mu \leq \big(\tfrac{1}{2}\big)^m \mu(X),\\ \int_E f_m\,d\mu - \int_E f_n\,d\mu &\leq \lambda(E) - \int_E f_n\,d\mu \leq \big(\tfrac{1}{2}\big)^n \mu(X),\end{aligned}$$

which implies

$$\left|\int_E (f_m - f_n)\,d\mu\right| \leq \big(\tfrac{1}{2}\big)^m \mu(X)$$

whenever $m \leq n$. Since this holds for all $E \in \mathcal{X}$, we get

$$\begin{aligned}\int |f_m - f_n|\,d\mu &= \int (f_m - f_n)^+\,d\mu + \int (f_m - f_n)^-\,d\mu \\ &= \int_{F^+_{m,n}} (f_m - f_n)\,d\mu - \int_{F^-_{m,n}} (f_m - f_n)\,d\mu \\ &= \left|\int_{F^+_{m,n}} (f_m - f_n)\,d\mu\right| + \left|\int_{F^-_{m,n}} (f_m - f_n)\,d\mu\right| \leq 2\big(\tfrac{1}{2}\big)^m \mu(X)\end{aligned}$$

for each m, n such that $m \leq n$, with $F^+_{m,n} = \{x \in X\colon (f_m - f_n)(x) \geq 0\}$ and $F^-_{m,n} = \{x \in X\colon (f_m - f_n)(x) \leq 0\}$. Since property (v) holds for all E in $\mathcal{X}$, λ is a finite measure, and f_n is a real-valued function in $\mathcal{M}^+$, it follows that each function f_n lies in $L^1(\mu) = L^1(X, \mathcal{X}, \mu)$. Thus, by the above inequality,

$$\|f_m - f_n\|_1 \leq \big(\tfrac{1}{2}\big)^{m-1} \mu(X)$$

if $1 \leq m \leq n$. Since μ is a finite measure, the above inequality says that $\{f_n\}$ is a Cauchy sequence in the Banach space $L^1(\mu)$, and so it converges in $L^1(\mu)$ to, say, $f \in L^1(\mu)$. Then the real-valued sequence $\{\int_E f_n\,d\mu\}$ converges in $\mathbb{R}$ to $\int_E f\,d\mu$ for every $E \in \mathcal{X}$. In fact (cf. Problem 5.16),

$$\left|\int_E f_n\,d\mu - \int_E f\,d\mu\right| \leq \int_E |f_n - f|\,d\mu \leq \int |f_n - f|\,d\mu = \|f_n - f\|_1 \to 0.$$

Recall that $f = \lim_n f_n \in L^1(\mu)$ is real-valued. Therefore, by property (v),

$$\lambda(E) = \lim_n \int_E f_n\,d\mu = \int_E f\,d\mu \quad \text{for every } E \in \mathcal{X}.$$

Note that we may take a nonnegative function f in the equivalence class $[f]$, that is, we may take $f \in \mathcal{M}^+$. Indeed, since $0 \leq \lambda(E) = \int_E f\,d\mu$ for every $E \in \mathcal{X}$, it follows by Problem 4.5(b) that $f \geq 0$ μ-a.e. (and also λ-a.e. because $\lambda \ll \mu$) for every $f \in [f]$. Moreover, such an f is μ-a.e. unique. In fact, if $g \in L^1(\mu)$ (and $g \in \mathcal{M}^+$) is such that $\lambda(E) = \int_E f\,d\mu = \int_E g\,d\mu$ for every $E \in \mathcal{X}$, then $f = g$ μ-a.e. by Problem 3.8(d) (or Problem 4.5(c)).

(b) Next assume that the measures λ and μ are σ-finite. Thus there are two sequences of $\mathcal{X}$-measurable sets, say $\{A_n\}$ and $\{B_n\}$, such that $\lambda(A_n) < \infty$ and $\mu(B_n) < \infty$ for every n, both covering X. Actually, by considering successive unions of them if necessary, we may assume that these sequences are increasing. Set $X_n = A_n \cap B_n$ so that $\{X_n\}$ is an increasing sequence of $\mathcal{X}$-measurable sets such that $\bigcup X_n = \bigcup A_n \cap \bigcup B_n = X$ and, for each n,

$$\lambda(X_n) < \infty \quad \text{and} \quad \mu(X_n) < \infty.$$

Take an arbitrary index n. Consider the σ-algebra $\mathcal{X}_n = \wp(X_n) \cap \mathcal{X}$. By part (a) there is a function $g_n \in \mathcal{M}(X_n, \mathcal{X}_n)^+$ such that $\lambda(E') = \int_{E'} g_n \, d\mu|_{X_n}$ for every $E' \in \mathcal{X}_n$ (cf. Problem 2.11). Let f_n be a function on X defined as follows: $f_n(x) = g_n(x)$ if $x \in X_n$, and $f_n(x) = 0$ if $x \in X \backslash X_n$. This function f_n lies in $\mathcal{M}^+ = \mathcal{M}(X, \mathcal{X})^+$ and, for every $E' \in \mathcal{X}_n \subseteq \mathcal{X}$,

$$\lambda(E') = \int_{E'} f_n \, d\mu.$$

Recall that $X_n \subseteq X_k$ for every $n \leq k$. Then $E' \in \mathcal{X}_k$ for every $k \geq n$ whenever $E' \in \mathcal{X}_n$. So the above identity holds for all $k \geq n$. That is, the sequence $\{f_n\}$ of functions in $\mathcal{M}^+$ is such that if $E' \in \mathcal{X}_n$ for some $n \geq 1$, then

$$\lambda(E') = \int_{E'} f_n \, d\mu = \int_{E'} f_k \, d\mu$$

for every $k \geq n$. Again from part (a), g_n is unique μ-a.e., then so is f_n, for each n. Thus the previous identity ensures that $f_k = f_n$ μ-a.e. on X_n for all $k \geq n$. (Same uniqueness argument: see Problem 3.8(d) or Problem 4.5(a) — nonnegative functions in $L^1(\mu)$.) Hence, since each f_n vanishes outside X_n, and since $\{X_n\}$ is an increasing sequence of sets, it follows that $\{f_n\}$ is an increasing sequence of functions in $\mathcal{M}^+$ such that

$$\lambda(E \cap X_n) = \int_{E \cap X_n} f_n \, d\mu = \int_E f_n \chi_{X_n} \, d\mu = \int_E f_n \, d\mu$$

for each $E \in \mathcal{X}$ (since $E \cap X_n \in \mathcal{X}_n$ and $f_n = f_n \chi_{X_n}$) and for each $n \geq 1$. Take an arbitrary set E in $\mathcal{X}$ and note that $\{E \cap X_n\}$ is an increasing sequence of sets in $\mathcal{X}$ that covers E (because $\{X_n\}$ is an increasing sequence of sets in $\mathcal{X}$ that covers X). Thus, by Proposition 2.2(c),

$$\lambda(E) = \lambda\Big(\bigcup_n (E \cap X_n)\Big) = \lim_n (E \cap X_n) = \lim_n \int_E f_n \, d\mu.$$

The Monotone Convergence Theorem completes the existence proof. Indeed, since $\{f_n\}$ is an increasing sequence of functions in $\mathcal{M}^+$, it converges pointwise to a function $\tilde{f}$ in $\mathcal{M}^+$, and so it follows by Theorem 3.4 that

$$\lambda(E) = \lim_n \int_E f_n \, d\mu = \lim_n \int f_n \chi_E \, d\mu = \int \lim_n f_n \chi_E \, d\mu = \int \tilde{f} \chi_E \, d\mu = \int_E \tilde{f} \, d\mu.$$

The function $\tilde{f} \in \mathcal{M}^+$ is the pointwise limit of an increasing sequence of functions in $\mathcal{M}^+$. Thus it may possibly be extended real-valued; but, it is μ-a.e. real-valued. Equivalently, the set $F_{+\infty} = \{x \in X \colon \tilde{f}(x) = +\infty\}$ has measure zero (i.e., $\mu(F_{+\infty}) = 0$). Indeed, recall that $\{X_n\}$ is an increasing sequence of sets that cover X, each f_n is null outside X_n, and $f_k = f_n$ μ-a.e. on X_n for all $k \geq n$. Thus, since $\tilde{f}(x) = \lim_n f_n(x)$ for every $x \in X$, it follows that $f_n = \tilde{f} \chi_{X_n}$ μ-a.e., and so $\mu(F_{+\infty} \cap X_n) = \mu(\{x \in X_n \colon f_n(x) = +\infty\})$, for each n. However $f_n \in L^1(X, \mathcal{X}, \mu)$, because $g_n = f_n \chi_{X_n} \in L^1(X_n, \mathcal{X}_n, \mu)$ and $f_n = 0$ on $X \backslash X_n$. Hence $\mu(F_{+\infty} \cap X_n) = 0$ for all n by Problem 3.9(b). Since $F_{+\infty} = F_{+\infty} \cap X = F_{+\infty} \cap \bigcup_n X_n = \bigcup_n (F_{+\infty} \cap X_n)$, it follows that $\mu(F_{+\infty}) \leq \sum_n \mu(F_{+\infty} \cap X_n) = 0$ (Problem 2.8(b)), and so $\mu(F_{+\infty}) = 0$. Set $f = \tilde{f}\chi_{X\backslash F_{+\infty}}$, a real-valued function in $\mathcal{M}^+$ such that (see Problem 3.8)

$$\lambda(E) = \int_E \tilde{f}\, d\mu = \int_{E\cap(X\backslash F_{+\infty})} \tilde{f}\, d\mu + \int_{E\cap F_{+\infty}} \tilde{f}\, d\mu = \int_E \tilde{f}\chi_{X\backslash F_{+\infty}}\, d\mu = \int_E f\, d\mu$$

for every $E \in \mathcal{X}$. Such an $f \in \mathcal{M}^+$ is unique μ-a.e. by Problem 3.8(d). $\square$

The real-valued function $f \in \mathcal{M}(X, \mathcal{X})^+$ in the statement of the Radon–Nikodým Theorem was not claimed to be integrable. Actually, f is μ-integrable (i.e., f lies in $\mathcal{L}(X, \mathcal{X}, \mu)$) if and only if λ is a finite measure. This function f is called the *Radon–Nikodým derivative* of λ with respect to μ, which it is often written as $f = \frac{d\lambda}{d\mu}$ (or $d\lambda = f\, d\mu$). As noticed in Problem 3.11, no independent meaning is assigned to the symbols $d\lambda$ and $d\mu$. So, if λ and μ are σ-finite measures such that $\lambda \ll \mu$, then there is a unique (μ-a.e.) real-valued function $\frac{d\lambda}{d\mu}$ in $\mathcal{M}(X, \mathcal{X})^+$ such that for every E in $\mathcal{X}$

$$\lambda(E) = \int_E \frac{d\lambda}{d\mu}\, d\mu.$$

Remark: A major applications of the Radon–Nikodým Theorem is the *Riesz Representation Theorem*. One of the versions of it say that *if $\Phi\colon L^p(\mu) \to \mathbb{R}$ is a bounded linear functional on the Banach space $L^p(\mu)$, then there is a unique $g \in L^q(\mu)$ such that $\Phi(f) = \int fg\, d\mu$ for every $f \in L^p(\mu)$ and $\|\Phi\| = \|g\|_q$* (where q is the Hölder conjugate of p; if $p = 1$ so that $q = \infty$, then μ is supposed to be σ-finite). See Proposition 12.A. The Riesz Representation Theorem holds in every Hilbert space, and so it can be proved for $p = 2$ without using the Radon–Nikodým Theorem and, perhaps surprisingly, this can be used to prove the Radon–Nikodým Theorem itself. Chapter 12 is entirely dedicated to the Riesz Representation Theorem.

7.3 The Lebesgue Decomposition Theorem

If λ is absolutely continuous with respect to μ, then they act synchronized in the sense that sets of small μ-measures have small λ-measures (Proposition 7.7). At the opposite end there are measures λ and μ which act complementary in the sense that sets of small μ-measure may have large λ-measure.

Definition 7.9. Take a measurable space $(X, \mathcal{X})$. A measure λ on $\mathcal{X}$ is *singular* with respect to a measure μ on $\mathcal{X}$ (notation: $\lambda \perp \mu$) if there exists a measurable partition $\{A, B\}$ of X such that

$$\lambda(A) = \mu(B) = 0.$$

The notion of singular measures means that λ and μ have *disjoint supports* (see Problem 2.13), which is also referred to by saying that λ is *concentrated* on a set of μ-measure zero. It is clear that $\perp$ is a symmetric relation on the collection of all measures on $\mathcal{X}$ (i.e., $\lambda \perp \mu$ if and only if $\mu \perp \lambda$). Thus we say that λ and μ are *mutually singular*, or simply *singular*, instead of λ is singular with respect to μ (or vice versa). Note that singularity may be equivalently restated as follows: $\lambda \perp \mu$ if there exists a partition $\{A, B\}$ of X such that $A \cap E$ and $B \cap E$ lie in $\mathcal{X}$ for every $E \in \mathcal{X}$ and

$$\lambda(A \cap E) = \mu(B \cap E) = 0.$$

Observe that the preceding two expressions are equivalent. In fact, A and B must be measurable (because $A \cap X$ and $B \cap X$ lie in $\mathcal{X}$ since X lies in $\mathcal{X}$), $A \cap E \subseteq A$, and $B \cap E \subseteq B$. The next result is another important consequence of the Radon–Nikodým Theorem.

Theorem 7.10. (Lebesgue Decomposition Theorem). *Consider a measurable space $(X, \mathcal{X})$. If λ and μ are σ-finite measures on $\mathcal{X}$, then there exists a unique pair of measures λ_a and λ_s on $\mathcal{X}$ such that $\lambda_a \ll \mu$, $\lambda_s \perp \mu$, and*

$$\lambda = \lambda_a + \lambda_s.$$

Proof. Let λ and μ be σ-finite measures on $\mathcal{X}$. Set $\nu = \mu + \lambda$, which is again σ-finite measure on $\mathcal{X}$ (Problem 2.14). Both μ and λ are absolutely continuous with respect to ν (i.e., $\nu(E) = 0$ implies $\mu(E) = \lambda(E) = 0$, and so $\mu \ll \nu$ and $\lambda \ll \nu$). Then the Radon–Nikodým Theorem says that there are real-valued functions f and g in $\mathcal{M}(X, \mathcal{X})^+$ such that for every $E \in \mathcal{X}$,

$$\mu(E) = \int_E f\,d\nu \quad \text{and} \quad \lambda(E) = \int_E g\,d\nu.$$

Take the measurable partition $\{F_0, F_+\}$ of X, with $F_0 = \{x \in X\colon f(x) = 0\}$ and $F_+ = \{x \in X\colon f(x) > 0\}$. Set $\lambda_s = \lambda_{F_0}$ and $\lambda_a = \lambda_{F_+}$, where the measures λ_{F_0} and λ_{F_+} on $\mathcal{X}$ are defined as in Problem 2.11, namely,

$$\lambda_s(E) = \lambda_{F_0}(E) = \lambda(E \cap F_0) \quad \text{and} \quad \lambda_a(E) = \lambda_{F_+}(E) = \lambda(E \cap F_+)$$

for every $E \in \mathcal{X}$. Since $\mu(F_0) = \int_{F_0} 0\,d\nu = 0$ and $\lambda_s(F_+) = \lambda(\varnothing) = 0$, we get

$$\lambda_s \perp \mu.$$

If $E \in \mathcal{X}$ is such that $\mu(E) = 0$, then $\int f\chi_E\,d\nu = \int_E f\,d\nu = 0$, and hence $f\chi_E = 0$ ν-a.e. (cf. Proposition 3.7(a)). That is, $f = 0$ ν-a.e. on E, and so $\nu(E \cap F_+) = 0$. Since $\lambda \ll \nu$, it follows that $\lambda_a(E) = \lambda(E \cap F_+) = 0$. Thus,

$$\lambda_a \ll \mu.$$

Now note that $\lambda(E) = \lambda((E \cap F_0) \cup (E \cap F_+)) = \lambda(E \cap F_0) + \lambda(E \cap F_+) = \lambda_s(E) + \lambda_a(E)$ for every $E \in \mathcal{X}$, and therefore

$$\lambda = \lambda_s + \lambda_a.$$

To prove uniqueness, suppose $\lambda = \lambda_1 + \lambda_2$, where λ_1 and λ_2 are (σ-finite) measures on $\mathcal{X}$ such that $\lambda_1 \perp \mu$ and $\lambda_2 \ll \mu$. Take the signed measures $\lambda_s - \lambda_1$ and $\lambda_a - \lambda_2$ on $\mathcal{X}$ so that $\lambda_s - \lambda_1 \perp \mu$ and $\lambda_a - \lambda_2 \ll \mu$ (cf. Problems 7.10 and 7.11). Since $\lambda_s + \lambda_a = \lambda_1 + \lambda_2$, it follows by Problem 7.12 that $\lambda_s - \lambda_1 = \lambda_a - \lambda_2 = 0$, and so $\lambda_1 = \lambda_s$ and $\lambda_2 = \lambda_a$. □

Remark: The signed measures $\lambda_s - \lambda_1$ and $\lambda_a - \lambda_2$ are well defined if we allow extended real-valued signed measures, and declare that $(\lambda_s - \lambda_1)(E) = 0$ if E in $\mathcal{X}$ is such that $\lambda_s(E) = \lambda_1(E) = +\infty$, and $(\lambda_a - \lambda_2)(E) = 0$ if E in $\mathcal{X}$ is such that $\lambda_a(E) = \lambda_2(E) = +\infty$. Also note that Problems 7.10, 7.11, and 7.12 are naturally extended to extended real-valued signed measures.

Theorem 7.10 decomposes every σ-finite measure λ into two parts: an absolute continuous and a singular, both with respect to a σ-finite *reference measure* μ (e.g., such a reference measure may be the Lebesgue measure in the particular case of $(X, \mathcal{X}) = (\mathbb{R}, \mathfrak{R})$). Next we refine this decomposition.

Definition 7.11. Take a measurable space $(X, \mathcal{X})$. A measure λ on $\mathcal{X}$ is *continuous* with respect to a measure μ on $\mathcal{X}$ if, for $\{x\} \in \mathcal{X}$,

$$\mu(\{x\}) = 0 \quad \text{implies} \quad \lambda(\{x\}) = 0$$

(i.e., $\lambda(\{x\}) = 0$ for every *measurable* singleton $\{x\}$ such that $\mu(\{x\}) = 0$).

Definition 7.12. Let $(X, \mathcal{X})$ be a measurable space. A measure λ on $\mathcal{X}$ is *discrete* with respect to measure μ on $\mathcal{X}$ if there exists a measurable partition $\{A, B\}$ of X such that (i) B is a countable set whose all subsets are measurable (equivalently, $B = \{b_n\}_{n \in I}$ with each singleton $\{b_n\}$ lying in $\mathcal{X}$, where the index set I is either finite or $I = \mathbb{N}$), and (ii)

$$\lambda(A) = \mu(B) = 0.$$

That is, λ is concentrated on a *countable* set of μ-measure zero; and so, if λ is discrete with respect μ, then λ and μ are singular — see Problems 7.13 and 7.14. In the very particular case where μ is the Lebesgue measure on the Borel algebra of subsets of the real line things get considerably simplified.

Proposition 7.13. *Consider a measurable space $(X, \mathcal{X})$. Suppose λ and μ are measures on $\mathcal{X}$. If λ is σ-finite and measurable singletons of X have μ-measure zero, then there is a unique pair of measures λ_c and λ_d on $\mathcal{X}$ such that λ_c is continuous and λ_d is discrete, both with respect to μ, and*

$$\lambda = \lambda_c + \lambda_d.$$

Proof. If λ is σ-finite, then there is a sequence $\{E_n\}$ of sets in $\mathcal{X}$ that cover X such that each E_n has finite μ-measure. For each integer $k \geq 1$ set

$$B_k(n) = \{x \in E_n \colon \{x\} \in \mathcal{X} \text{ and } \lambda(\{x\}) \geq \tfrac{1}{k}\}.$$

Suppose $B_k(n)$ is an infinite set. Thus it has a countably infinite subset, say $C_k(n) = \bigcup_m \{b_m\} \subseteq B_k(n)$, consisting of distinct points b_m of $B_k(n)$. Since each singleton $\{b_m\}$ is $\mathcal{X}$-measurable, $C_k(n)$ also lies in $\mathcal{X}$, and hence $\lambda(C_k(n)) = \sum_m \lambda(\{b_m\}) = \infty$ because $\lambda(\{b_m\}) \geq \frac{1}{k}$ for all m. But this contradicts the fact that $\lambda(C_k(n)) \leq \lambda(E_n) < \infty$. Therefore, each $B_k(n)$ is a finite set in $\mathcal{X}$. Thus, since $X = \bigcup_n E_n$,

$$B_k = \bigcup_n B_k(n) = \{x \in X \colon \{x\} \in \mathcal{X} \text{ and } \lambda(\{x\}) \geq \tfrac{1}{k}\}$$

is a countable set in $\mathcal{X}$ for every $k \geq 1$, and so

$$B = \bigcup_k B_k = \{x \in X \colon \{x\} \in \mathcal{X} \text{ and } \lambda(\{x\}) \neq 0\}$$

is again a countable set (recall: a countable union of countable sets is countable) in $\mathcal{X}$. Indeed, B is measurable (because, after all, B is a countable union of measurable singletons). Take the measurable partition $\{A, B\}$ of X so that $A = X \backslash B$. Set $\lambda_c = \lambda_A$ and $\lambda_d = \lambda_B$, where the measures λ_A and λ_B on $\mathcal{X}$ are defined as in Problem 2.11. That is, for each $E \in \mathcal{X}$,

$$\lambda_d(E) = \lambda_B(E) = \lambda(E \cap B) \quad \text{and} \quad \lambda_c(E) = \lambda_A(E) = \lambda(E \cap A).$$

Since $B = \{b_n\}_{n \in I}$ is a countable set consisting of measurable singletons, and since measurable singletons have μ-measure zero, we get

$$\lambda_d(A) = \lambda(A \cap B) = 0 \quad \text{and} \quad \mu(B) = \mu\Big(\bigcup_n \{b_n\}\Big) = \sum_n \mu(\{b_n\}) = 0,$$

and so λ_d is discrete with respect to μ. If $\{x\}$ is an $\mathcal{X}$-measurable singleton of X, then either $\lambda(\{x\}) = 0$ or $\lambda(\{x\}) \neq 0$. In the former case, $\{x\} \subseteq A$ so that $\lambda_c(\{x\}) = \lambda(\{x\}) = 0$. In the latter case, $\{x\} \subseteq B$ so that $\lambda_c(\{x\}) =$

$\lambda(\varnothing) = 0$. Thus $\lambda_c(\{x\}) = 0$ for all $\{x\} \in \mathcal{X}$, and hence λ_c is continuous with respect to μ (because $\mu(\{x\}) = 0$ for all $\{x\} \in \mathcal{X}$). Now observe that

$$\lambda(E) = \lambda\big((E \cap A) \cup (E \cap B)\big) = \lambda(E \cap A) + \lambda(E \cap B) = \lambda_c(E) + \lambda_d(E)$$

for every $E \in \mathcal{X}$, and therefore

$$\lambda = \lambda_c + \lambda_d.$$

To prove uniqueness, suppose $\lambda = \lambda_1 + \lambda_2$, where λ_1 and λ_2 are measures on $\mathcal{X}$ such that λ_1 is continuous and λ_2 is discrete, both with respect to μ. Take a singleton $\{x\}$ in $\mathcal{X}$. Since $\mu(\{x\}) = 0$, it follows that $\lambda_1(\{x\}) = \lambda_c(\{x\}) = 0$, and so $\lambda_2(\{x\}) = \lambda(\{x\}) = \lambda_d(\{x\})$. Thus $\lambda_2 = \lambda_d$ by Problem 7.14. If $\lambda_1 \neq \lambda_c$, then there is a measurable set $E \subseteq A$ such that $\lambda_1(E) \neq \lambda_c(E)$. But $\lambda_2(E) = \lambda_d(E) = 0$ because $E \subseteq A$ and $\lambda_2 = \lambda_d$, and hence $\lambda(E) = \lambda_1(E) = \lambda_c(E)$, which is a contradiction. Then $\lambda_1 = \lambda_c$. □

Corollary 7.14. *Consider a measurable space $(X, \mathcal{X})$. If μ is a σ-finite measure on $\mathcal{X}$ such that measurable singletons of X have μ-measure zero, then every σ-finite measure λ on $\mathcal{X}$ has a unique decomposition*

$$\lambda = \lambda_a + \lambda_{sc} + \lambda_{sd},$$

where the measures λ_a, λ_{sc}, λ_{sd} on $\mathcal{X}$ are absolutely continuous, singular and continuous, singular and discrete, respectively, all with respect to μ.

Proof. By Theorem 7.10, $\lambda = \lambda_a + \lambda_s$, where λ_a is absolutely continuous and λ_s is singular, with respect μ. If λ is σ-finite, then so is λ_s (the same countable covering of X that makes λ σ-finite, works for λ_s). Thus $\lambda_s = \lambda_{sc} + \lambda_{sd}$ by Proposition 7.13, where λ_{sc} is continuous and λ_{sd} is discrete, with respect to μ. Since $\lambda_s \perp \mu$, there exists a measurable partition $\{A, B\}$ of X such that $\lambda_s(A) = \lambda_{sc}(A) + \lambda_{sd}(A) = 0 = \mu(B)$, and so $\lambda_{sc}(A) = \lambda_{sd}(A) = 0$. Hence λ_{sc} and λ_{sd} also are singular with respect to μ. □

The identity $\lambda = \lambda_a + \lambda_{sc} + \lambda_{sd}$ is called the *canonical decomposition* of λ with respect to a reference measure μ, and the measures λ_{sc} and λ_{sd} are called *singular-continuous*, and *singular-discrete* (with respect to μ).

7.4 Problems

Problem 7.1. Consider a signed measure $\nu\colon \mathcal{X} \to \mathbb{R}$ on $\mathcal{X}$. If A and B are $\mathcal{X}$-measurable sets and $\{E_n\}$ is a sequence of sets in $\mathcal{X}$, then show that

(a) $\nu(B \backslash A) = \nu(B) - \nu(A)$ if $A \subseteq B$,

(b) $\nu\big(\bigcup_n E_n\big) = \lim_n \nu(E_n)$ if $\{E_n\}$ is increasing,

(c) $\nu\big(\bigcap_n E_n\big) = \lim_n \nu(E_n)$ if $\{E_n\}$ is decreasing.

Hint: Proof of Proposition 2.2 and Definition 2.3.

Problem 7.2. Prove the assertions. (a) A measurable subset of a positive set is positive. (b) A countable union of positive sets is a positive set.

Hint: (b) Let $\{A_n\}$ be a sequence of sets. Set $A'_{n+1} = A_{n+1}\backslash\big(\bigcup_{i=1}^{n} A_i\big)$ with $A'_1 = A_1$. This $\{A'_n\}$ is a sequence of disjoint sets such that $\bigcup_n A'_n = \bigcup_n A_n$ (i.e., $\{A'_n\}$ is a *disjointification* of $\{A_n\}$). Suppose each A_n is a positive set. Each A'_n is a measurable subset of A_n and so is a positive set by (a). Thus, $\nu(E \cap \bigcup_n A_n) = \nu(E \cap \bigcup_n A'_n) = \nu\big(\bigcup_n (E \cap A'_n)\big) = \sum_n \nu(E \cap A'_n) \geq 0$.

Problem 7.3. Take a signed measure ν on a σ-algebra $\mathcal{X}$. If A and B lie in $\mathcal{X}$ and B is positive with respect to ν, then

$$A \subseteq B \quad \text{implies} \quad 0 \leq \nu(A) \leq \nu(B).$$

Hint: Problems 7.1(a) and 7.2(a).

Problem 7.4. This is the signed-measure version of Problem 2.8. Suppose $\nu\colon \mathcal{X} \to \mathbb{R}$ is a signed measure on a σ-algebra $\mathcal{X}$, and let $\{E_n\}$ be a sequence of $\mathcal{X}$-measurable sets. Show that

(a) $\nu\big(\bigcup_n E_n\big) = \lim_n \nu\big(\bigcup_{i=1}^{n} E_i\big)$.

If each E_n is positive with respect to ν, then

(b) $\nu\big(\bigcup_n E_n\big) \leq \sum_n \nu(E_n)$.

Hints: (a) Problem 7.1(b). (b) Disjointification and Problem 7.3.

Problem 7.5. Take a signed measure ν on $\mathcal{X}$. For an arbitrary $E \in \mathcal{X}$, set $\mathcal{E} = \wp(E) \cap \mathcal{X}$ (the σ-algebra of all measurable subsets of E). Show that

$$\nu^+(E) = \sup_{F \in \mathcal{E}} \nu(F) \quad \text{and} \quad \nu^-(E) = -\inf_{F \in \mathcal{E}} \nu(F).$$

Hint: Theorem 7.4: $\nu(F) = \nu^+(F) - \nu^-(F) \leq \nu^+(F) \leq \nu^+(E) = \nu(A^+ \cap E)$.

Problem 7.6. Again, take a signed measure ν on $\mathcal{X}$. Prove that

$$N \in \mathcal{X} \text{ is a null set with respect to } \nu \text{ if and only if } |\nu|(N) = 0.$$

Hint: Definition of $|\nu|$ on the one hand; Problem 7.5 on the other hand.

Problem 7.7. Consider a measurable space $(X, \mathcal{X})$. Suppose ν, ν_1, and ν_2 are signed measures on $\mathcal{X}$. Let α be any real number. Verify that $\nu_1 + \nu_2$ and $\alpha\nu$ are again signed measures on $\mathcal{X}$, where for each $E \in \mathcal{X}$,

$$(\alpha\nu)(E) = \alpha\nu(E) \quad \text{and} \quad (\nu_1 + \nu_2)(E) = \nu_1(E) + \nu_2(E).$$

Now let $\mathcal{S} = \mathcal{S}(X, \mathcal{X}, \mathbb{R})$ stand for the collection of all signed measures on $\mathcal{X}$. Since addition and scalar multiplication of signed measures are again signed measures, it follows that $\mathcal{S}$ is a (real) linear space (in fact, a linear manifold of the real linear space $\mathbb{R}^{\mathcal{X}}$ of all real-valued set functions on $\mathcal{X}$). Consider the total variation of signed measures in $\mathcal{S}$. Prove that

(a) $|\alpha\nu| = |\alpha||\nu|$

and

(b) $|\nu_1 + \nu_2| \leq |\nu_1| + |\nu_2|$.

Hint: Take a Hahn decomposition $\{A_\nu^+, A_\nu^-\}$ of X with respect to ν. Show that $\{A_{\alpha\nu}^+, A_{\alpha\nu}^-\}$ is a Hahn decomposition of X with respect to $\alpha\nu$, where $A_{\alpha\nu}^+ = A_\nu^+$ and $A_{\alpha\nu}^- = A_\nu^-$ if $\alpha \geq 0$ or $A_{\alpha\nu}^+ = A_\nu^-$ and $A_{\alpha\nu}^- = A_\nu^+$ if $\alpha \leq 0$. Then show that $(\alpha\nu)^+ = \alpha\nu^+$ and $(\alpha\nu)^- = \alpha\nu^-$ if $\alpha \geq 0$ or $(\alpha\nu)^+ = -\alpha\nu^-$ and $(\alpha\nu)^- = -\alpha\nu^+$ if $\alpha \leq 0$. Thus conclude the identity in (a): $|\alpha\nu| = (\alpha\nu)^+ + (\alpha\nu)^- = |\alpha|(\nu^+ + \nu^-) = |\alpha||\nu|$. To verify the inequality in (b) note that $\nu_1 + \nu_2 = (\nu_1^+ + \nu_2^+) - (\nu_1^- + \nu_2^-)$, apply Theorem 7.4 again to show that $(\nu_1 + \nu_2)^+ \leq \nu_1^+ + \nu_2^+$ and $(\nu_1 + \nu_2)^- \leq \nu_1^- + \nu_2^-$, and hence $|\nu_1 + \nu_2| = (\nu_1 + \nu_2)^+ + (\nu_1 + \nu_2)^- \leq (\nu_1^+ + \nu_1^-) + (\nu_1^+ + \nu_2^-) = |\nu_1| + |\nu_2|$.

Next consider the function $\| \, \| : \mathcal{S} \to \mathbb{R}$ defined by

$$\|\nu\| = |\nu|(X)$$

for every $\nu \in \mathcal{S}$. This is a norm on $\mathcal{S}$. In other words, show that

(c) $(\mathcal{S}, \| \, \|)$ is a normed space.

Hint: Use (a) and (b) to verify axioms (iii) and (iv) of Definition 5.1.

Also show that this normed space is complete. That is, show that

(d) $(\mathcal{S}, \| \, \|)$ is a Banach space.

Hint: Consider the following well-known result from elementary functional analysis. *A normed space is a Banach space if and only if every absolutely summable sequence is summable* (cf. Suggested Readings for Chapter 5). Observe that if $\{\nu_n\}$ is a sequence of signed measures in $\mathcal{S}$, then

$$\max\{\nu_n^+(E), \nu_n^-(E)\} \leq \nu_n^+(E) + \nu_n^-(E) \leq \nu_n^+(X) + \nu_n^-(X) = |\nu_n|(X),$$

where $0 \leq \min\{\nu_n^+(E), \nu_n^-(E)\}$, so that $\nu_n(E) = \nu_n^+(E) - \nu_n^-(E)$ makes a summable sequence for every E in $\mathcal{X}$ (i.e., $\{\nu_n\}$ is summable in $\mathcal{S}$) whenever $\{\|\nu_n\|\}$ is a summable in $\mathbb{R}$ (i.e., whenever $\{\nu_n\}$ is absolutely summable).

Problem 7.8. Take a measurable space $(X, \mathcal{X})$. Show that absolute continuity $\ll$ is a reflexive and transitive but not a symmetric relation on the collection of all measures on $\mathcal{X}$. That is, if λ, μ, and ν are measures on $\mathcal{X}$, then show that $\mu \ll \mu$ (reflexivity), and that $\lambda \ll \mu$ and $\mu \ll \nu$ imply $\lambda \ll \nu$ (transitivity), but $\lambda \ll \mu$ does not imply $\mu \ll \lambda$. If $\lambda \ll \mu$ and $\mu \ll \lambda$, then λ and μ are called *equivalent measures* (common notation: $\lambda \equiv \mu$ or $\lambda \sim \mu$).

Problem 7.9. Consider a measure space $(X, \mathcal{X})$. Let λ, μ, and ν be σ-finite measures on $\mathcal{X}$. Prove the following propositions.

(a) If $\lambda \ll \mu$ and $g \in \mathcal{M}(X, \mathcal{X})^+$, then $\int_E g\, d\lambda = \int_E g \frac{d\lambda}{d\mu}\, d\mu$ for every $E \in \mathcal{X}$.

Hint: Theorem 7.8 and Problem 3.11 (recall: $g\chi_E \in \mathcal{M}(X, \mathcal{X})^+$).

(b) If $\lambda \ll \nu$ and $\mu \ll \nu$, then $\frac{d(\lambda+\mu)}{d\nu} = \frac{d\lambda}{d\nu} + \frac{d\mu}{d\nu}$ ν-almost everywhere.

Hint: Theorem 7.8, Proposition 3.5(b), and Problem 3.8(d).

(c) If $\lambda \ll \mu \ll \nu$, then $\frac{d\lambda}{d\nu} = \frac{d\lambda}{d\mu}\frac{d\mu}{d\nu}$ ν-almost everywhere.

Hint: Recall that $\lambda \ll \nu$. Apply Theorem 7.8 for each relation $\ll$ followed by Problem 3.11 as in part (a). Then use Problem 3.8(d).

(d) If $\lambda \ll \mu$ and $\mu \ll \lambda$, then $\frac{d\lambda}{d\mu} = (\frac{d\mu}{d\lambda})^{-1}$ almost everywhere.

Note: $\lambda \ll \mu \ll \lambda$ means $\lambda \equiv \mu$ (i.e., λ and μ are equivalent measures) so that μ-almost everywhere is equivalent to λ-almost everywhere.

Hint: $\frac{d\lambda}{d\lambda}$ is the identity. Use part (a) with $\nu = \lambda$. Swap λ and μ.

Problem 7.10. Let ν and μ be signed measures on a σ-algebra $\mathcal{X}$. The signed measure ν is *absolutely continuous* with respect to μ if, for $E \in \mathcal{X}$,

$$|\mu|(E) = 0 \quad \text{implies} \quad \nu(E) = 0$$

(i.e., $\nu(E) = 0$ for every $E \in \mathcal{X}$ such that $|\mu|(E) = 0$). Same notation as for measures: $\nu \ll \mu$. Show that the following assertions are pairwise equivalent.

(a) $\nu \ll \mu$.

(b) $\nu^+ \ll \mu$ and $\nu^- \ll \mu$.

(c) $|\nu| \ll |\mu|$.

Hint: Take a Hahn decomposition $\{A_\nu^+, A_\nu^-\}$ of X with respect to ν. Verify that $|\mu|(E) = 0$ implies $|\mu|(A^+ \cap E) = |\mu|(A^- \cap E) = 0$ and, if (a) holds, this implies $\nu^+(E) = \nu^-(E) = 0$. Thus conclude that (a) implies (b). That (b) implies (a) follows from the fact that $\nu = \nu^+ - \nu^-$ (Theorem 7.4). Similarly, verify that (b) and (c) are equivalent because $|\nu| = \nu^+ + \nu^-$.

Consider a third signed measure λ on $\mathcal{X}$ and show that

(d) $\lambda \ll \mu$ and $\nu \ll \mu$ imply $(\lambda + \nu) \ll \mu$.

Hint: $|\lambda + \nu| \leq |\lambda| + |\nu|$ according to Problem 7.7(b).

Problem 7.11. Let ν and μ be signed measures on a σ-algebra $\mathcal{X}$. One is *singular* with respect to the other (or *mutually singular*, or simply *singular*) if their total variations $|\nu|$ and $|\mu|$ are singular measures on $\mathcal{X}$ (according to Definition 7.9). Same notation as for measures: $\nu \perp \mu$. That is,

$$\nu \perp \mu \quad \text{if and only if} \quad |\nu| \perp |\mu|.$$

Since $|\nu| = \nu^+ + \nu^-$, show that

(a) $\nu \perp \mu$ implies $\nu^+ \perp \mu$ and $\nu^- \perp \mu$.

Consider a third signed measure λ on $\mathcal{X}$ and show that

(b) $\lambda \perp \mu$ and $\nu \perp \mu$ imply $(\lambda + \nu) \perp \mu$.

Hint: If $|\lambda|(A) = |\mu|(B) = 0$ and $|\nu|(C) = |\mu|(D)$, where $\{A, B\}$ and $\{C, D\}$ are measurable partitions of X, then $\{E, F\}$ forms another measurable partition of X, with $E = (A \cap C)$ and $F = (B \cap C) \cup (A \cap D) \cup (B \cap D)$, such that $|\lambda|(E) = |\nu|(E) = |\mu|(F) = 0$. Now recall from Problem 7.7(b) that $|\lambda + \nu| \leq |\lambda| + |\nu|$, and so $|\lambda + \nu|(E) = \mu(F) = 0$.

Problem 7.12. Let $(X, \mathcal{X})$ be a measurable space. If λ and μ are measures (or signed measures) on $\mathcal{X}$ such that $\lambda \ll \mu$ and $\lambda \perp \mu$, then $\lambda = 0$.

Problem 7.13. Consider a measurable space $(X, \mathcal{X})$. Let λ and μ be measures on $\mathcal{X}$. Prove the following propositions.

(a) If $\lambda \ll \mu$, then λ is continuous with respect μ.

(b) If λ is discrete with respect to μ, then $\lambda \perp \mu$.

(c) If λ is continuous and discrete with respect to μ, then $\lambda = 0$.

Note that the converses are not true. In particular, although *discrete* implies *singular*; *singular* does not imply *discrete*. See examples in Problem 7.15.

Hint: Definitions 7.6, 7.9, and 7.11. (c) If λ is discrete and continuous with respect to μ, then $\lambda(X)=\lambda(A)+\lambda(B)=\lambda(B)=\sum_n \lambda(b_n)=0$, since each $\{b_n\}$ is measurable and $\mu(\{b_n\})\le\mu(B)=0$, so $\lambda(\{b_n\})=0$.

Problem 7.14. Let λ and μ be measures on a σ-algebra $\mathcal{X}$. For each $E\in\mathcal{X}$ take the σ-algebra $\mathcal{E}=\wp(E)\cap\mathcal{X}$. If λ is discrete with respect to μ, then

$$\lambda(E)=\sum_{\{x\}\in\mathcal{E}}\lambda(\{x\}) \quad \text{for every} \quad E\in\mathcal{X},$$

and $\mu(\{x\})=0$ whenever $\{x\}$ in $\mathcal{X}$ is such that $\lambda(\{x\})\neq 0$.

Hint: $\lambda(E)=\lambda(E\cap B)$, where B is a countable set of measurable singletons.

Problem 7.15. Let $\mu,\lambda,\nu\colon\mathfrak{R}\to\overline{\mathbb{R}}$ be measures on the Borel algebra $\mathfrak{R}$ of subsets of $\mathbb{R}$, where μ is the Lebesgue measure (Example 2C, Problem 2.7). Consider the components of the canonical decomposition in Corollary 7.14.

(a) Set $F=\chi_{[x,\infty)}\colon\mathbb{R}\to\{0,1\}$, the characteristic function of $[x,\infty)$ for some $x\in\mathbb{R}$ (Example 1B). Set $\lambda=\mu_F=\delta_x\colon\mathfrak{R}\to\{0,1\}$, the Borel–Stieltjes measure generated by F (Example 2D), which is the Dirac measure at x (Example 2A). Show that $\lambda\perp\mu$. Actually, show that $\lambda=\delta_x$ is *singular-discrete with respect to Lebesgue measure* (and so is $\sum_{q\in\mathbb{Q}}\delta_q$).

(b) Let $f,F\colon\mathbb{R}\to\mathbb{R}$ be measurable functions. Define $\lambda,\nu\colon\mathfrak{R}\to\overline{\mathbb{R}}$ as follows. $\lambda(E)=\int_E f\,d\mu$ (if f is nonnegative) and $\nu(E)=\mu_F(E)=\int_E\frac{dF}{dx}\,d\mu(x)$, the Borel–Stieltjes measure generated by F (if F is nondecreasing and continuously differentiable), for each $E\in\mathfrak{R}$. Prove: λ and ν are *absolutely continuous with respect to Lebesgue measure* μ, and σ-finite (if f is locally L^1). (*Hint*: Example 2D, Propositions 3.5 and 3.7, Theorem 7.8.)

(c) Consider the Cantor set $C\subset[0,1]$ obtained by successive removal of the central open third of $[0,1]$ (Problem 2.9). Writing each point of $[0,1]$ in its ternary (i.e., base 3) expansion, it can be shown that x lies in C if and only if it has only 0's and 2's, and no 1's, in its ternary expansion. Moreover, it can be verified that the map $\Phi\colon C\to[0,1]$ that changes the 2's into 1's, and interprets the result in its binary expansion (i.e., as a base 2 number), is a one-to-one correspondence between C and $[0,1]$ (i.e., Φ is injective and surjective, which shows that C is uncountable). It can also be verified that Φ is increasing, uniformly continuous, and assumes the same values at the end points of every bounded open interval in the complement $[0,1]\backslash C$ of the Cantor set. Thus the function $\Phi\colon C\to[0,1]$ has a unique extension $F\colon[0,1]\to[0,1]$ over the closed interval $[0,1]$ (i.e., $F|_C=\Phi$) such that F is piecewise constant (constant on each successively removed open third remaining in $[0,1]\backslash C$ — e.g.,

$F(x) = \frac{1}{2}$ for all x in the central open third $(\frac{1}{3}, \frac{2}{3})$ of $[0,1]$), continuous, increasing ($F(0) = 0,\ F(1) = 1$), and differentiable with $\frac{dF}{dx} = 0$ μ-a.e. (since $\mu(C) = 0$). This $F\colon [0,1] \to [0,1]$ is the *Cantor function* (or *Cantor–Lebesgue function*, or *Lebesgue singular function* — whose graph is sometimes referred to as *Devil's Staircase*). Now set $\lambda = \mu_F$, the Borel–Stieltjes measure generated by F (Example 2D). Recalling that every singleton in $\mathbb{R}$ is $\mathfrak{R}$-measurable (Problem 2.7), and that F is continuous, show that $\lambda(\{x]\}) = 0$ for every singleton $\{x\} \subset [0,1]$, and so verify that λ is continuous with respect to Lebesgue measure μ (Definition 7.11). Also show that λ is concentrated on C; that is, $\lambda(C) = \lambda([0.1])$, and hence verify that λ and μ are singular (i.e., $\lambda \perp \mu$). So conclude: $\lambda = \mu_F$ is *singular-continuous with respect to Lebesgue measure.*

Suggested Reading

Bartle [4], Berberian [7], Halmos [18], Kelley and Srinivasan [22], Royden [35], Rudin [36], Shilov and Gurevich [38]. See also [29, Section 6.8]. For construction of the Cantor function (Problem 7.15(c)) see [4], [9], [32], [37].

8

Extension of Measures

8.1 Measure on an Algebra and Outer Measure

In Chapter 2 (see Example 2C) we considered the Lebesgue measure λ on the Borel algebra $\Re$, which is the σ-algebra of subsets of the real line $\mathbb{R}$ generated by the collection of all open intervals; and we have been using the notion of Lebesgue measure since then, although it has not been properly constructed so far. Indeed, in Example 2C we promised to prove existence and uniqueness of the Lebesgue measure $\lambda\colon \Re \to \overline{\mathbb{R}}$ in Chapter 8. We will comply with that promise in Section 8.3, as a special case of the following program. (1) First we introduce the concept of a measure μ on an algebra $\mathcal{A}$ (rather than on a σ-algebra) of subsets of set X. (2) Then we consider the notion of an outer measure μ^* generated by that measure μ on an algebra $\mathcal{A}$, which is a set function on the power set $\wp(X)$. (3) Finally, we show that this outer measure μ^* induces a σ-algebra $\mathcal{A}^*$ of subsets of X (such that $\mathcal{A} \subseteq \mathcal{A}^*$) upon which the restriction $\mu^*|_{\mathcal{A}^*}$ is a measure on the σ-algebra $\mathcal{A}^*$. This is the Carathéodory Extension Theorem, which is the central result of this chapter, whose applications go as far as Chapters 9, 11, and 13.

The difference between an algebra and a σ-algebra of subsets of a set X is that in an algebra $\mathcal{A}$ any *finite union* of sets in $\mathcal{A}$ is required to remain in $\mathcal{A}$, while in a σ-algebra $\mathcal{X}$ it is imposed, in addition, that any *countable union* of sets in $\mathcal{X}$ must remain in $\mathcal{X}$ (see Definition 1.1). We will now define the notion of a measure μ on an algebra $\mathcal{A}$. Since a countable union of sets in $\mathcal{A}$ is not necessarily in $\mathcal{A}$, countable additivity for μ will be restricted to countable families of sets in $\mathcal{A}$ whose union still lies in $\mathcal{A}$.

C.S. Kubrusly, *Essentials of Measure Theory*,
DOI 10.1007/978-3-319-22506-7_8

Definition 8.1. A *measure on an algebra* is an extended real-valued set function μ on an algebra $\mathcal{A}$ of subsets of a set X,

$$\mu\colon \mathcal{A} \to \overline{\mathbb{R}},$$

that fulfills the following axioms.

(a) $\mu(\varnothing) = 0$,

(b) $\mu(E) \geq 0 \quad$ for every $\quad E \in \mathcal{A}$,

(c) $\mu\big(\bigcup_n E_n\big) = \sum_n \mu(E_n)$
for every countable family $\{E_n\}$ of pairwise disjoint sets in $\mathcal{A}$ for which $\bigcup_n E_n$ lies in $\mathcal{A}$.

Remarks: Similarly to the remarks that follow Definitions 2.1 and 2.3, if the countable set $\{\mu(E_n)\}$ of nonnegative (extended) real numbers in (c) is infinite (countably infinite), then the (infinite) series $\sum_n \mu(E_n)$ either converges unconditionally to a real number (i.e., the real value of the sum does not depend on the order of the summands) or diverges to infinity. Properties of measures on a σ-algebra are naturally transferred to measures on an algebra up to the assumption $\bigcup_n E_n \in \mathcal{A}$ in axiom (c), which is not necessary for a measure on a σ-algebra.

Let $\mathcal{A}$ be any algebra of subsets of a set X. A measure μ on $\mathcal{A}$ generates a set function μ^* on the power set $\wp(X)$ as follows.

Definition 8.2. Suppose $\mu\colon \mathcal{A} \to \overline{\mathbb{R}}$ is a measure on an algebra $\mathcal{A}$ of subsets of a set X. For each subset S of X (i.e., for each $S \in \wp(X)$) consider the collection $\mathcal{C}_S$ of all countable families $\{E_n\}$ of sets in $\mathcal{A}$ that cover S,

$$\mathcal{C}_S = \big\{\{E_n\}\colon\, E_n \in \mathcal{A} \text{ and } S \subseteq \textstyle\bigcup_n E_n\big\}.$$

The extended real-valued set function μ^* on the power set $\wp(X)$,

$$\mu^*\colon \wp(X) \to \overline{\mathbb{R}},$$

defined for each $S \in \wp(X)$ by

$$\mu^*(S) = \inf_{\{E_n\} \in \mathcal{C}_S} \sum\nolimits_n \mu(E_n),$$

is the *outer measure* generated by the measure μ on the algebra $\mathcal{A}$.

In spite of the terminology, an *outer measure* μ^* may be far from being a measure (in the sense of Definitions 2.1 or 8.1) since it is not necessarily additive. Actually, it may happen that $\mu^*(A \cup B) \neq \mu^*(A) + \mu^*(B)$ for some disjoint sets A and B in $\wp(X)$ — we will comment on this in Problems 8.16

and 8.22. Additivity is weakened to *subadditivity*. In fact, outer measures are *countably subadditive* (Proposition 8.3(e) below). On the other hand, μ^* has the advantage of being defined for every subset of X and inherits some properties of a measure, such as $\mu^*(\varnothing) = 0$ and $\mu^*(S) \geq 0$ for every subset S of X and, what is more important, $\mu^*(E) = \mu(E)$ for every $E \in \mathcal{A}$.

Proposition 8.3. *Let $\mu^*\colon \wp(X) \to \overline{\mathbb{R}}$ be the outer measure generated by a measure $\mu\colon \mathcal{A} \to \overline{\mathbb{R}}$ on an algebra $\mathcal{A}$ of subsets of a set X. Then*

(a) $\mu^*(\varnothing) = 0$,

(b) $\mu^*(S) \geq 0$ *for every* $S \in \wp(X)$,

(c) $\mu^*(S_1) \leq \mu^*(S_2)$ *whenever* $S_1 \subseteq S_2 \subseteq X$,

(d) $\mu^*(E) = \mu(E)$ *for every* $E \in \mathcal{A}$ (i.e., $\mu^*|_{\mathcal{A}} = \mu$),

(e) $\mu^*\big(\bigcup_n E_n\big) \leq \sum_n \mu^*(E_n)$
for every countable family $\{E_n\}$ *of subsets of* X.

Proof. Observe that properties (a), (b), and (c) are trivially verified by the definition of outer measure (Definition 8.2). To verify property (d) take an arbitrary $E \in \mathcal{A}$, and let $\{E'_n\}$ be a sequence of subsets of $\mathcal{A}$ such that $E'_1 = E$ and $E'_n = \varnothing$ for all $n \neq 1$. Since $\{E'_n\} \in \mathcal{C}_E$, we get

$$\mu^*(E) = \inf_{\{E_n\} \in \mathcal{C}_E} \sum\nolimits_n \mu(E_n) \leq \sum\nolimits_n \mu(E'_n) = \mu(E).$$

To verify the reverse inequality proceed as follows. If $\{E_n\}$ is any countable family in $\mathcal{C}_E$, then $\{E \cap E_n\}$ is again a countable family in $\mathcal{C}_E$ such that $E = \bigcup_n (E \cap E_n)$, and so (cf. Problem 2.8(b) and Proposition 2.2(a))

$$\mu(E) \leq \sum\nolimits_n \mu(E \cap E_n) \leq \sum\nolimits_n \mu(E_n).$$

Since this holds for all $\{E_n\} \in \mathcal{C}_E$, we get

$$\mu(E) \leq \inf_{\{E_n\} \in \mathcal{C}_E} \sum\nolimits_n \mu(E_n) = \mu^*(E),$$

completing the proof of (d): the restriction $\mu^*|_{\mathcal{A}}$ of the outer measure μ^* to the algebra $\mathcal{A}$ coincides with measure μ on the algebra $\mathcal{A}$. Finally we verify countable subadditivity. Take an arbitrary $\varepsilon > 0$. Let $\{E_n\}$ be an arbitrary countable family of subsets of X, and recall that for each E_n,

$$\mu^*(E_n) = \inf_{\{F_k\}\in\mathcal{C}_{E_n}} \sum\nolimits_k \mu(F_k).$$

Then for each $n \geq 1$ there is a countable family $\{F'_{n,k}\}$ in $\mathcal{C}_{E_n}$ for which

$$\sum\nolimits_k \mu(F'_{n,k}) \leq \mu^*(E_n) + \tfrac{\varepsilon}{n^2}.$$

Note that $\{F'_{n,k}\} = \bigcup_n \bigcup_k F'_{n,k}$ is a countable family of sets in $\mathcal{A}$ covering $\bigcup_n E_n$, which means that $\{F'_{n,k}\}$ lies in $\mathcal{C}_{\bigcup_n E_n}$. Since

$$\mu^*\Big(\bigcup\nolimits_n E_n\Big) = \inf_{\{F_{n,k}\}\in\mathcal{C}_{\bigcup_n E_n}} \sum\nolimits_{n,k} \mu(F_{n,k}),$$

it follows that

$$\mu^*\Big(\bigcup\nolimits_n E_n\Big) \leq \sum\nolimits_{n,k} \mu(F'_{n,k}) = \sum\nolimits_n \sum\nolimits_k \mu(F'_{n,k}) \leq \sum\nolimits_n \mu^*(E_n) + \varepsilon \sum\nolimits_n \tfrac{1}{n^2},$$

and so, since this holds for every $\varepsilon > 0$, and since $\sum_n \frac{1}{n^2} < \infty$,

$$\mu^*\Big(\bigcup\nolimits_n E_n\Big) \leq \inf_{\varepsilon>0}\Big(\sum\nolimits_n \mu^*(E_n) + \varepsilon \sum\nolimits_n \tfrac{1}{n^2}\Big) = \sum\nolimits_n \mu^*(E_n),$$

which completes the proof of (e). □

8.2 The Carathéodory Extension Theorem

Consider the outer measure μ^* generated by a measure μ on an algebra $\mathcal{A}$ of subsets of a set X. A set $E \in \wp(X)$ is said to be *μ^*-measurable* (or satisfies the *Carathéodory condition*) if

$$\mu^*(S) = \mu^*(S \cap E) + \mu^*(S \backslash E)$$

for every $S \in \wp(X)$. This means that μ^* behaves additively on E. Let

$$\mathcal{A}^* = \big\{E \in \wp(X)\colon\ E \text{ is } \mu^*\text{-measurable}\big\}$$

be the collection of all μ^*-measurable subsets of X. The next result is a decisive one for constructing measures out of set functions that are not measures on σ-algebras. It says that *this $\mathcal{A}^*$ is a σ-algebra such that $\mathcal{A} \subseteq \mathcal{A}^*$, and the restriction of the outer measure μ^* to $\mathcal{A}^*$ is a measure (on the σ-algebra $\mathcal{A}^*$) that extends the measure μ (on the algebra $\mathcal{A}$) over $\mathcal{A}^*$.*

Theorem 8.4. (Carathéodory Extension Theorem). *$\mathcal{A}^*$ is a σ-algebra that includes the algebra $\mathcal{A}$, and the restriction of μ^* to $\mathcal{A}^*$ is a measure on $\mathcal{A}^*$.*

Proof. Note that the empty set $\varnothing$ and the whole set X clearly lie in $\mathcal{A}^*$ (i.e., they are μ^*-measurable). Also, since $S\cap E = S\backslash(X\backslash E)$ for every pair of sets E and S in $\wp(X)$, it follows that the complement of every set in $\mathcal{A}^*$ is again in $\mathcal{A}^*$. Next take an arbitrary $S\in\wp(X)$. If $F\in\mathcal{A}^*$, then

$$\mu^*(S) = \mu^*(S\cap F) + \mu^*(S\backslash F),$$

and if $E\in\mathcal{A}^*$, then

$$\mu^*(S\cap F) = \mu^*(S\cap F\cap E) + \mu^*((S\cap F)\backslash E).$$

Now observe that $[S\backslash(E\cap F)]\cap F = (S\cap F)\backslash(E\cap F) = (S\cap F)\backslash E$, and also that $[S\backslash(E\cap F)]\backslash F = (S\backslash F)\backslash(E\cap F) = S\backslash F$. Thus, if $F\in\mathcal{A}^*$,

$$\begin{aligned}\mu^*(S\backslash(E\cap F)) &= \mu^*([S\backslash(E\cap F)]\cap F) + \mu^*([S\backslash(E\cap F)]\backslash F)\\ &= \mu^*((S\cap F)\backslash E) + \mu^*(S\backslash F).\end{aligned}$$

The above three displayed identities ensure that if E and F lie in $\mathcal{A}^*$, then

$$\begin{aligned}\mu^*(S) &= \mu^*(S\cap F\cap E) + \mu^*((S\cap F)\backslash E) + \mu^*(S\backslash F)\\ &= \mu^*(S\cap E\cap F) + \mu^*(S\backslash(E\cap F)),\end{aligned}$$

and so $E\cap F$ lies in $\mathcal{A}^*$. Therefore, since intersection of sets in $\mathcal{A}^*$ and complements of sets in $\mathcal{A}^*$ are both again in $\mathcal{A}^*$, it follows that union of sets in $\mathcal{A}^*$ also lie in $\mathcal{A}^*$ (because $E\cup F = X\backslash[(X\backslash E)\cap(X\backslash F)]$ — De Morgan Laws). Then a trivial induction ensures that any finite union of sets in $\mathcal{A}^*$ remains in $\mathcal{A}^*$, and hence $\mathcal{A}^*$ is an algebra. That is,

(a) $\bigcup_{i=1}^n F_i$ lies in $\mathcal{A}^*$ for every finite family $\{F_i\}_{i=1}^n$ of sets in $\mathcal{A}^*$.

Furthermore, if S and F are sets in $\wp(X)$, if E is a set $\mathcal{A}^*$, and if E and F are disjoint (so that $S\cap(E\cup F)\backslash E = S\cap F$), then

$$\begin{aligned}\mu^*(S\cap(E\cup F)) &= \mu^*(S\cap(E\cup F)\cap E) + \mu^*(S\cap(E\cup F)\backslash E)\\ &= \mu^*(S\cap E) + \mu^*(S\cap F),\end{aligned}$$

and so μ^* acts additively on the intersection of any set in $\wp(X)$ with every pair of disjoint sets in $\mathcal{A}^*$. Thus another trivial induction ensures that

(b) $\mu^*\big(S\cap\big(\bigcup_{i=1}^n E_i\big)\big) = \sum_{i=1}^n \mu^*(S\cap E_i)$ for every finite family $\{E_i\}_{i=1}^n$ of pairwise disjoint sets in $\mathcal{A}^*$ and every set S in $\wp(X)$.

In particular, for $S = X$, this shows that μ^* is *finitely additive* on $\mathcal{A}^*$:

(c) $\mu^*\big(\bigcup_{i=1}^n E_i\big) = \sum_{i=1}^n \mu^*(E_i)$ for every finite family $\{E_i\}_{i=1}^n$ of pairwise disjoint sets in $\mathcal{A}^*$.

Next these results on finite families are extended to countably infinite families, which shows that $\mathcal{A}^*$ is a σ-algebra and μ^* is countably additive on $\mathcal{A}^*$, and hence the restriction of μ^* to $\mathcal{A}^*$ is a measure (by Definition 2.1). Thus take any (infinite) sequence $\{F_n\}$ of sets in $\mathcal{A}^*$, and consider the disjointification $\{E_n\}$ of $\{F_n\}$ given by $E_1 = F_1$ and $E_{n+1} = F_{n+1}\backslash\big(\bigcup_{i=1}^n F_i\big)$ (see Hints to Problems 2.3 and 7.2). Clearly, $\{E_n\}$ is a sequence of pairwise disjoint sets in $\mathcal{A}^*$ (each E_n lies in $\mathcal{A}^*$ because finite union and difference of sets in an algebra remain in the algebra). For each integer $n \geq 1$ set $G_n = \bigcup_{i=1}^n E_i = \bigcup_{i=1}^n F_i$, which lies in $\mathcal{A}^*$ (finite union of sets in $\mathcal{A}^*$), and set $G = \bigcup_{n=1}^\infty E_n = \bigcup_{n=1}^\infty F_n$ in $\wp(X)$. Take an arbitrary S in $\wp(X)$. Note by the countable subadditivity of Proposition 8.3(e) that

$$\mu^*(S \cap G) = \mu^*\Big(S \cap \bigcup_{i=1}^\infty E_i\Big) = \mu^*\Big(\bigcup_{i=1}^\infty (S \cap E_i)\Big) \leq \sum_{i=1}^\infty \mu^*(S \cap E_i),$$

and hence, since $S = (S \cap G) \cup (S\backslash G)$, subadditivity ensures again that

$$\mu^*(S) \leq \mu^*(S \cap G) + \mu^*(S\backslash G) \leq \sum_{i=1}^\infty \mu^*(S \cap E_i) + \mu^*(S\backslash G).$$

On the other hand, since each $G_n = \bigcup_{i=1}^n E_i$ lies in $\mathcal{A}^*$, and since the sequence $\{E_n\}$ consists of pairwise disjoint sets in $\mathcal{A}^*$, it follows by (b) that

$$\mu^*(S) = \mu^*(S \cap G_n) + \mu^*(S\backslash G_n) = \sum_{i=1}^n \mu^*(S \cap E_i) + \mu^*(S\backslash G_n)$$

for all n. Since $G_n \subseteq G$, we get $S\backslash G \subseteq S\backslash G_n$, so $\mu^*(S\backslash G) \leq \mu^*(S\backslash G_n)$ by Proposition 8.3(c). Therefore,

$$\sum_{i=1}^n \mu^*(S \cap E_i) + \mu^*(S\backslash G) \leq \mu^*(S)$$

for all n, which implies that

$$\sum_{i=1}^\infty \mu^*(S \cap E_i) + \mu^*(S\backslash G) \leq \mu^*(S).$$

Hence, for every $S \in \wp(X)$,

$$\mu^*(S) = \mu^*(S \cap G) + \mu^*(S\backslash G) = \sum_{i=1}^\infty \mu^*(S \cap E_i) + \mu^*(S\backslash G).$$

The first identity in the above expression says that G lies in $\mathcal{A}^*$, that is,

(a′) $\bigcup_n F_n$ lies in $\mathcal{A}^*$ for every countable family $\{F_n\}$ (not necessarily pairwise disjoint) of sets in $\mathcal{A}^*$, proving that the algebra $\mathcal{A}^*$ is, in fact, a σ-algebra. Furthermore, taking $S = G$ (so that $\mu^*(S\backslash G) = 0$ and $S \cap E_i = E_i$ for every $i \geq 1$), we also get

$$\mu^*(G) = \sum_{i=1}^{\infty} \mu^*(E_i).$$

If $\{F_n\}$ is a sequence of pairwise disjoint sets, then $E_n = F_n$ for each n, and so the above identity ensures that μ^* is countably additive on $\mathcal{A}^*$:

(c′) $\mu^*\big(\bigcup_n E_n\big) = \sum_n \mu^*(E_n)$ for every countably infinite family $\{E_n\}$ of pairwise disjoint sets in $\mathcal{A}^*$.

Properties (a′) and (c′) ensure that the restriction of μ^* to $\mathcal{A}^*$ is a measure on the σ-algebra $\mathcal{A}^*$. Finally we verify the inclusion $\mathcal{A} \subseteq \mathcal{A}^*$. Take an arbitrary $A \in \mathcal{A}$, an arbitrary $S \in \wp(X)$, and an arbitrary $\varepsilon > 0$. According to Definition 8.2, there exists a sequence $\{A_n\}$ of sets in $\mathcal{A}$ such that

$$S \subseteq \bigcup_n A_n \quad \text{and} \quad \sum_n \mu(A_n) \leq \mu^*(S) + \varepsilon.$$

Note that $\{(A_n \cap A), (A_n\backslash A)\}$ is a partition of A_n for every n, which is made up of sets in the algebra $\mathcal{A}$. Thus, according to Proposition 8.3(c,e,d),

$$\mu^*(S \cap A) \leq \mu^*\Big(\bigcup_n (A_n \cap A)\Big) \leq \sum_n \mu^*(A_n \cap A) = \sum_n \mu(A_n \cap A),$$

$$\mu^*(S\backslash A) \leq \mu^*\Big(\bigcup_n (A_n\backslash A)\Big) \leq \sum_n \mu^*(A_n\backslash A) = \sum_n \mu(A_n\backslash A),$$

and so, by using the additivity of Definition 8.1(c),

$$\begin{aligned}\mu^*(S \cap A) + \mu^*(S\backslash A) &\leq \sum_n \big(\mu(A_n \cap A) + \mu(A_n\backslash A)\big) \\ &= \sum_n \mu((A_n \cap A) \cup (A_n\backslash A)) = \sum_n \mu(A_n) \leq \mu^*(S) + \varepsilon.\end{aligned}$$

Since $\varepsilon > 0$ is arbitrary, this implies that

$$\mu^*(S \cap A) + \mu^*(S\backslash A) \leq \mu^*(S).$$

On the other hand, since $S = (S \cap A) \cup (S\backslash A)$,

$$\mu^*(S) \leq \mu^*(S \cap A) + \mu^*(S\backslash A)$$

by the subadditivity of Proposition 8.3(e). Hence,

$$\mu^*(S) = \mu^*(S \cap A) + \mu^*(S\backslash A)$$

for every $S \in \wp(X)$, which means that A lies in $\mathcal{A}^*$. Thus $\mathcal{A} \subseteq \mathcal{A}^*$. □

The Carathéodory Extension Theorem is sometimes referred to as the Carathéodory–Fréchet–Hahn–Kolmogorov Extension Theorem. Since the restriction of the outer measure μ^* to the algebra $\mathcal{A}$ coincides with the measure μ on $\mathcal{A}$ (i.e., $\mu^*|_{\mathcal{A}} = \mu$), since the restriction $\mu^*|_{\mathcal{A}^*}$ of the outer measure μ^* to the σ-algebra $\mathcal{A}^*$ is a measure, and since $\mathcal{A} \subseteq \mathcal{A}^* \subseteq \wp(X)$, it follows that the measure $\mu^*|_{\mathcal{A}^*}$ on the σ-algebra $\mathcal{A}^*$ extends the measure μ on the algebra $\mathcal{A}$ over $\mathcal{A}^*$. This measure $\mu^*|_{\mathcal{A}^*}$ is complete in the sense of Section 2.3. Equivalently, the σ-algebra $\mathcal{A}^*$ is complete with respect to it. This means that if $N \in \mathcal{A}^*$ and $\mu^*(N) = 0$, then every subset E of N lies in $\mathcal{A}^*$ (and so $\mu^*(E) = 0$ according to Proposition 8.3(b,c)). More is true: *all sets with outer measure zero are* μ^**-measurable* (and so $\mathcal{A}^*$ is complete).

Proposition 8.5. *Consider the outer measure* μ^* *on* $\wp(X)$ *(induced by a measure on* $\mathcal{A}$*). Let* $\mathcal{A}^*$ *be the* σ*-algebra of all* μ^**-measurable sets in* $\wp(X)$*.*

(a) *If* $N \in \wp(X)$ *is such that* $\mu^*(N) = 0$*, then* $N \in \mathcal{A}^*$*.*

(b) *If* $E \subseteq N \in \wp(X)$ *and* $\mu^*(N) = 0$*, then* $E \in \mathcal{A}^*$ *and* $\mu(E) = 0$*.*

Proof. Assertion (b) follows from assertion (a) and Proposition 8.3(b,c). Assertion (a) says that every set with outer measure zero is μ^*-measurable. To verify this take arbitrary sets S and N in $\wp(X)$ such that $\mu^*(N) = 0$. Since $S \cup N = (S \cap N) \cup (S \backslash N)$, we get from Proposition 8.3(c,e) that

$$\mu^*(S) \le \mu^*(S \cup N) \le \mu^*(S \cap N) + \mu^*(S \backslash N) \le \mu^*(N) + \mu^*(S) = \mu^*(S),$$

and so $\mu^*(S) = \mu^*(S \cap N) + \mu^*(S \backslash N)$; that is, $N \in \mathcal{A}^*$. □

A measure on an algebra inherits most of the attributes of a measure on a σ-algebra. For instance, a measure μ on an algebra $\mathcal{A}$ of subsets of a set X is *finite* if $\mu(X) < \infty$. Similarly, μ is σ*-finite* if X is covered by a countable family of sets in $\mathcal{A}$ of finite measure, that is, if there exists a sequence $\{A_n\}$ of sets in $\mathcal{A}$ such that $\mu(A_n) < \infty$ for every n and $X = \bigcup_n A_n$. Finiteness and σ-finiteness are naturally extended from a measure μ on an algebra $\mathcal{A}$ to the associated outer measure μ^* on the power set $\wp(X)$ by Proposition 8.3(d) since $X \in \mathcal{A} \subseteq \wp(X)$ — just replace μ with μ^*. The next result says that if μ is σ-finite, then its extension over the σ-algebra $\mathcal{A}^*$ is unique.

Theorem 8.6. (Hahn Extension Theorem). *If a measure* μ *on the algebra* $\mathcal{A}$ *is* σ*-finite, then its extension to a measure on the* σ*-algebra* $\mathcal{A}^*$ *is unique.*

Proof. Consider the outer measure μ^* generated by a measure μ on an algebra $\mathcal{A}$ of subsets of a set X. The collection $\mathcal{A}^*$ of all μ^*-measurable subsets of X is a σ-algebra such that $\mathcal{A} \subseteq \mathcal{A}^*$, and the restriction of μ^* to $\mathcal{A}^*$ is a measure that extends μ over A^* (Theorem 8.5). Let ν is a measure on $\mathcal{A}^*$ that extends μ over $\mathcal{A}^*$ (i.e., suppose ν is a measure on $\mathcal{A}^*$ such that $\nu(E) = \mu(E)$ for every $E \in \mathcal{A}$). The proof will be split into two parts. The claimed result is proved supposing μ is a finite measure in part (a). This is extended to the case when μ is σ-finite in part (b).

(a) Assume that μ is a finite measure on the algebra $\mathcal{A}$. Thus any extension ν of μ to $\mathcal{A}^*$ is again finite (because $X \in \mathcal{A} \subseteq \mathcal{A}^* \subseteq \wp(X)$ implies $\mu^*(X) = \nu(X) = \mu(X) < \infty$). Take an arbitrary set E in $\mathcal{A}^*$, and an arbitrary sequence $\{A_n\}$ in $\mathcal{C}_E$. That is, let $\{A_n\}$ be any sequence of sets in $\mathcal{A}$ such that $E \subseteq \bigcup_n A_n$. (These sequences do exist; trivial example: $A_1 = X$.) Thus

$$\nu(E) \leq \nu\Big(\bigcup\nolimits_n A_n\Big) \leq \sum\nolimits_n \nu(A_n) = \sum\nolimits_n \mu(A_n).$$

(See Proposition 2.2(a) and Problem 2.8(b).) Since this holds for every $\{A_n\} \in \mathcal{C}_E$, it follows by the definition of the outer measure μ^* induced by the measure μ on the algebra $\mathcal{A}$ (Definition 8.2) that

$$\nu(E) \leq \inf_{\{E_n\} \in \mathcal{C}_E} \sum\nolimits_n \mu(E_n) = \mu^*(E).$$

Recall that $\mu^*|_{\mathcal{A}^*}$ (μ^* restricted to $\mathcal{A}^*$) and ν are measures on the σ-algebra $\mathcal{A}^*$, so that they are additive by Definition 2.1(c). Moreover, also recall that $\mu^*(X) = \nu(X) = \mu(X) < \infty$. Thus the above inequality leads to

$$\mu^*(E) = \mu^*(X) - \mu^*(X \backslash E) = \nu(X) - \mu^*(X \backslash E) \leq \nu(X) - \nu(X \backslash E) = \nu(E).$$

Hence $\nu(E) = \mu^*(E)$ for all $E \in \mathcal{A}^*$, which proves uniqueness if μ is finite.

(b) Assume that μ is σ-finite. Thus any extension ν of μ to $\mathcal{A}^*$ is again σ-finite. In fact, if μ is σ-finite, then there exists a sequence $\{A_n\}$ of sets in $\mathcal{A} \subseteq \mathcal{A}^* \subseteq \wp(X)$ for which $\mu^*(A_n) = \nu(A_n) = \mu(A_n) < \infty$ for every $n \geq 1$ and $X = \bigcup_n A_n$, and so μ^* and ν are σ-finite as well. Set $A'_n = \bigcup_{i=1}^n A_n$ so that $\{A'_n\}$ is an *increasing* sequence of sets of finite measure that cover X. Take an arbitrary E in $\mathcal{A}^*$. Since $\{E \cap A'_n\}$ is a sequence of sets in $\mathcal{A}^*$ (intersection and finite union of sets in $\mathcal{A}^*$ remain in $\mathcal{A}^*$) of finite measure (since $\mu^*(E \cap A'_n) \leq \mu^*(A'_n) < \infty$ and $\nu(E \cap A'_n) \leq \nu(A'_n) < \infty$), it follows from part (a) that for every $n \geq 1$,

$$\mu^*(E \cap A'_n) = \nu(E \cap A'_n).$$

Thus, since $\{E \cap A'_n\}$ is an increasing sequence of sets in $\mathcal{A}^*$ such that $E = \bigcup_n (E \cap A'_n)$, and since both ν and the restriction $\mu^*|_{\mathcal{A}^*}$ of μ^* to the σ-algebra $\mathcal{A}^*$ are measures on $\mathcal{A}^*$, we get by Proposition 2.2(c) that

$$\mu^*(E) = \lim_n \mu^*(E \cap A'_n) = \lim_n \nu(E \cap A'_n) = \nu(E).$$

Hence $\nu(E) = \mu^*(E)$ for all $E \in \mathcal{A}^*$, proving uniqueness if μ is σ-finite. □

Let λ^* denote the measure on the σ-algebra $\mathcal{A}^*$ obtained by the restriction of the outer measure $\mu^*\colon \wp(X) \to \overline{\mathbb{R}}$ to $\mathcal{A}^*$. That is,

$$\lambda^* = \mu^*|_{\mathcal{A}^*}\colon \mathcal{A}^* \to \overline{\mathbb{R}}.$$

The previous results are summarized as follows. *If $\mu\colon \mathcal{A} \to \overline{\mathbb{R}}$ is a σ-finite measure defined on an algebra $\mathcal{A}$, then there exist a σ-algebra $\mathcal{A}^*$ including $\mathcal{A}$ and a unique extension of μ to a measure over the σ-algebra $\mathcal{A}^*$. This measure on $\mathcal{A}^*$ is σ-finite and, by uniqueness, coincides with $\lambda^*\colon \mathcal{A}^* \to \overline{\mathbb{R}}$.* That is, there exists a unique measure λ^* on the σ-algebra $\mathcal{A}^*$ such that

$$\mu = \lambda^*|_{\mathcal{A}}\colon \mathcal{A} \to \overline{\mathbb{R}};$$

equivalently, such that $\lambda^*(E) = \mu(E)$ for every $E \in \mathcal{A}$. Moreover, *the measure space $(X, \mathcal{A}^*, \lambda^*)$ is complete*. This means that the σ-algebra $\mathcal{A}^*$ is complete with respect to the measure λ^* or, equivalently, the measure λ^* is complete on the σ-algebra $\mathcal{A}^*$.

8.3 Construction of Lebesgue Measure

Consider the following classes of (left-open) intervals of the real line.

Class $\mathcal{C}_1$: $\{(\alpha, \beta] \subseteq \mathbb{R}\colon\ \alpha, \beta \in \mathbb{R},\ \alpha \leq \beta\}$.

Class $\mathcal{C}_2$: $\{(-\infty, \beta] \subseteq \mathbb{R}\colon\ \beta \in \mathbb{R}\}$.

Class $\mathcal{C}_3$: $\{(\alpha, +\infty) \subseteq \mathbb{R}\colon\ \alpha \in \mathbb{R}\}$.

Class $\mathcal{C}_4$: $\{(-\infty, +\infty)\}$.

These four classes are exhaustive. Let $\mathcal{I}$ be the family of all (left-open) intervals of the real line; each of them belonging to one of the above classes. The empty set $\varnothing$ is of class $\mathcal{C}_1$ (for the case of $\alpha = \beta$), and the only interval of class $\mathcal{C}_4$ is $\mathbb{R}$ itself. It is clear that the intersection of any two sets in $\mathcal{I}$ is again a set in $\mathcal{I}$ and that the complement of any set in $\mathcal{I}$ is a finite union of disjoint sets in $\mathcal{I}$. This means that $\mathcal{I}$ is a *semialgebra*. But the finite union of sets in $\mathcal{I}$ is not necessarily a set in $\mathcal{I}$, and so $\mathcal{I}$ is not an algebra. However, the collection $\mathfrak{I}$ of all finite unions of sets in $\mathcal{I}$ is an algebra (Problem 8.3).

Definition 8.7. Consider the collection $\Im \subseteq \wp(\mathbb{R})$ consisting of all *finite* unions of (left-open) intervals of the real line (i.e., of all *finite* unions of intervals from the semialgebra $\mathcal{I}$),

$$\Im = \{F \subseteq \mathbb{R}\colon F \text{ is a finite union of intervals from } \mathcal{I}\}.$$

Let $\ell\colon \Im \to \overline{\mathbb{R}}$ an extended real-valued set function with following properties.

(a) $\ell((\alpha,\beta]) = \beta - \alpha$ for any $\alpha, \beta \in \mathbb{R}$ such that $\alpha \leq \beta$,

(b) $\ell((-\infty,\beta]) = \ell((\alpha,+\infty)) = \ell((-\infty,+\infty)) = \infty$,

(c) $\ell(\bigcup_i I_i) = \sum_i \ell(I_i)$
for every *finite* family $\{I_i\}$ of pairwise disjoint intervals in $\mathcal{I}$.

A function with these properties is referred to as the *length function* on $\Im$.

We show that properties (a), (b), and (c) are enough to make the length function $\ell\colon \Im \to \overline{\mathbb{R}}$ well defined (and so unique) on the collection $\Im$, which is an algebra, and ℓ is a measure on it. For the prove we proceed as follows. First we check that ℓ is well defined at every set in $\Im$ (Problem 8.2). Next we show that $\Im$ is an algebra (Problem 8.3), and so we conclude that the set function ℓ is increasing (Problem 8.4). This is applied to prove an auxiliary result in Proposition 8.8 that is used to prove in Lemma 8.9 that the length function ℓ is countably additive, and so it is a measure on the algebra $\Im$.

Proposition 8.8. *Let $(a,b]$ be an interval of class C_1. If $\{(\alpha_k,\beta_k]\}$ is a countable family of disjoint intervals of class C_1, then*

$$\text{(a)}\qquad (a,b] = \bigcup_k (\alpha_k,\beta_k] \quad\text{implies}\quad \ell((a,b]) = \sum_k \ell((\alpha_k,\beta_k]).$$

If $\{(a_i,b_i]\}$ is a finite set of disjoint intervals of class C_1, then

$$\text{(b)}\qquad \bigcup_i (a_i,b_i] = \bigcup_k (\alpha_k,\beta_k] \quad\text{implies}\quad \ell\Big(\bigcup_i (a_i,b_i]\Big) = \sum_k \ell((\alpha_k,\beta_k]).$$

Proof. First note that if the countable family $\{(\alpha_k,\beta_k]\}$ of disjoint intervals is finite, then the results in (a) and (b) are trivially verified by Definition 8.7(c). Thus suppose the countable family $\{(\alpha_k,\beta_k]\}$ is infinite. To avoid trivialities, assume that all intervals $\{(\alpha_k,\beta_k]\}$ are *nonempty*, which means that $\alpha_k < \beta_k$ for every k.

(a) Suppose $(a,b] = \bigcup_k (\alpha_k,\beta_k]$. Take an arbitrary finite subfamily $\{(\alpha_i,\beta_i]\}$ of the infinite family $\{(\alpha_k,\beta_k]\}$. Since $\{(\alpha_i,\beta_i]\}$ has a finite number of subintervals of $(a,b]$, it follows that $a \leq \min\{\alpha_i\}$ and $\max\{\beta_i\} \leq b$, and so

$$\sum_i \ell((\alpha_i,\beta_i]) = \sum_i (\beta_i - \alpha_i) \leq b - a = \ell((a,b]).$$

This holds for all finite subfamilies $\{(\alpha_i, \beta_i]\}$ of $\{(\alpha_k, \beta_k]\}$. By taking the supremum over all finite subsets of the countably infinite set $\{\ell((\alpha_k, \beta_k])\}$ of positive numbers we get

$$\text{(a}_1\text{)} \qquad \sum\nolimits_k \ell((\alpha_k, \beta_k]) = \sup \sum\nolimits_i \ell((\alpha_i, \beta_i]) \leq \ell((a, b]).$$

To verify the reverse inequality, proceed as follows. Take an arbitrary $\varepsilon > 0$ and let $\{\varepsilon_k\}$ be any sequence of *positive* numbers such that $\sum_k \varepsilon_k \leq \varepsilon$. Since $(a, b] = \bigcup_k (\alpha_k, \beta_k]$, it follows that $b \in (\alpha_{k_1}, \beta_{k_1}]$ for some index k_1, and so $b = \beta_{k_1}$. Furthermore, it also follows that $a = \inf_k \{\alpha_k\}$, which ensures the existence of an index $k_0 \neq k_1$ such that $\alpha_{k_0} < a + \varepsilon_0$. Hence the following infinite family of *open* intervals

$$\big\{(\alpha_{k_0} - \varepsilon_0, \beta_{k_0} + \varepsilon_0)\,, (\alpha_k, \beta_k + \varepsilon_k) \;\text{ for every }\; k \neq k_0 \;\text{ with }\; k_0 \neq k_1\big\}$$

covers the closed and bounded interval $[a, b]$ (i.e., it covers the compact interval $[a, b]$ — by the Heine–Borel Theorem). The definition of compactness (cf. Definition 11.1(c)) says that this family of open intervals has a *finite* subfamily of open intervals that still covers $[a, b]$,

$$\big\{(\alpha_{k_0} - \varepsilon_0, \beta_{k_0} + \varepsilon_0)\,, (\alpha_j, \beta_j + \varepsilon_j) \;\text{ for every }\; j \in J\big\},$$

with J being a finite index set such that $k_0 \notin J$ and $k_1 \in J$. If the intervals in $\{(\alpha_k, \beta_k]\}$ are pairwise disjoint, then we may assume that $\alpha_{k_0} < \alpha_j < \alpha_{k_1}$ for all $j \in J \backslash \{k_1\}$. Note that J is not empty ($k_1 \in J$). Let $n \in \mathbb{N}$ be the cardinality of the finite set J (i.e., the number of elements of J), and relabel this finite family of open intervals with nonnegative integers $i \in [0, 1, \ldots, n]$. Since the intervals in $\{(\alpha_k, \beta_k]\}$ are disjoint, we can order the endpoints of the intervals that appear in the above finite family as $\alpha_i < \beta_i \leq \alpha_{i+1} < \beta_{i+1}$ for each $i \in [0, 1, \ldots, n{-}1]$, identifying k_0 with 0 and k_1 with n, so that

$$a_0 - \varepsilon_0 < a \leq \alpha_0 < \beta_0 \leq \alpha_i < \beta_i \leq \alpha_{i+1} < \beta_{i+1} < \alpha_n < \beta_n = b < \beta_n + \varepsilon_n$$

for $i \in [1, \ldots, n{-}2]$. This finite family of open intervals is then rewritten as

$$\big\{(\alpha_0 - \varepsilon_0, \beta_0 + \varepsilon_0)\,, (\alpha_i, \beta_i + \varepsilon_i) \;\text{ for every }\; i \in [1, \ldots, n]\big\},$$

which covers $[a, b]$, and so it covers the interval $(a, b]$ of class $\mathcal{C}_1$:

$$\begin{aligned}(a, b] \subseteq [a, b] \;&\subseteq\; (\alpha_0 - \varepsilon_0, \beta_0 + \varepsilon_0) \cup \bigcup_{i=1}^{n} (\alpha_i, \beta_i + \varepsilon_i) \\ &\subseteq (\alpha_0 - \varepsilon_0, \alpha_0] \cup (\alpha_0, \beta_0] \cup (\beta_0, \beta_0 + \varepsilon_0] \cup \bigcup_{i=1}^{n} (\alpha_i, \beta_i] \cup \bigcup_{i=1}^{n} (\beta_i, \beta_i + \varepsilon_i].\end{aligned}$$

Let i be an arbitrary index in $[0,\dots,n{-}1]$. Since $\{(\alpha_k,\beta_k]\}$ consists of disjoint intervals, it follows that $\beta_i \le \alpha_{i+1}$. Also, since the above union of intervals of class $\mathcal{C}_1$ covers $[a,b]$, it follows that $\alpha_{i+1} \le \beta_i + \varepsilon_i$. Then consider a finite sequence $\{\delta_i\}$ of nonnegative numbers given by

$$0 \le \delta_i = \alpha_{i+1} - \beta_i \le \varepsilon_i \;\text{ for each }\; i \in [0,\dots,n{-}1] \quad\text{and}\quad 0 < \delta_n = \varepsilon_n.$$

Replacing $\{\varepsilon_i\}$ with $\{\delta_i\}$ in the intervals of the form $(\beta_i, \beta_i + \varepsilon_i]$ we get a *finite* covering for $(a,b]$ consisting of *disjoint* intervals of class $\mathcal{C}_1$, viz.,

$$(\alpha_0 - \varepsilon_0, \alpha_0] \cup (\alpha_0, \beta_0] \cup (\beta_0, \beta_0 + \delta_0] \cup \bigcup_{i=1}^{n}(\alpha_i, \beta_i] \cup \bigcup_{i=1}^{n}(\beta_i, \beta_i + \delta_i].$$

Hence, according to Definition 8.7(a,c) and Problem 8.4,

$$\ell((a,b]) \le \varepsilon_0 + \sum_{i=0}^{n} \ell((\alpha_i,\beta_i]) + \sum_{i=0}^{n} \delta_i \le \sum\nolimits_k \ell((\alpha_k,\beta_k]) + 2\varepsilon$$

for $\sum_{i=0}^{n}\delta_i < \sum_k \varepsilon_k \le \varepsilon$. Since this holds for an arbitrary $\varepsilon > 0$, we get

(a$_2$) $$\ell((a,b]) \le \sum\nolimits_k \ell((\alpha_k,\beta_k]).$$

Therefore the result in (a) follows by (a$_1$) and (a$_2$).

(b) The disjointness assumption on both $\{(\alpha_k,\beta_k]\}$ and $\{(a_i,b_i]\}$ ensures that, if $\bigcup_i (a_i,b_i] = \bigcup_k (\alpha_k,\beta_k]$, then $(a_i,b_i] = \bigcup_j (\alpha_{i,j},\beta_{i,j}]$ for each i, where $\{(\alpha_{i,j},\beta_{i,j}]\} = \{(\alpha_k,\beta_k]\}$. Thus, applying Definition 8.7(c) and the result in item (a), we get

$$\ell\Big(\bigcup\nolimits_k (\alpha_k,\beta_k]\Big) = \ell\Big(\bigcup\nolimits_i (a_i,b_i]\Big) = \sum\nolimits_i \ell((a_i,b_i]) = \sum\nolimits_i \sum\nolimits_j \ell((\alpha_{i,j},\beta_{i,j}]).$$

Since the summands $\{\ell((\alpha_{i,j},\beta_{i,j}])\}$ are nonnegative real numbers, the doubly indexed sum is unconditionally convergent, and hence

$$\sum\nolimits_i \sum\nolimits_j \ell((\alpha_{i,j},\beta_{i,j}]) = \sum\nolimits_k \ell((\alpha_k,\beta_k]). \qquad \square$$

Lemma 8.9. *The collection $\Im$ of all finite unions of (left-open) intervals of the real line is an algebra of subsets of $\mathbb{R}$, and the length function $\ell\colon \Im \to \overline{\mathbb{R}}$ is a measure defined on the algebra $\Im$.*

Proof. Take the collection $\Im$ of all finite unions of intervals from $\mathcal{I}$ as in Definition 8.7. It is readily verified that $\Im$ is an algebra of subsets of $\mathbb{R}$ (cf. Problem 8.3). We claim that the length function ℓ is a measure on the algebra $\Im$. Note that $\ell(\varnothing) = 0$ by Definition 8.7(a). Next note that every set in $\Im$ can be expressed as a finite union of *disjoint* intervals from $\mathcal{I}$ so that

$\ell(F) \geq 0$ for every $F \in \Im$ by Definition 8.7(a,b,c). Indeed, every countable family of intervals in $\mathcal{I}$ admits a disjointification *consisting of intervals in* $\mathcal{I}$ (see Hints to Problems 2.3 and 7.2). To complete the proof, it remains to verify axiom (c) of Definition 8.1, viz.,

(c) $\ell\big(\bigcup_n F_n\big) = \sum_n \ell(F_n)$

for every countable family $\{F_n\}$ of pairwise disjoint sets in $\Im$ for which $\bigcup_n F_n$ lies in $\Im$.

Recall that every set in the algebra $\Im$ can be expressed as a finite union of *disjoint* intervals from $\mathcal{I}$. Hence the identity in property in (c) is readily verified by Definition 8.7(c) if $\{F_i\}$ is a *finite* family of pairwise disjoint sets in $\Im$. Thus suppose $\{F_n\}$ is a countably infinite family (equivalently, an infinite sequence) of pairwise disjoint sets in $\Im$ such that $\bigcup_n F_n$ lies in $\Im$. Take the extended nonnegative-valued sequence $\{\ell(F_n)\}$. If $\sum_n \ell(F_n) < \infty$, then $\ell\big(\bigcup_i F_i\big) = \sum_i \ell(F_i) < \sum_n \ell(F_n) < \infty$ for every finite subunion $\bigcup_i F_i$ of $\bigcup_n F_n$ because (c) holds for finite families of disjoint sets in $\Im$. So

$$\ell\Big(\bigcup\nolimits_n F_n\Big) = \sup \ell\Big(\bigcup\nolimits_i F_i\Big) \leq \sum\nolimits_n \ell(F_n) < \infty,$$

where the supremum is taken over all *finite subunions* of the countably infinite union $\bigcup_n F_n$. Hence, if $\ell\big(\bigcup_n F_n\big) = \infty$, then axiom (c) holds:

$$\ell\Big(\bigcup\nolimits_n F_n\Big) = \sum\nolimits_n \ell(F_n) = \infty.$$

Thus suppose $\ell\big(\bigcup_n F_n\big) < \infty$ so that $\ell(F_n) < \infty$ for each n by Problem 8.4, and recall that each F_n and also $\bigcup_n F_n$ are sets in $\Im$. Then each set F_n is a finite union of disjoint intervals of class $\mathcal{C}_1$,

$$F_n = \bigcup\nolimits_j (\alpha_{n,j}, \beta_{n,j}].$$

Since $\{F_n\}$ is a sequence of disjoint sets, this implies that

$$\bigcup\nolimits_n F_n = \bigcup\nolimits_{n,j} (\alpha_{n,j}, \beta_{n,j}] = \bigcup\nolimits_k (\alpha_k, \beta_k],$$

where $\{(\alpha_{n,j}, \beta_{n,j}]\} = \{(\alpha_k, \beta_k]\}$ is an infinite family of disjoint intervals of class $\mathcal{C}_1$, which also is a finite union of disjoint intervals of class $\mathcal{C}_1$,

$$\bigcup\nolimits_n F_n = \bigcup\nolimits_i (a_i, b_i].$$

Therefore, $\bigcup_i (a_i, b_i] = \bigcup_k (\alpha_k, \beta_k]$. Using Proposition 8.8(b) and recalling the unconditional convergence argument that closed that proof, we get

$$\ell\Big(\bigcup\nolimits_n F_n\Big) = \ell\Big(\bigcup\nolimits_i (a_i, b_i]\Big) = \sum\nolimits_k \ell((\alpha_k, \beta_k]) = \sum\nolimits_{n,j} \ell((\alpha_{n,j}, \beta_{n,j}])$$
$$= \sum\nolimits_n \sum\nolimits_j \ell((\alpha_{n,j}, \beta_{n,j}]) = \sum\nolimits_n \ell\Big(\bigcup\nolimits_j (\alpha_{n,j}, \beta_{n,j}]\Big) = \sum\nolimits_n \ell(F_n)$$

by Definition 8.7(c), so that (c) holds, thus completing the proof. □

Now we can to apply the Carathéodory Extension Theorem of Section 8.2 to build up the Lebesgue measure.

Theorem 8.10. *There exists a σ-algebra $\Im^*$ of subsets of $\mathbb{R}$ that includes the algebra $\Im$ and extends the length function $\ell\colon \Im \to \overline{\mathbb{R}}$ uniquely to a measure $\lambda^*\colon \Im^* \to \overline{\mathbb{R}}$, which is σ-finite and complete on the σ-algebra $\Im^*$.*

Proof. Lemma 8.9 says that $\ell\colon \Im \to \overline{\mathbb{R}}$ is a measure on the algebra $\Im$. Let $\ell^*\colon \wp(\mathbb{R}) \to \overline{\mathbb{R}}$ be the outer measure generated by ℓ (as in Definition 8.2), and let $\Im^*$ be the collection of all sets in $\wp(\mathbb{R})$ that are ℓ^*-measurable:

$$\Im^* = \big\{F \in \wp(\mathbb{R})\colon \ell^*(S) = \ell^*(S \cap F) + \ell^*(S\backslash F) \text{ for every } S \in \wp(\mathbb{R})\big\}.$$

According to Theorem 8.4, $\Im^*$ is a σ-algebra of subsets of $\mathbb{R}$ that includes the algebra $\Im$, and the restriction of ℓ^* to $\Im^*$ is a measure. In other words, $\Im \subseteq \Im^* \subseteq \wp(\mathbb{R})$ and $\ell^*|_{\Im^*}$ is a measure on $\Im^*$. Thus this measure $\ell^*|_{\Im^*}$ extends ℓ over $\Im^*$ according to Proposition 8.3(d). Also, it is readily verified that the measure ℓ on the algebra $\Im$ is σ-finite. Indeed, the real line $\mathbb{R}$ is covered by the countably infinite family of intervals $\{(q_k - \varepsilon, q_k + \varepsilon]\}$ of class $\mathcal{C}_1$ of length 2ε for any $\varepsilon > 0$, where $\{q_k\}$ (with k running over all integers $\mathbb{Z}$) is an enumeration of the rational numbers $\mathbb{Q}$ (see Example 2C). Since ℓ is σ-finite, it Theorem 8.6 ensures that there is a unique measure on $\Im^*$, say $\lambda^*\colon \Im^* \to \overline{\mathbb{R}}$, that extends ℓ over $\Im^*$, which is again σ-finite. By uniqueness, this extension of ℓ over $\Im^*$ coincides with the restriction of ℓ^* to $\Im^*$; that is, $\lambda^* = \ell^*|_{\Im^*}$. Thus Proposition 8.5 ensures that λ^* is a complete measure on the σ-algebra $\Im^*$ (i.e., the measure space $(\mathbb{R}, \Im^*, \lambda^*)$ is complete). □

The σ-algebra $\Im^*$ of Theorem 8.10 is referred to as the *Lebesgue algebra*. Sets in $\Im^*$ are called *Lebesgue sets* (or $\Im^*$-measurable). The measure λ^* on $\Im^*$ of Theorem 8.10 is also called *Lebesgue measure*. We close the section by considering a collection of a few basic properties of Lebesgue measure.

Consider the Borel algebra $\Re$, consisting of the Borel sets (i.e., $\Re$-measurable sets), which is the σ-algebra generated by the open intervals of the real line $\mathbb{R}$ or, equivalently, by the left-open intervals in $\mathcal{I}$, which means that $\Re$ is the intersection of all σ-algebras of subsets of $\mathbb{R}$ that include the family $\mathcal{I}$ (so $\Re$ is the smallest σ-algebra including $\mathcal{I}$ — see also the remark following Problem 1.14). Since any σ-algebra that includes the family $\mathcal{I}$ necessarily includes the algebra $\Im$ (finite union of intervals from $\mathcal{I}$), it follows that

(P$_1$) $\quad\mathfrak{R}$ is the smallest σ-algebra including the algebra $\mathfrak{I}$.

Thus $\mathfrak{R}$ properly includes the algebra $\mathfrak{I}$ (because $\mathfrak{I}$ is not a σ-algebra — Problem 8.3). Moreover, $\mathfrak{R}$ is included in the σ-algebra $\mathfrak{I}^*$ (since $\mathfrak{I}^*$ includes $\mathfrak{I}$). This leads to the following chain if inclusions:

$$\mathfrak{I} \subset \mathfrak{R} \subseteq \mathfrak{I}^* \subseteq \wp(\mathbb{R}).$$

It is clear that the restriction of the Lebesgue measure $\lambda\colon \mathfrak{R} \to \overline{\mathbb{R}}$ of Example 2C to the algebra $\mathfrak{I}$ is the length function $\ell\colon \mathfrak{I} \to \overline{\mathbb{R}}$ (Problem 2.7(c)). Recall that the restriction of the outer measure $\ell^*\colon \wp(\mathbb{R}) \to \overline{\mathbb{R}}$ to the Lebesgue algebra $\mathfrak{I}^*$ is the Lebesgue measure $\lambda^*\colon \mathfrak{I}^* \to \overline{\mathbb{R}}$ (Proof of Theorem 8.10). So

$$\lambda|_{\mathfrak{I}} = \ell \quad \text{and} \quad \ell^*|_{\mathfrak{I}^*} = \lambda^*.$$

Since the length function ℓ on $\mathfrak{I}$ extends to the measure λ^* on $\mathfrak{I}^*$ so that $\lambda^*|_{\mathfrak{I}} = \ell = \lambda|_{\mathfrak{I}}$, and since this extension is *unique* (Theorem 8.10), we can infer that the restriction of the Lebesgue measure $\lambda^*\colon \mathfrak{I}^* \to \overline{\mathbb{R}}$ to the Borel algebra $\mathfrak{R}$ is the Lebesgue measure $\lambda\colon \mathfrak{R} \to \overline{\mathbb{R}}$ of Example 2C. That is,

(P$_2$) $$\lambda^*|_{\mathfrak{R}} = \lambda.$$

Summing up,

$$\ell = \lambda|_{\mathfrak{I}} = \lambda^*|_{\mathfrak{I}} = \ell^*|_{\mathfrak{I}}, \qquad \lambda = \lambda^*|_{\mathfrak{R}} = \ell^*|_{\mathfrak{R}}, \qquad \lambda^* = \ell^*|_{\mathfrak{I}^*}.$$

Recall that λ is a Borel measure in the sense of Problem 2.13 (which implies that all bounded sets in $\mathfrak{R}$ have finite measure). Then so is its extension λ^* over $\mathfrak{I}^*$. Since $\mathfrak{R}$-measurable sets are $\mathfrak{I}^*$-measurable, all the Borel sets of Problem 2.7 are Lebesgue sets, and their λ^*-measures (as Lebesgue sets) coincide with their λ-measures (as Borel sets), which in turn coincide with their ℓ^*-outer measures. In particular, the Cantor set in $\mathfrak{R}$ of Problem 2.9 is an uncountable set with Lebesgue measure zero. Also, the measure space $(\mathbb{R}, \mathfrak{I}^*, \lambda^*)$ *is the completion of the measure space* $(\mathbb{R}, \mathfrak{R}, \lambda)$. Equivalently, the σ-algebra $\mathfrak{I}^*$ *is the completion of the σ-algebra* $\mathfrak{R}$ with respect to λ, or the measure λ^* *is the completion of the measure* λ on $\mathfrak{R}$. That is,

(P$_3$) $$\mathfrak{I}^* = \overline{\mathfrak{R}} \quad \text{and} \quad \lambda^* = \overline{\lambda}$$

(see the remark that closes Section 2.3). In fact, since $(\mathbb{R}, \mathfrak{I}^*, \lambda^*)$ is a complete measure space, and according to Problem 8.6(b), it follows that

$$\mathfrak{I}^* = \big\{ \overline{E} \subseteq \mathbb{R} : \overline{E} = E \cup A, \text{ with } E \in \mathfrak{R},\ A \subseteq N \in \mathfrak{R} \text{ and } \lambda(N) = 0 \big\}.$$

This is precisely the definition of $\overline{\mathfrak{R}}$. So, according of Problem 2.15, *every* $\mathfrak{I}^*$*-measurable function is a.e. equal to an* $\mathfrak{R}$*-measurable functions*. That is, if $f: \mathbb{R} \to \overline{\mathbb{R}}$ is an $\Im^*$-measurable (or *Lebesgue measurable*) function, then there is an $\Re$-measurable (or *Borel measurable*) function $g: \mathbb{R} \to \overline{\mathbb{R}}$ such that $g = f$ λ-a.e. Observe that the Borel–Stieltjes measure on $\Re$ of Example 2D can be naturally extended to a complete *Lebesgue–Stieltjes measure* on $\Im^*$. The above properties are easily verified. However, $\Im^*$ *is neither the smallest nor the largest* σ*-algebra including* $\Im$, so the inclusions below are proper:

$$(\mathrm{P}_4) \qquad \Re \subset \Im^* \subset \wp(\mathbb{R}).$$

These *proper* inclusions are all but trivial. They have been quite critical in many aspects (including historical aspects). Completeness of the measure space $(\mathbb{R}, \Im^*, \lambda^*)$ can be used to give an existential proof for the first proper inclusion. Indeed, let $\mathcal{I}_{\mathbb{Q}}$ be the subfamily of $\mathcal{I}$ consisting of those intervals from $\mathcal{I}$ with rational endpoints. It is readily verified that this is a countably infinite set ($\#\mathcal{I}_{\mathbb{Q}} = \#\mathbb{N}$, where $\#$ stands for cardinality), and also that the smallest σ-algebra of subsets of $\mathbb{R}$ that includes $\mathcal{I}_{\mathbb{Q}}$ is the smallest σ-algebra that includes $\mathcal{I}$, and so it is the smallest σ-algebra that includes the algebra $\Im$, which is precisely the Borel algebra $\Re$. Then $\Re$ is the σ-algebra generated by $\mathcal{I}_{\mathbb{Q}}$. Since $\#\mathcal{I}_{\mathbb{Q}} = \#\mathbb{N} < \#\mathbb{R}$, it follows that $\#\Re \leq \#\mathbb{R}$ [18, Problem 9(c), p. 22]. But $\#\mathbb{R} \leq \#\Re$ trivially (for each $x \in \mathbb{R}$, $(x, x+1) \in \Re$). Hence,

$$\#\Re = \#\mathbb{R}.$$

Now recall that the Cantor set $C \in \Re \subseteq \Im^*$ of Problem 2.9 is uncountable (actually, $\#C = \#\mathbb{R}$) and has measure zero. Since λ^* is a complete measure on $\Im^*$ and $\lambda^*(C) = 0$, it follows that all subsets of C are measurable; that is, $\wp(C) \subseteq \Im^*$. Thus, recalling that $\#X < \#\wp(X)$ for every set X, we get

$$\#\mathbb{R} = \#C < \#\wp(C) \leq \#\Im^*.$$

Therefore $\#\Re < \#\Im^*$ and $\Re \subseteq \Im^*$ so that

$$\Re \subset \Im^*.$$

This proves the first proper inclusion in P_4 and, *en passant*, it also proves that $\Re$ is not complete (as anticipated in the remark the closes Section 2.3) since $\Im^*$ is the completion of $\Re$. As for the second proper inclusion,

$$\Im^* \subset \wp(\mathbb{R}),$$

a proof of it is worked out in Problem 8.14 based on translation invariance (an important property that we discuss later). It is also worth noticing that by Properties P_3 and P_4 the σ-algebra $\Re$ is not complete with respect to λ, and this implies that there are Lebesgue sets of measure zero that are not Borel. However, *every Lebesgue set of measure zero is a subset of a Borel set of measure zero*:

(P$_5$) $\quad N \in \Im^*$ and $\lambda^*(N) = 0$ imply $N \subseteq G \in \Re$ and $\lambda(G) = 0$.

In fact, since $N \in \Im^*$ and $\lambda^*(N) = 0$ imply $N \in \wp(\mathbb{R})$ and $\ell^*(N) = 0$, Property P$_5$ is an immediate consequence of Problem 8.8(c), which also says that the Borel set G of measure zero actually is a G_δ. Next we focus on the *translation invariance property*. For every $\alpha \in \mathbb{R}$ and every $S \subseteq \mathbb{R}$, set

$$S + \alpha = \{\xi + \alpha \in \mathbb{R}\colon \xi \in S\} \subseteq \mathbb{R}.$$

If $E \in \Im^*$, then $E + \alpha \in \Im^*$ and

(P$_6$) $$\lambda^*(E + \alpha) = \lambda^*(E)$$

(cf. Problem 8.12). Translation invariance plays a central role when we set about to build up a *nonmeasurable set* (as we do in Problem 8.14). Another important consequence of it reads as follows.

(P$_7$) Sets with positive outer measure include nonmeasurable subsets.

(See Problem 8.18.) Special case: *every Lebesgue set of positive measure includes a nonmeasurable set*. Translation invariance is the central topic of Chapter 13 (see, in particular, Proposition 13.F).

Remarks: In the first paragraph of Chapter 1 we observed that "the power set is too large a set to be the domain" of some measures. Now (and only now) are we ready to offer a proper explanation of that assertion. In fact, we have the following problem. It is not possible to construct a set function with the following four properties: (1) defined on the whole $\wp(\mathbb{R})$, (2) assigning to each interval the value of its length, (3) countably additive, and (4) translation invariant, as we can infer from Problem 8.14. Indeed, properties (2), (3), and (4) were all we needed in the proof of Problem 8.14. Weakening property (1) is a possible approach to face this problem (as we have, in fact, done in Chapters 2 and 8), supplying a measure λ^* defined on a proper σ-algebra $\Im^*$, which retains the useful properties (2), (3), and (4). However, there are other approaches (e.g., if we keep (1), (2), and (4) but replace (3) with subadditivity, then we get the outer measure ℓ^* on $\wp(\mathbb{R})$). Assuming the *Continuum Hypothesis* (i.e., the hypothesis that every uncountable subset of $\mathbb{R}$ has the same cardinality as $\mathbb{R}$ itself), then it can be shown that there is no set function satisfying properties (1), (2), and (3) only. (See the references in the Suggested Reading section.)

8.4 Problems

Problem 8.1. Axiom (c) of Definition 8.1 is referred to as *countable additivity* (same as in axiom (c) of Definition 2.1 for a measure on σ-algebra). Let $\mu\colon \mathcal{A} \to \overline{\mathbb{R}}$ be a set function on an algebra $\mathcal{A}$ and consider the assertion:

∘ $\mu\big(\bigcup_i E_i\big) = \sum_i \mu(E_i)$ for every finite family $\{E_i\}$ of pairwise disjoint sets in $\mathcal{A}$.

This is referred to as *finite additivity.* It is clear that countable additivity implies finite additivity. Verify that the converse fails. (*Hint:* Problem 2.5.) Suppose the set function μ is such that

(i) $\mu(E) \ge 0$ for every $E \in \mathcal{A}$

and, if $\{A_n\}$ is a sequence of sets in $\mathcal{A}$,

(ii) $\lim_k \mu\big(\bigcup_{n=1}^k A_n\big) = \mu\big(\bigcup_{n=1}^\infty A_n\big)$

whenever $\bigcup_{n=1}^\infty A_n$ lies in $\mathcal{A}$. Prove the following proposition. *If μ is finitely additive, then it is countably additive.* In other words, take an arbitrary sequence $\{E_n\}$ of pairwise disjoint sets in $\mathcal{A}$. If

$$\mu\Big(\bigcup_{n=1}^m E_n\Big) = \sum_{n=1}^m \mu(E_n)$$

for every $m \ge 1$ (i.e., *finite additivity*) and (i) and (ii) hold, then show that

$$\mu\Big(\bigcup_{n=1}^\infty E_n\Big) = \sum_{n=1}^\infty \mu(E_n)$$

whenever $\bigcup_{n=1}^\infty E_n$ lies in $\mathcal{A}$ (i.e., *countable additivity*).

Hint: First use the proof of Proposition 2.2(a), recalling that μ is finitely additive now acting on an algebra $\mathcal{A}$, to show that if (i) holds, then

(i′) $A, B \in \mathcal{A}$ and $A \subseteq B$ imply $\mu(A) \le \mu(B)$.

Now take an arbitrary sequence $\{E_n\}$ of pairwise disjoint sets in $\mathcal{A}$ such that $\bigcup_{n=1}^\infty E_n \in \mathcal{A}$. Use finite additivity, the fact that $\big\{\bigcup_{n=1}^m E_n\big\}$ is an increasing sequence of sets in $\mathcal{A}$, and the result in (i′) to check that

$$\sum_{n=1}^m \mu(E_n) = \mu\Big(\bigcup_{n=1}^m E_n\Big) \le \mu\Big(\bigcup_{n=1}^\infty E_n\Big) \quad \text{for all} \quad m \ge 1.$$

Next apply the same argument, recalling from (i′) that $\big\{\mu\big(\bigcup_{n=1}^m E_n\big)\big\}$ is an increasing sequence of nonnegative elements from $\overline{\mathbb{R}}$, to show that

$$\mu\Big(\bigcup_{n=1}^k E_n\Big) \le \lim_m \mu\Big(\bigcup_{n=1}^m E_n\Big) = \lim_m \sum_{n=1}^m \mu(E_n) = \sum_{n=1}^\infty \mu(E_n) \quad \text{for all} \quad k \ge 1.$$

Problem 8.2. The notion of length function $\ell\colon \Im \to \overline{\mathbb{R}}$ was introduced in Definition 8.7. Show that properties (a), (b), and (c) in Definition 8.7 are enough to ensure that $\ell\colon \Im \to \overline{\mathbb{R}}$ is defined on the whole collection $\Im$.

Hint: Every countable family of intervals in $\mathcal{I}$ admits a disjointification *consisting of intervals in* $\mathcal{I}$ (use the Hints to Problems 2.3 and 7.2). Use Definition 8.7(c) to verify that a length is defined for every set in $\Im$.

Problem 8.3. Consider again the setup of Definition 8.7.

(a) First show that $\Im$ is an algebra of subsets of the real line $\mathbb{R}$.

Hint: In order to show that axiom (b) of Definition 1.1 holds true; that is, to check that *the complement of a set in* $\Im$ *belongs to* $\Im$, proceed as follows. Verify that (i) if I is an interval in $\mathcal{I}$, then $\mathbb{R}\backslash I$ is the union of no more than two intervals in $\mathcal{I}$, and therefore it is a set in $\Im$, and (ii) if E and F are sets in $\Im$, then $E \cap F$ is again a set in $\Im$ (since the intersection of any pair of intervals in $\mathcal{I}$ is again an interval in $\mathcal{I}$). According to De Morgan laws, $\mathbb{R}\backslash\bigcup_i I_i = \bigcap_i(\mathbb{R}\backslash I_i)$ lies in $\Im$ for every *finite* union $\bigcup_i I_i$ of intervals I_i from $\mathcal{I}$.

(b) Now verify that $\Im$ is not a σ-algebra of subsets of $\mathbb{R}$.

Hint: $\bigcup_{n=1}^{\infty}(2n-1, 2n] = (1,2] \cup (3,4] \cup \ldots$ is not a set in $\Im$.

Problem 8.4. Let $\Im$ be the algebra of Problem 8.3(a). Verify that

$$E \subseteq F \quad \text{with} \quad E, F \in \Im \quad \text{implies} \quad \ell(E) \leq \ell(F).$$

Hint: $F = E \cup (F\backslash E)$ and $F\backslash E \in \Im$. Recall the Hint to Problem 8.2.

Problem 8.5. Consider the union in the Hint to Problem 8.3(b). Observe that it suggests the following infinite union in $\Im$ (which in fact is in class $\mathcal{C}_3$) of disjoint intervals of class $\mathcal{C}_1$,

$$\bigcup_{k=1}^{\infty}(k, k+1] \;=\; (1,2] \cup (2,3] \cup (3,4] \cup \ldots \;=\; (1,+\infty),$$

which has an infinite subunion $\bigcup_{n=1}^{\infty}(2n-1, 2n]$ not in $\Im$. Show that this can happen even if the original infinite union is of class $\mathcal{C}_1$ (thus having finite length) by exhibiting a sequence $\{(\alpha_k, \beta_k]\}$ of intervals of class $\mathcal{C}_1$ such that

$$\bigcup_{k=1}^{\infty}(\alpha_k, \beta_k] = (0,1] \quad \text{but} \quad \bigcup_{n=1}^{\infty}(\alpha_n, \beta_n] \notin \Im$$

for some subsequence $\{(\alpha_n, \beta_n]\}$ of $\{(\alpha_k, \beta_k]\}$. *Hint:* $\left\{\left(\frac{1}{2^k}, \frac{1}{2^{k-1}}\right]\right\}$.

Problem 8.6. Take the Borel algebra $\Re$ and the Lebesgue algebra $\Im^*$, which are σ-algebras of subsets of $\mathbb{R}$ with $\Re \subseteq \Im^*$. Prove the following assertions.

(a) If $F \in \Im^*$, then there exists $E \in \Re$ such that $\lambda^*(F \backslash E) = 0$.

Hint: Since $\lambda^*(F) = \lambda^*(F \cap E) + \lambda^*(F \backslash E)$, it follows that, if $\lambda^*(F \backslash E) > 0$ for every $E \in \Re$, then $\lambda^*(F \cap E) < \lambda^*(F)$ (whenever $\lambda^*(F) < \infty$) for every $E \in \Re$, which is a contradiction since $\mathbb{R} \in \Re$.

(b) If $F \in \Im^*$, then there exist $E \in \Re$ and $N \in \Im^*$ such that $F = E \cup N$ and $E \cap N = \varnothing$, where $\lambda^*(N) = 0$ and so $\lambda^*(F) = \lambda(E)$.

Problem 8.7. Take the outer measure $\ell^*: \wp(\mathbb{R}) \to \overline{\mathbb{R}}$ generated by the length function $\ell: \Im \to \overline{\mathbb{R}}$, which is a measure on the algebra $\Im$ (by Lemma 8.6). Apply Definitions 8.2 and 8.7 to verify that ℓ^* is written as

(a)
$$\ell^*(S) = \inf_{\{I_n\} \in \mathcal{I}_S} \sum\nolimits_n \ell(I_n)$$

for every $S \in \wp(\mathbb{R})$, where

$$\mathcal{I}_S = \big\{\{I_n\}: I_n \in \mathcal{I} \text{ and } S \subseteq \textstyle\bigcup_n I_n\big\}$$

is the collection of all countable families $\{I_n\}$ of intervals in $\mathcal{I}$ that cover S. Let $I \in \mathcal{I}$ be an arbitrary interval, and let I° and I^- denote interior and closure of I, respectively (with respect to the usual topology of the $\mathbb{R}$). Suppose $\lambda: \Re \to \overline{\mathbb{R}}$ is the Lebesgue measure on the σ-algebra $\Re$. Recall that $\Re$ includes the algebra $\Im$ (i.e., $\Im \subset \Re$). Verify that $I^\circ \in \Re$ and $I^- \in \Re$, and

(b)
$$\lambda(I^\circ) = \lambda(I) = \lambda(I^-)$$

for every $I \in \mathcal{I}$. (*Hint:* Problem 2.7.) Recall that $I^\circ \subseteq I \subseteq I^-$ and show that

(c)
$$\ell^*(S) = \inf_{\{I_n\} \in \mathcal{I}_S^\circ} \sum\nolimits_n \lambda(I_n^\circ) = \inf_{\{I_n\} \in \mathcal{I}_S^-} \sum\nolimits_n \lambda(I_n^-),$$

where the infimum is taken over all countable coverings of S consisting either of *open* intervals (equivalently, of the interior of intervals in $\mathcal{I}$),

$$\mathcal{I}_S^\circ = \big\{\{I_n\}: I_n \in \mathcal{I} \text{ and } S \subseteq \textstyle\bigcup_n I_n^\circ\big\},$$

or of *closed* intervals (equivalently, of the closure of intervals in $\mathcal{I}$),

$$\mathcal{I}_S^- = \big\{\{I_n\}: I_n \in \mathcal{I} \text{ and } S \subseteq \textstyle\bigcup_n I_n^-\big\}.$$

Hint: Covering by closed intervals is a consequence of item (b) since $I \subseteq I^-$ and $\ell(I) = \lambda(I)$. For the case of covering by open intervals, take an arbitrary $\varepsilon > 0$ and observe that for each $I_n \in \mathcal{I}$ there exists an open interval $J_n \in \Re$ such that $I_n \subseteq J_n$ and $\lambda(J_n) \leq \ell(I_n) + \frac{\varepsilon}{2^n}$. Thus $\bigcup_n I_n \subseteq \bigcup_n J_n$ and $\sum_n \lambda(J_n) \leq \sum_n \ell(I_n) + \varepsilon$ (if n runs over $\mathbb{N}$).

Problem 8.8. Take a measure $\ell\colon \Im \to \overline{\mathbb{R}}$ on an algebra $\Im$, let $\ell^*\colon \wp(\mathbb{R}) \to \overline{\mathbb{R}}$ be the outer measure generated by ℓ (Definition 8.2), let $\lambda\colon \Re \to \overline{\mathbb{R}}$ be the Lebesgue measure on the σ-algebra $\Re$, and let $\lambda^*\colon \Im^* \to \overline{\mathbb{R}}$ be the Lebesgue measure on the σ-algebra $\Im^*$. Recall that $\Re \subseteq \Im^*$ and $\lambda(E) = \lambda^*(E)$ for every $E \in \Re$, and also that open sets are measurable (open sets are Borel sets by Problem 2.7(d), and so they are Lebesgue sets). Take any $S \in \wp(\mathbb{R})$ and an arbitrary $\varepsilon > 0$. Show that there is an open set $U_\varepsilon \subseteq \mathbb{R}$ such that

(a) $$S \subseteq U_\varepsilon \quad \text{and} \quad \lambda(U_\varepsilon) \leq \ell^*(S) + \varepsilon.$$

Then, since $\ell^*(S) \leq \lambda(U_\varepsilon)$ and infimum is the maximum of all lower bounds,

(b) $$\ell^*(S) = \inf_{U \in \mathcal{T}} \{\lambda(U)\colon S \subseteq U\},$$

where $\mathcal{T}$ is the topology (i.e., the collection of all open sets) of $\mathbb{R}$.

Hint: Take the first identity in Problem 8.7(c) with the infimum taken over $\mathcal{I}_S^\circ$. Thus there is $\{J_n\} \in \mathcal{I}_S^\circ$ such that $\sum_n \lambda(J_n^\circ) \leq \ell^*(S) + \varepsilon$. Since every union of open sets is again an open set, set $U = \bigcup_n J_n^\circ$ in $\Re$ and verify that $\lambda(U) = \lambda^*(U) = \ell^*(U) \leq \sum_n \ell^*(J_n^\circ) = \sum_n \lambda(J_n^\circ)$ — Proposition 8.3(e).

Finite intersections of open sets is an open set, but a countable intersection of open sets, which is referred to as a G_δ, is not necessarily an open set, although always measurable (i.e., a G_δ is a Borel set). Show that there exists a G_δ, say $G \in \Re$, such that

(c) $$S \subseteq G \quad \text{and} \quad \ell^*(S) = \lambda(G).$$

Hint: Item (a) ensure the existence of a sequence $\{U_n\}$ of open sets such that $S \subseteq U_n$ and $\lambda(U_n) \leq \ell^*(S) + \frac{1}{n}$ for all n. Set $G = \bigcap_n U_n$ in $\Re$ and verify that $S \subseteq G \subseteq U_n$, and so $\ell^*(S) \leq \ell^*(G) = \lambda(G) \leq \lambda(U_n)$ for all n.

Problem 8.9. Apply Problem 8.8(c) and the fact that the measure λ^* is complete to show that *every set with outer measure zero is a Lebesgue set*:

(a) $N \in \wp(\mathbb{R})$ and $\ell^*(N) = 0$ imply $N \in \Im^*$ and $\lambda^*(N) = 0$; and

(b) $E, N \in \wp(\mathbb{R})$, $E \subseteq N$ and $\ell^*(N) = 0$ imply $E \in \Im^*$ and $\lambda^*(E) = 0$.

Take an arbitrary set $S \in \wp(\mathbb{R})$. According to Problem 8.8(c), $S \subseteq G$ with $\ell^*(S) = \lambda(G)$ for some $G \in \Re$. Since $G = S \cup (G\backslash S)$ and $S = G\backslash(G\backslash S)$, use item (a) to prove the following assertion.

(c) If $\ell^*(S) < \infty$, then $S \in \Im^*$ if and only if $\ell^*(G\backslash S) = 0$.

Problem 8.10. Consider the setup of Problem 8.8. Prove that

$$\lambda^*(E) = \inf_{U\in\mathfrak{R}} \{\lambda(U)\colon U \text{ is open and } E \subseteq U\} \quad \text{for every} \quad E \in \Im^*.$$

This means that λ^* is *outer regular* (see Section 11.2).

Hint: Problem 8.8(b) and Property P_2.

Problem 8.11. Consider again the setup of Problem 8.8. Prove that

$$\lambda^*(E) = \sup_{F\in\mathfrak{R}} \{\lambda(F)\colon F \text{ is closed and } F \subseteq E\} \quad \text{for every} \quad E \in \Im^*.$$

This implies that λ^* is *inner regular* (see Section 11.2).

Hint: Take an arbitrary $E \in \Im^*$ and an arbitrary $\varepsilon > 0$. Set $E' = \mathbb{R}\backslash E \in \Im^*$ (the complement of E lies in $\Im^*$). According to Problem 8.8(a), there is an open set U_ε such that $E' \subseteq U_\varepsilon$ and $\lambda(U_\varepsilon) \le \lambda^*(E') + \varepsilon$. First prove that

$$\lambda^*(U_\varepsilon\backslash E') \le \varepsilon.$$

If $\lambda^*(E') < \infty$, then the above inequality is an immediate consequence of Proposition 2.2(b). If $\lambda^*(E') = \infty$, then proceed as follows. Since λ^* is a σ-finite measure on $\Im^*$, every set in $\Im^*$ is σ-finite, and so there is a countable covering of E', say $\{E'_n\}$, made up of measurable subsets of E' of finite measure. Since $\lambda^*(E'_n) < \infty$, there is an open set $U_{\varepsilon,n}$ such that $E'_n \subseteq U_{\varepsilon,n}$ and $\lambda^*(U_{\varepsilon,n}\backslash E'_n) \le \frac{\varepsilon}{2^n}$. Take the open set $U_\varepsilon = \bigcup_n U_{\varepsilon,n} \supseteq \bigcup_n E'_n = E'$. Show that $\lambda^*(U_\varepsilon\backslash E') \le \sum_n \lambda^*(U_{\varepsilon,n}\backslash E'_n) \le \varepsilon$. This proves the claimed inequality. Now take the closed set $F_\varepsilon = \mathbb{R}\backslash U_\varepsilon \subseteq E$, verify that $E\backslash F_\varepsilon = U_\varepsilon\backslash E'$, and hence $\lambda^*(E\backslash F_\varepsilon) \le \varepsilon$. Thus $\lambda^*(E) = \lambda^*(E \cap F_\varepsilon) + \lambda^*(E\backslash F_\varepsilon) \le \lambda^*(F_\varepsilon) + \varepsilon$. But $\lambda^*(F_\varepsilon) \le \lambda^*(E)$ (and supremum is the minimum of all upper bounds).

Problem 8.12. Prove the translation invariance property (i.e., prove P_6).

Hint: Take an arbitrary real number $\alpha \in \mathbb{R}$. Verify that

$$\ell(I + \alpha) = \ell(I)$$

for every $I \in \mathcal{I}$. Apply Problem 8.7(a) to show that

$$\ell^*(S + \alpha) = \ell^*(S)$$

for every $S \in \wp(\mathbb{R})$. Recall that $E \in \Im^*$ if and only if

$$\ell^*(S) = \ell^*(S \cap E) + \ell^*(S\backslash E) \quad \text{for every} \quad S \in \wp(\mathbb{R}).$$

Take arbitrary $A, B \in \wp(\mathbb{R})$. Verify that $(A + \alpha) \cap B = (A \cap (B - \alpha)) + \alpha$ and $(\mathbb{R}\backslash B) + \alpha = \mathbb{R}\backslash(B + \alpha)$, and so $(A + \alpha)\backslash B = (A\backslash(B - \alpha)) + \alpha$. Thus show that if $E \in \Im^*$, then

$$\begin{aligned}\ell^*(S) &= \ell(S-\alpha) = \ell^*\big((S-\alpha)\cap E\big) + \ell^*\big((S-\alpha)\backslash E\big)\\ &= \ell^*\big((S\cap(E+\alpha)-\alpha\big) + \ell^*\big((S\backslash(E+\alpha)-\alpha\big)\\ &= \ell^*\big(S\cap(E+\alpha)\big) + \ell^*\big(S\backslash(E+\alpha)\big).\end{aligned}$$

Hence $E+\alpha$ also lies in $\Im^*$. Since $\lambda^* = \ell^*|_{\Im^*}$, conclude the result in P_6:

$$E \in \Im^* \quad \text{implies} \quad E+\alpha \in \Im^* \;\text{ and }\; \lambda^*(E+\alpha) = \lambda^*(E).$$

Problem 8.13. Consider a binary operation $\dot{+}\colon [0,1)\times[0,1) \to [0,1)$ defined as follows. If α and β lie in the interval $[0,1)$, then

$$\alpha \dot{+} \beta = \begin{cases} \alpha+\beta, & \alpha+\beta<1,\\ \alpha+\beta-1, & \alpha+\beta\ge 1.\end{cases}$$

This is called *sum modulo* 1. For every $\alpha \in [0,1)$ and every $S \subseteq [0,1)$, set

$$S \dot{+} \alpha = \{\xi \dot{+} \alpha \in \mathbb{R}\colon \xi \in S\} \;\subseteq\; [0,1).$$

This is called *translation modulo* 1, Prove *translation invariance modulo* 1; that is, prove translation invariance with respect to sum modulo 1.

(a) If $S \subseteq [0,1)$, then $\ell^*(S \dot{+} \alpha) = \ell^*(S)$.

(b) If $E \subseteq [0,1)$ and $E \in \Im^*$, then $E \dot{+} \alpha \in \Im^*$ and $\lambda^*(E \dot{+} \alpha) = \lambda^*(E)$.

Hint: Take an arbitrary real number $\alpha \in [0,1)$, and take an arbitrary set $E \subseteq [0,1)$ such that $E \in \Im^*$. Consider the sets $E_1 = E \cap [0,1-\alpha)$ and $E_2 = E \cap [1-\alpha,1)$. Show that E_1 and E_2 lie in $\Im^*$. Since $\xi+\alpha<1$ for every $\xi \in E_1$, and $\xi+\alpha \ge 1$ for every $\xi \in E_2$, verify that $E_1 \dot{+} \alpha = E_1 + \alpha$ and $E_2 \dot{+} \alpha = E_2 + (\alpha-1)$. Therefore, conclude that $E_1 \dot{+} \alpha$ and $E_2 \dot{+} \alpha$ lie in $\Im^*$. Finally, using translation invariance again (for ordinary sums, as in Problem 8.12), show that

$$\begin{aligned}\lambda^*(E \dot{+} \alpha) &= \lambda^*(E_1 \dot{+} \alpha) + \lambda^*(E_2 \dot{+} \alpha)\\ &= \lambda^*(E_1+\alpha) + \lambda^*(E_2+\alpha) = \lambda^*(E_1) + \lambda^*(E_2) = \lambda^*(E).\end{aligned}$$

Problem 8.14. Let $\sim$ be a relation on the interval $[0,1)$ defined by

$$\alpha \sim \beta \quad \text{if and only if} \quad \alpha - \beta \in \mathbb{Q}.$$

In other words, α and β in $[0,1)$ are related if their difference is a rational number. This is an equivalence relation. That is, $\sim$ is reflexive (i.e., $\alpha \sim \alpha$), transitive (i.e., $\alpha \sim \beta$ and $\beta \sim \gamma$ imply $\alpha \sim \gamma$), and symmetric (i.e., $\alpha \sim \beta$ implies $\beta \sim \alpha$). Thus $\sim$ induces a partition of $[0,1)$ into equivalence classes

$$[\alpha] = \{\alpha' \in [0,1) \colon \alpha' \sim \alpha\} = \{\alpha' \in [0,1) \colon \alpha' - \alpha \in \mathbb{Q}\}.$$

The *Axiom of Choice* ensures the existence of sets consisting of elements from every equivalence class (one and just one from each equivalence class). These are the *Vitali sets*. Let $V \subseteq [0,1)$ be a Vitali set and let $\{q_n\}$ be an enumeration of $\mathbb{Q} \cap [0,1)$. For each n set $V_n = V \dotplus q_n \subseteq [0,1)$; rational translations modulo 1 of V. Show that $\{V_n\}$ forms a partition of $[0,1)$.

(a) $V_m \cap V_n = \varnothing$ whenever $m \neq n$.

Hint: Take an arbitrary $\xi \in V_n \cap V_m$. Show that $\xi = \alpha \dotplus q_n = \beta \dotplus q_m$, with $\alpha, \beta \in V$, $\alpha \dotplus q_n \in V_n$, and $\beta \dotplus q_m \in V_m$. Then $\alpha - \beta \in \mathbb{Q}$ (i.e., $\alpha \sim \beta$), so α and β come from the same equivalence class. Since V has only one element from each equivalence class, $\alpha = \beta$. Hence $m = n$.

(b) $\bigcup_n V_n = [0,1)$.

Hint: Take an arbitrary $\alpha \in [0,1)$. Thus α is in some equivalence class (since these classes form a partition of $[0,1)$), and so $\alpha \sim \beta$ for some $\beta \in V \subseteq [0,1)$ (since V has one element from each class), which means $\alpha - \beta \in \mathbb{Q}$ so that $\alpha = \beta + q$ for some q in $\mathbb{Q}$. First show that if $\beta \leq \alpha$, then $\alpha = \beta + q_n$ for q_n in $\mathbb{Q} \cap [0,1)$ so that $\alpha = \beta \dotplus q_n$, and therefore $\alpha \in V_n$ for some n. Next verify that if $\alpha < \beta$, then $\alpha = \beta - q_m$ for q_m in $\mathbb{Q} \cap (0,1)$ so that $\alpha = \beta \dotplus p_m$ with $p_m = 1 - q_m$ in $\mathbb{Q} \cap (0,1)$, and so $\alpha \in V_m$ for some m. Then conclude that $[0,1) \subseteq \bigcup_n V_n$.

Now use Problem 8.13 to prove that

$$V \notin \Im^*.$$

Hint: If $V \in \Im^*$, then $\lambda^*([0,1)) = \sum_n \lambda^*(V_n) = \sum_n \lambda^*(V) \neq 1$.

Outcome: *Vitali sets are not measurable*, thus completing the proof of P_4. This was the first example of a nonmeasurable set, given by Giuseppe Vitali in 1905. The use of the Axiom of Choice is essential here.

Problem 8.15. Exhibit a sequence $\{V_n\}$ of disjoint sets in $\wp(\mathbb{R})$ such that

$$\ell^*\Big(\bigcup_n V_n\Big) < \sum_n \ell^*(V_n).$$

Problem 8.16. Exhibit a pair of disjoint sets A and B in $\wp(\mathbb{R})$ such that

$$\ell^*(A \cup B) \neq \ell^*(A) + \ell^*(B).$$

Hint: According to Propositions 8.3(c) and 8.5(a), the outer measure of a Vitali set V is such that $0 < \ell^*(V) \leq 1$. Take the sequence $\{V_n\}$ of Problem 8.14. Set $q_0 = 0$ so that $V_0 = V$. Note that the equivalence class containing zero is $[0] = \mathbb{Q} \cap [0,1)$. Use the same argument as in the Hint to Problem 8.14(b) and show that $\bigcup_n V_n \backslash V_0 = (0,1)$. Set $A = V_0$ and $B = \bigcup_n V_n \backslash V_0$.

Problem 8.17. Show that measurable subsets of Vitali sets have measure zero. In other words, prove that

$$E \in \Im^* \text{ and } E \subset V \quad \text{imply} \quad \lambda^*(E) = 0.$$

Hint: Consider the setup of Problem 8.14. Set $E_n = E + q_n \subset V_n$. Show that $\{E_n\}$ is a sequence of disjoint sets in $\Im^*$ with $\lambda^*(E_n) = \lambda^*(E)$ and

$$\sum_n \lambda^*(E_n) = \lambda^*\Big(\bigcup_n E_n\Big) \leq \ell^*\Big(\bigcup_n V_n\Big) = \ell^*([0,1)) = 1.$$

Problem 8.18. Prove Property P_7.

If $S \in \wp(\mathbb{R})$ and $\ell^*(S) > 0$, then there exists $S_0 \subseteq S$ such that $S_0 \notin \Im^*$.

Hint: Suppose $\ell^*(S) > 0$. Then use Problem 8.7(c) to show that there is a translation of S, say S', such that $\ell^*(S' \cap (0,1)) > 0$. Set $A = S' \cap (0,1)$ and $A_n = A \cap V_n$, with $\{V_n\}$ as in Problem 8.14. If all subsets of S are measurable, then show that A_n is measurable. Since $A_n \subseteq V$, Problem 8.17 ensures that $\lambda^*(A_n) = 0$. Thus verify the following contradiction:

$$0 < \ell^*(A) \leq \ell^*\Big(A \cap \bigcup_n V_n\Big) = \ell^*\Big(\bigcup_n A_n\Big) \leq \sum_n \ell^*(A_n) = \sum_n \lambda^*(A_n) = 0.$$

Problem 8.19. Take $E \in \Im^*$ and $S \in \wp(\mathbb{R})$ arbitrary. Show that

(a) $\ell^*(E \cup S) + \ell^*(E \cap S) = \lambda^*(E) + \ell^*(S)$,

(b) $E \subseteq S$ and $\lambda^*(E) < \infty$ imply $\ell^*(S \backslash E) = \ell^*(S) - \lambda^*(E)$.

Hint: $\ell^*(E \cup S) = \ell^*((E \cup S) \cap E) + \ell^*((E \cup S) \backslash E) = \lambda^*(E) + \ell^*(S \backslash E)$ and $\ell^*(E \cap S) + \ell^*(S \backslash E) = \ell^*(S)$, proving (a). Replace S with $S \backslash E$ in (a).

Problem 8.20. If $S \in \wp(\mathbb{R})$ is such that $\ell^*(S) < \infty$, then show that

(a) $S \in \Im^*$ if and only if $\lambda^*(E) = \ell^*(S)$ for some $E \in \Im^*$ such that $E \subseteq S$,

(b) $S \in \Im^*$ if and only if $\ell^*(S \backslash E) = 0$ for some $E \in \Im^*$ such that $E \subseteq S$.

Hint: For the nontrivial part of (a): $\ell^*(S \backslash E) = \ell^*(S) - \lambda(E) = 0$ by Problem 8.19(b); use Problem 8.9(a) to conclude that $S = (S \backslash E) \cup E \in \Im^*$.

Problem 8.21. Prove that if $E \in \Im^*$, $\lambda^*(E) < \infty$, and $S \subseteq E$, then

$$S \in \Im^* \quad \text{if and only if} \quad \lambda^*(E) = \ell^*(S) + \ell^*(E\backslash S).$$

This is a special application of the Carathéodory condition for measurability. It gives a necessary and sufficient condition for a subset of a Lebesgue measurable set of finite measure to be Lebesgue measurable.

Hint: If $S \in \Im^*$, then the claimed equation follows from the Carathéodory condition. Conversely, note that $E\backslash S \subseteq G$ with $\ell^*(E\backslash S) = \lambda(G)$ for some $G \in \Re$ by Problem 8.8(c). Thus show that $\ell^*(E\backslash S) \leq \lambda^*(E \cap G) \leq \lambda(G) = \ell^*(E\backslash S)$, and hence $\lambda^*(E) = \lambda^*(E \cap G) + \lambda^*(E\backslash G) = \ell^*(E\backslash S) + \lambda^*(E\backslash G)$. If the claimed equation holds, then $\lambda^*(E\backslash G) = \ell^*(S) < \infty$, since $\lambda^*(E) < \infty$. Verify that $E\backslash G \subseteq S$, apply Problem 8.20(a), and conclude that $S \in \Im^*$.

Problem 8.22. Prove the following proposition. Every measurable $E \in \Im^*$ with $0 < \lambda^*(E) < \infty$ has a nonmeasurable partition, A and B in $\wp(\mathbb{R})\backslash\Im^*$ with $A \cup B = E$ and $A \cap B = \varnothing$, such that

$$\lambda^*(E) = \lambda^*(A \cup B) < \ell^*(A) + \ell^*(B).$$

Hint: According to Problem 8.18, there is a nonmeasurable $A \subset E$. Thus $\{A, B\}$ with $B = E\backslash A$ is a nonmeasurable partition of E so that, by subadditivity, $\lambda^*(E) \leq \ell^*(A) + \ell^*(B)$. The inequality is strict by Problem 8.21.

Problem 8.23. Now prove the converse. Take any nonmeasurable set with finite outer measure, say $A \in \wp(\mathbb{R})\backslash\Im^*$ with $\ell^*(A) < \infty$, and let $G \in \Re$ be such that $A \subset G$ and $\ell^*(A) = \lambda(G)$ as in Problem 8.8(c). Show that $B = G\backslash A \in \wp(\Re)\backslash\Im^*$ is such that $\ell^*(B) > 0$ (by Problem 8.9), and hence

$$\lambda(G) = \lambda(A \cup B) < \ell^*(A) + \ell^*(B).$$

Suggested Reading

Bartle [4], Cohn [10], Halmos [18], Royden [35]. See also [29, Part Two].

9

Product Measures

9.1 Construction of Product Measure

Let $X \times Y$ denote the *Cartesian product* of two sets X and Y, which is the set of all ordered pairs (x, y) where $x \in X$ and $y \in Y$. Consider two measure spaces $(X, \mathcal{X}, \mu)$ and $(Y, \mathcal{Y}, \nu)$. In this section we construct a σ-algebra of subsets of the Cartesian product $X \times Y$, denoted by $\mathcal{X} \times \mathcal{Y}$, which is induced by the σ-algebras $\mathcal{X}$ and $\mathcal{Y}$, such that a measure π on $\mathcal{X} \times \mathcal{Y}$ is given by the product of the measures μ on $\mathcal{X}$ and ν on $\mathcal{Y}$. Since we will be dealing with the product of measures, we must consider the problem of defining the product "zero times infinity" because these are possible values for extended real-valued measures. Therefore we declare again (see Sections 1.3 and 3.1) that $0 \cdot +\infty = +\infty \cdot 0 = 0$.

Consider an arbitrary pair of subsets of X and Y, $A \subseteq X$ and $B \subseteq Y$. The Cartesian product $A \times B \subseteq X \times Y$ is called a *a rectangle* from $X \times Y$. If $(X, \mathcal{X})$ and $(Y, \mathcal{Y})$ are measurable spaces, and if E and F are measurable subsets of X and Y, respectively (i.e., if $E \in \mathcal{X}$ and $F \in \mathcal{Y}$), then $E \times F$ is referred to as a *measurable rectangle* from $X \times Y$ (i.e., a rectangle $E \times F$ from $X \times Y$ consisting of an $\mathcal{X}$-measurable set $E \subseteq X$ and a $\mathcal{Y}$-measurable set $F \subseteq Y$). Let $\mathcal{X} \times \mathcal{Y}$ denote the σ-algebra generated by the measurable rectangles from $X \times Y$ (i.e., the smallest σ-algebra of subsets of $X \times Y$ containing all measurable rectangles $E \times F$ with $E \in \mathcal{X}$ and $F \in \mathcal{Y}$), and consider the measurable space $(X \times Y,\ \mathcal{X} \times \mathcal{Y})$, which is referred to as the *Cartesian*

C.S. Kubrusly, *Essentials of Measure Theory*,
DOI 10.1007/978-3-319-22506-7_9

product of the measurable spaces $(X,\mathcal{X})$ and $(Y,\mathcal{Y})$. In particular, it is clear that $E\times F$ is a measurable rectangle from $X\times Y$ if and only if $E\times F\in\mathcal{X}\times\mathcal{Y}$.

Example 9A. Take the Borel algebra $\mathfrak{R}$ of subsets of $\mathbb{R}$ and consider the Cartesian product $(\mathbb{R}\times\mathbb{R},\mathfrak{R}\times\mathfrak{R})$ of two copies of the same measurable space $(\mathbb{R},\mathfrak{R})$. Note that open subsets of $\mathbb{R}^2=\mathbb{R}\times\mathbb{R}$ lie in $\mathfrak{R}\times\mathfrak{R}$. In fact, since $\mathbb{R}^2$ is a separable space, it follows by Problem 1.14 that open sets of $\mathbb{R}^2$ are countably covered by the topological base consisting of open rectangles made up of open intervals. Indeed, it can be shown that the σ-algebra $\mathfrak{R}\times\mathfrak{R}$ coincides with the σ-algebra generated by the open sets of $\mathbb{R}^2$, and so $\mathfrak{R}\times\mathfrak{R}$ is the *Borel algebra* of subsets of $\mathbb{R}^2$ (see the remark that follows Problem 1.14).

Proposition 9.1. *The collection $\mathcal{P}$ of all finite unions of measurable rectangles is an algebra of subsets of $X\times Y$, included in $\mathcal{X}\times\mathcal{Y}$.*

Proof. Every finite union of measurable rectangles admits a disjointification (cf. Hints to Problems 2.3 and 7.2) consisting of measurable rectangles. Thus finite unions of sets in $\mathcal{P}$ are in $\mathcal{P}$. It is readily verified that the complement of a measurable rectangle lies in $\mathcal{P}$, and that a finite intersection of sets in $\mathcal{P}$ is a set in $\mathcal{P}$. Applying De Morgan Laws we conclude that the complement of a set in $\mathcal{P}$ lies in $\mathcal{P}$. So $\mathcal{P}$ is an algebra. It is clear that $\mathcal{P}$ is included in the σ-algebra $\mathcal{X}\times\mathcal{Y}$ generated by measurable rectangles. □

As we have observed in the previous proof, any set in $\mathcal{P}$ admits a disjointification consisting of measurable rectangles. This means that any set in $\mathcal{P}$ can be expressed as a *finite* union of *disjoint* measurable rectangles.

Definition 9.2. Take an arbitrary set $P=\bigcup_i E_i\times F_i$ in $\mathcal{P}$, where $\{E_i\times F_i\}$ is an arbitrary *finite* partition of P consisting of measurable rectangles from $X\times Y$. Let $\mu\colon\mathcal{X}\to\overline{\mathbb{R}}$ and $\nu\colon\mathcal{Y}\to\overline{\mathbb{R}}$ be measures on $\mathcal{X}$ and $\mathcal{Y}$, respectively, and define a set function $\varpi\colon\mathcal{P}\to\overline{\mathbb{R}}$ by the following *finite* sum: for $P\in\mathcal{P}$,

$$\text{(a)}\qquad \varpi(P)=\varpi\Big(\bigcup\nolimits_i E_i\times F_i\Big)=\sum\nolimits_i\mu(E_i)\,\nu(F_i)=\sum\nolimits_i\varpi(E_i\times F_i).$$

In particular, for each measurable rectangle $E\times F$ from $X\times Y$,

$$\text{(b)}\qquad \varpi(E\times F)=\mu(E)\,\nu(F).$$

In (a) we have three identities. The first identity just reminds us that $\{E_i\times F_i\}$ is a partition of P, the second identity is the definition of the set function ϖ on $\mathcal{P}$, and the third identity is a consequence of the second, according to the particular case in (b). By additivity of the measures μ and ν, it is easy to verify that the sums in (a) are "partition invariant": they remain the same for every finite partition of P made up of measurable rectangles. We show that the set function ϖ is a measure on the algebra $\mathcal{P}$, and then we extend this to a measure on the σ-algebra $\mathcal{X}\times\mathcal{Y}$. But first we need the following auxiliary result to prove that ϖ is countably additive.

Proposition 9.3. *Let $\{E_k\times F_k\}$ be a countable family of disjoint rectangles in $\mathcal{X}\times\mathcal{Y}$, and let $A\times B$ be a rectangle in $\mathcal{X}\times\mathcal{Y}$. Then*

(a) $$A\times B = \bigcup_k E_k\times F_k \quad \text{implies} \quad \varpi(A\times B) = \sum_k \varpi(E_k\times F_k).$$

If $\{A_i\times B_i\}$ is a finite *set of disjoint rectangles in $\mathcal{X}\times\mathcal{Y}$, then*

(b) $$\bigcup_i A_i\times B_i = \bigcup_k E_k\times F_k \quad \text{implies} \quad \varpi\Big(\bigcup_i A_i\times B_i\Big) = \sum_k \varpi(E_k\times F_k).$$

Proof. Let $A\times B$ be any measurable rectangle with $A\in\mathcal{X}$ and $B\in\mathcal{Y}$, and let $\{E_k\times F_k\}$ be an arbitrary countable partition of $A\times B$ with E_k in $\mathcal{X}$ and F_k in $\mathcal{Y}$ so that $\{E_k\}$ and $\{F_k\}$ are countable partitions of A and B, respectively. If $\{E_k\times F_k\}$ is a finite partition, then the results in (a) and (b) are trivially verified by Definition 9.2(a). Thus suppose it is countably infinite.

(a) As usual, let χ_S denote the characteristic function of a set S. Then

$$\begin{aligned}\chi_A(x)\,\chi_B(y) = \chi_{A\times B}\big((x,y)\big) &= \sum_k \chi_{E_k\times F_k}\big((x,y)\big)\\ &= \sum_k \chi_{E_k}(x)\,\chi_{F_k}(y),\end{aligned}$$

for every pair $(x,y)\in X\times Y$. (The second identity holds since $\{E_k\times F_k\}$ is a family of disjoint sets that cover $A\times B$.) Fix x, integrate with respect to ν, and apply the Monotone Convergence Theorem (cf. Problem 3.7) to get

$$\chi_A(x)\,\nu(B) = \sum_k \chi_{E_k}(x)\,\nu(F_k).$$

Next, integrate with respect to μ, repeating the same argument, to get

$$\mu(A)\,\nu(B) = \sum_k \mu(E_k)\,\nu(F_k).$$

Thus we get the result in (a) by Definition 9.2(b), namely,

$$\varpi(A\times B) = \sum_k \varpi(E_k\times F_k).$$

(b) The disjointness assumption on both $\{E_k\times F_k\}$ and $\{A_i\times B_i\}$ ensures that if $\bigcup_i A_i\times B_i = \bigcup_k E_k\times B_k$, then $A_i\times B_i = \bigcup_j E_{i,j}\times F_{i,j}$ for each i, where $\{E_{i,j}\times F_{i,j}\} = \{E_k\times F_k\}$. Hence, using Definition 9.2(a) and the result in (a), we get

$$\varpi\Big(\bigcup_k E_k \times F_k\Big) = \varpi\Big(\bigcup_i A_i \times B_i\Big) = \sum_i \varpi(A_i \times B_i)$$
$$= \sum_i \sum_j \varpi(E_{i,j} \times E_{i,j}).$$

If $\varpi(A_i \times B_i) = \infty$ for some i, then $\varpi\big(\bigcup_i A_i \times B_i\big) = \sum_k \varpi(E_k \times F_k) = \infty$ and the result in (b) holds. Suppose $\varpi(A_i \times B_i) < \infty$ for every i. Then the summands $\{\varpi(E_{i,j} \times E_{i,j})\}$ are nonnegative real numbers by (a), and so the doubly indexed sum is unconditionally convergent. Therefore,

$$\sum_i \sum_j \varpi(E_{i,j} \times E_{i,j}) = \sum_k \varpi(E_k \times F_k),$$

and the result in (b) still holds. (Compare with Proposition 8.8.) □

Lemma 9.4. *The set function ϖ is a measure on the algebra $\mathcal{P}$.*

Proof. Observe that $\varpi(\varnothing) = 0$ and $\varpi(P) \geq 0$ for every $P \in \mathcal{P}$, trivially. Consider an arbitrary countable family $\{P_n\}$ of pairwise disjoint sets in $\mathcal{P}$ for which $\bigcup_n P_n$ lies in $\mathcal{P}$. Then, as we have seen before, each set P_n is a finite union of disjoint measurable rectangles,

$$P_n = \bigcup_j E_{n,j} \times F_{n,j}.$$

Since $\{P_n\}$ is a sequence of disjoint sets, this implies that

$$\bigcup_n P_n = \bigcup_{n,j} E_{n,j} \times F_{n,j} = \bigcup_k E_k \times F_k,$$

where $\{E_{n,j} \times F_{n,j}\} = \{E_k \times F_k\}$ is a countable family of disjoint measurable rectangles, and also a finite union of disjoint measurable rectangles,

$$\bigcup_n P_n = \bigcup_i A_i \times B_i.$$

Therefore, $\bigcup_i A_i \times B_i = \bigcup_k E_k \times F_k$. Using Proposition 9.3(b), and recalling the unconditional convergence argument that closed that proof, we get

$$\varpi\Big(\bigcup_n P_n\Big) = \varpi\Big(\bigcup_i A_i \times B_i\Big) = \sum_k \varpi(E_k \times F_k) = \sum_{n,j} \varpi(E_{n,j} \times F_{n,j})$$
$$= \sum_n \sum_j \varpi(E_{n,j} \times F_{n,j}) = \sum_n \varpi\Big(\bigcup_j E_{n,j} \times F_{n,j}\Big) = \sum_n \varpi(P_n)$$

by Definition 9.2(a). This shows that ϖ is countably additive. □

Theorem 9.5. (Product Measure Theorem). *Let $(X, \mathcal{X}, \mu)$ and $(Y, \mathcal{Y}, \nu)$ be measure spaces. Then there exists a measure $\pi\colon \mathcal{X} \times \mathcal{Y} \to \overline{\mathbb{R}}$ such that*

$$\pi(E \times F) = \mu(E)\, \nu(F)$$

for every measurable rectangle $E\times F$ with $E \in \mathcal{X}$ and $F \in \mathcal{Y}$. Furthermore, if μ and ν are σ-finite, then the measure π is unique and σ-finite as well.

Proof. Recall that ϖ is a measure on the algebra $\mathcal{P}$ (Lemma 9.4). Then it follows by the Carathéodory Extension Theorem (Theorem 8.4) that there exists an extension of ϖ to a measure π^* on an σ-algebra $\mathcal{P}^*$ that includes the algebra $\mathcal{P}$. Furthermore, it is readily verified that if μ and ν are σ-finite on $\mathcal{X}$ and $\mathcal{Y}$, then ϖ is σ-finite on $\mathcal{P}$, and so the Hahn Extension Theorem (Theorem 8.6) says that π^* is unique on $\mathcal{P}^*$ and σ-finite. Recall that $\mathcal{X}\times\mathcal{Y}$ is the smallest σ-algebra including $\mathcal{P}$. Let π be the restriction of π^* to $\mathcal{X}\times\mathcal{Y}$, so that π is a σ-finite measure on $\mathcal{X}\times\mathcal{Y}$ (since π^* is σ-finite), which (by the uniqueness of π^*) must be the extension of ϖ over $\mathcal{X}\times\mathcal{Y}$. Summing up:

$$\mathcal{P} \subseteq \mathcal{X}\times\mathcal{Y} \subseteq \mathcal{P}^*, \qquad \pi = \pi^*|_{\mathcal{X}\times\mathcal{Y}}, \qquad \text{and} \qquad \varpi = \pi|_{\mathcal{P}} = \pi^*|_{\mathcal{P}}.$$

Observe from Definition 9.2(b) that $\varpi(E\times F) = \mu(E)\,\nu(F)$ for each measurable rectangle $E\times F$. Since $\varpi = \pi|_{\mathcal{P}}$, it follows that $\varpi(E\times F) = \pi(E\times F)$, and therefore the measure $\pi\colon \mathcal{X}\times\mathcal{Y} \to \mathbb{R}$ is such that

$$\pi(E\times F) = \mu(E)\,\nu(F) \quad \text{for every} \quad E\times F \in \mathcal{X}\times\mathcal{Y}. \qquad \square$$

The value of π at each Cartesian product $E\times F$ in $\mathcal{X}\times\mathcal{Y}$ is the product of the values of μ at E in $\mathcal{X}$ and ν at F in $\mathcal{Y}$. This motivates the notation

$$\pi = \mu \times \nu,$$

which is referred to as the *product measure* (or as the *product* of the measures μ and ν). Accordingly, $(X\times Y,\ \mathcal{X}\times\mathcal{Y},\ \mu\times\nu)$ is the (Cartesian) *product space* of the measure spaces $(X, \mathcal{X}, \mu)$ and $(Y, \mathcal{Y}, \nu)$.

9.2 Sections of Sets and Functions

Let S be an arbitrary subset of the Cartesian product $X\times Y$. Associated to each point $x \in X$, consider the set

$$S_x = \{y \in Y\colon\ (x, y) \in S\},$$

which is called the *x-section* of S. Similarly, for each $y \in Y$, consider the set

$$S^y = \{x \in X\colon\ (x, y) \in S\},$$

which is called the *y-section* of S. The reason for this notation with subscript and superscript is to distinguish x-sections (subsets of Y for each $x \in X$) from y-sections (subsets of X for each $y \in Y$). Observe that sections are not "slices", which means that $S_x \subseteq Y$ (or $S^y \subseteq X$) and $\{x\}\times S_x \subseteq X\times Y$ (or $S^y\times\{y\} \subseteq X\times Y$) are, in general, different sets. Also note that sections of a rectangle $A\times B \subseteq X\times Y$ are either empty or "sides" of the rectangle:

$$(A\times B)_x = \begin{cases} B, & x \in A, \\ \varnothing, & x \notin A, \end{cases} \qquad (A\times B)^y = \begin{cases} A, & y \in B, \\ \varnothing, & y \notin B. \end{cases}$$

In particular,

$$(X\times Y)_x = Y \quad \text{and} \quad (X\times Y)^y = X.$$

Now let $f\colon X\times Y \to \overline{\mathbb{R}}$ be an arbitrary extended real-valued function on $X\times Y$. For each $x \in X$ consider the function $f_x\colon Y \to \overline{\mathbb{R}}$ defined by

$$f_x(y) = f(x,y) \quad \text{for every} \quad y \in (X\times Y)_x = Y,$$

which is called the *x-section* of f. Similarly, for each $y \in Y$ consider the function $f^y\colon X \to \overline{\mathbb{R}}$ defined by

$$f^y(x) = f(x,y) \quad \text{for every} \quad x \in (X\times Y)^y = X,$$

which is called the *y-section* of f. If $f\colon A\times B \to \overline{\mathbb{R}}$ is defined on a rectangle $A\times B \subseteq X\times Y$, then its x-sections and y-sections are defined on $B \subseteq Y$ and $A \subseteq X$, respectively: $f_x\colon B \to \overline{\mathbb{R}}$ and $f^y\colon A \to \overline{\mathbb{R}}$.

Proposition 9.6. *Every section of a measurable set is measurable:*

(a) *$E \in \mathcal{X}\times\mathcal{Y}$ implies $E^y \in \mathcal{X}$ and $E_x \in \mathcal{Y}$ for every $x \in X$ and $y \in Y$.* (i.e., *if $E \in X\times Y$ is $\mathcal{X}\times\mathcal{Y}$-measurable, then $E^y \in X$ is $\mathcal{X}$-measurable and $E_x \in Y$ is $\mathcal{Y}$-measurable*).

Every section of a measurable function is measurable:

(b) *If $f\colon X\times Y \to \overline{\mathbb{R}}$ is $\mathcal{X}\times\mathcal{Y}$-measurable, then $f^y\colon X \to \overline{\mathbb{R}}$ is $\mathcal{X}$-measurable and $f_x\colon Y \to \overline{\mathbb{R}}$ is $\mathcal{Y}$-measurable for every $x \in X$ and $y \in Y$.*

Proof. Take a pair of measurable spaces $(X,\mathcal{X})$ and $(Y,\mathcal{Y})$, and consider their Cartesian product $(X\times Y,\ \mathcal{X}\times\mathcal{Y})$.

(a) The assertion in (a) is an immediate consequence of Problem 9.5. In fact, $\mathcal{X}\times\mathcal{Y}$ is included in the collection $(\mathcal{X}\times\mathcal{Y})_X$ of all subsets of $X\times Y$ for which all x-sections are $\mathcal{Y}$-measurable. Hence, if $E \in \mathcal{X}\times\mathcal{Y}$, then $E_x \in \mathcal{Y}$. Similarly, $E \in \mathcal{X}\times\mathcal{Y}$ also implies $E^y \in \mathcal{X}$. (See Problem 9.5.)

(b) To prove assertion (b) proceed as follows. Take any $\alpha \in \mathbb{R}$. If a function $f\colon X\times Y \to \overline{\mathbb{R}}$ is $\mathcal{X}\times\mathcal{Y}$-measurable, then the set $\{(x,y) \in X\times Y\colon f(x,y) > \alpha\}$ is $\mathcal{X}\times\mathcal{Y}$-measurable, and so every x-section $\{(x,y) \in X\times Y\colon f(x,y) > \alpha\}_x$ is $\mathcal{Y}$-measurable by item (a). Note that for each $x \in X$,

$$\begin{aligned}\{(x,y)\in X\times Y\colon f(x,y)>\alpha\}_x &= \{y\in Y\colon f(x,y)>\alpha\}\\ &= \{y\in Y\colon f_x(y)>\alpha\}.\end{aligned}$$

Therefore the set $\{y\in Y\colon f_x(y)>\alpha\}$ is $\mathcal{Y}$-measurable. This means that the function $f_x\colon Y\to\overline{\mathbb{R}}$ is measurable. Similarly, using the same argument, the function $f^y\colon X\to\overline{\mathbb{R}}$ is measurable. □

Next we apply the Monotone Class Lemma (cf. Problems 1.18 and 1.19) to prove an important result that will play a crucial role in Section 9.3.

Lemma 9.7. *Let $(X,\mathcal{X},\mu)$ and $(Y,\mathcal{Y},\nu)$ be measure spaces. For each measurable set E in $\mathcal{X}\times\mathcal{Y}$ consider the nonnegative functions $f_E\colon X\to\overline{\mathbb{R}}$ and $g_E\colon Y\to\overline{\mathbb{R}}$ defined by*

$$f_E(x)=\nu(E_x)\quad\text{and}\quad g_E(y)=\mu(E^y)$$

for every $x\in X$ and every $y\in Y$, respectively. If μ and ν are σ-finite measures, then $f_E\colon X\to\overline{\mathbb{R}}$ and $g_E\colon Y\to\overline{\mathbb{R}}$ are measurable functions such that

$$\int_X f_E\,d\mu=\pi(E)=\int_Y g_E\,d\nu.$$

Proof. Consider the collection $\mathcal{K}$ of all sets E in $\mathcal{X}\times\mathcal{Y}$ such that

$$\Big\{f_E\in\mathcal{M}(X,\mathcal{X})^+,\;\; g_E\in\mathcal{M}(Y,\mathcal{Y})^+,\;\;\text{and}\;\; \textstyle\int_X f_E\,d\mu=\pi(E)=\int_Y g_E\,d\nu\Big\}.$$

Thus $\mathcal{K}$ is the subcollection of $\mathcal{X}\times\mathcal{Y}$ for which the conclusion of the lemma holds true. We split the proof into two parts. In part (a) we apply the Monotone Class Lemma to prove that if μ and ν are finite measures, then

$$\mathcal{K}=\mathcal{X}\times\mathcal{Y}$$

(i.e., the stated assertion holds true for finite measures). In part (b) we apply the Monotone Convergence Theorem to extend it to σ-finite measures.

(a) Let $A\times B$ in $\mathcal{X}\times\mathcal{Y}$ be any measurable rectangle. Recall that x-sections $(A\times B)_x$ are B if $x\in A$ and empty otherwise, and y-sections $(A\times B)^y$ are A if $y\in B$ and empty otherwise. Thus, for each $x\in X$ and each $y\in Y$ set

$$\begin{aligned}f_{A\times B}(x)&=\nu((A\times B)_x)=\nu(B)\,\chi_A(x),\\ g_{A\times B}(y)&=\mu((A\times B)^y)=\mu(A)\,\chi_B(y).\end{aligned}$$

If μ and ν are finite measure, then these define *real-valued* functions on X and on Y such that (cf. Example 1B and Proposition 1.5) $f_{A\times B}=\nu(B)\,\chi_A$ is in $\mathcal{M}(X,\mathcal{X})^+$, $g_{A\times B}=\mu(A)\,\chi_B$ is in $\mathcal{M}(Y,\mathcal{Y})^+$, and (cf. Problem 3.3.(a))

$$\int_X f_{A\times B}\,d\mu = \nu(B)\,\mu(A) = \pi(A\times B) = \mu(A)\,\nu(B) = \int_Y g_{A\times B}\,d\nu.$$

Consider the algebra $\mathcal{P}$ of Proposition 9.1. Let P be an arbitrary set in $\mathcal{P}$. Since $P = \bigcup_{i=1}^n A_i\times B_i$ is a finite union of *disjoint* measurable rectangles $\{A_i\times B_i\}$, the above results ensure that $P \in \mathcal{K}$, and so $\mathcal{P} \subseteq \mathcal{K}$. Indeed,

$$\begin{aligned} f_P(x) &= \nu(P_x) = \nu\big((\textstyle\bigcup_{i=1}^n A_i\times B_i)_x\big) = \nu\big(\textstyle\bigcup_{i=1}^n (A_i\times B_i)_x\big) \\ &= \textstyle\sum_{i=1}^n \nu((A_i\times B_i)_x) = \sum_{i=1}^n \nu(B_i)\,\chi_{A_i}(x) = \sum_{i=1}^n f_{A_i\times B_i}(x), \end{aligned}$$

$$\begin{aligned} g_P(y) &= \mu(P^y) = \mu\big((\textstyle\bigcup_{i=1}^n A_i\times B_i)^y\big) = \mu\big(\textstyle\bigcup_{i=1}^n (A_i\times B_i)^y\big) \\ &= \textstyle\sum_{i=1}^n \mu((A_i\times B_i)^y) = \sum_{i=1}^n \mu(A_i)\,\chi_{B_i}(y) = \sum_{i=1}^n g_{A_i\times B_i}(y), \end{aligned}$$

for every $x \in X$ and $y \in Y$. Hence (cf. Proposition 1.5) $f_P = \sum_{i=1}^n f_{A_i\times B_i}$ is in $\mathcal{M}(X,\mathcal{X})^+$, $g_P = \sum_{i=1}^n g_{A_i\times B_i}$ is in $\mathcal{M}(Y,\mathcal{Y})^+$, and (cf. Problem 3.7(a))

$$\begin{aligned} \int_X f_P\,d\mu &= \sum_{i=1}^n \int_X f_{A_i\times B_i}\,d\mu \\ &= \sum_{i=1}^n \mu(A_i)\,\nu(B_i) = \sum_{i=1}^n \pi(A_i\times B_i) = \pi\Big(\bigcup_{i=1}^n A_i\times B_i\Big) = \pi(P) \\ &= \sum_{i=1}^n \int_Y g_{A_i\times B_i}\,d\nu = \int_Y g_P\,d\nu. \end{aligned}$$

Therefore, $P \in \mathcal{K}$ and so

(i) $$\mathcal{P} \subseteq \mathcal{K}.$$

First suppose $\{E_n\}$ is an increasing sequence of sets in $\mathcal{K}$. For each n set

$$f_{E_n}(x) = \nu((E_n)_x) \quad\text{and}\quad g_{E_n}(y) = \mu((E_n)^y)$$

for every $x \in X$ and every $y \in Y$. Since $E_n \in \mathcal{K}$, this defines two sequences $\{f_{E_n}\}$ and $\{g_{E_n}\}$ of functions in $\mathcal{M}(X,\mathcal{X})^+$ and in $\mathcal{M}(Y,\mathcal{Y})^+$ such that

$$\int_X f_{E_n}\,d\mu = \pi(E_n) = \int_Y g_{E_n}\,d\nu.$$

But $\{f_{E_n}\}$ and $\{g_{E_n}\}$ are increasing sequences of extended real-valued functions, because $\{E_n\}$ is an increasing sequence of sets. Thus these sequences of functions converge pointwise. Set $E = \bigcup_n E_n$, which lies in the σ-algebra $\mathcal{X}\times\mathcal{Y}$. Since the sequences $\{(E_n)_x\}$ and $\{(E_n)^y\}$ of sections in $\mathcal{Y}$ and $\mathcal{X}$ are increasing with $\bigcup_n (E_n)_x = E_x$ in $\mathcal{Y}$ and $\bigcup_n (E_n)^y = E^y$ in $\mathcal{X}$ for each $x \in X$ and each $y \in Y$, it follows by Proposition 2.2(c) that the functions $f_E\colon X \to \overline{\mathbb{R}}$ and $g_E\colon Y \to \overline{\mathbb{R}}$ are the pointwise limits of $\{f_{E_n}\}$ and $\{g_{E_n}\}$:

$$\lim_n f_{E_n}(x) = \lim_n \nu((E_n)_x) = \nu\Big(\bigcup\nolimits_n (E_n)_x\Big) = \nu(E_x) = f_E(x),$$

$$\lim_n g_{E_n}(y) = \lim_n \mu((E_n)^y) = \mu\Big(\bigcup\nolimits_n (E_n)^y\Big) = \mu(E^y) = g_E(y),$$

for every $x \in X$ and $y \in Y$. Still by Proposition 2.2.(c) we get

$$\lim_n \pi(E_n) = \pi\Big(\bigcup\nolimits_n E_n\Big) = \pi(E).$$

Note from Proposition 1.8 that $f_E \in \mathcal{M}(X, \mathcal{X})^+$ and $g_E \in \mathcal{M}(Y, \mathcal{Y})^+$. Use the Monotone Convergence Theorem (Theorem 3.4) to conclude that

$$\int_X f_E \, d\mu = \lim_n \int_X f_{E_n} \, d\mu = \pi(E) = \lim_n \int_Y g_{E_n} \, d\nu = \int_Y g_E \, d\nu.$$

This implies that

$$E \in \mathcal{K}.$$

Next suppose $\{F_n\}$ be a decreasing sequence of sets in $\mathcal{K}$ and set $F = \bigcap_n F_n$ in $\mathcal{X}\times\mathcal{Y}$. Proceeding as before, consider the similarly defined functions f_{F_n} in $\mathcal{M}(X, \mathcal{X})^+$ and g_{F_n} in $\mathcal{M}(Y, \mathcal{Y})^+$ such that

$$\int_X f_{F_n} \, d\mu = \pi(F_n) = \int_Y g_{F_n} \, d\nu$$

for each n. Still under the assumption that μ and ν are finite measures, which implies that the product measure $\pi = \mu\times\nu$ is finite as well, we may apply Proposition 2.2(d) for the finite measure π (instead of Proposition 2.2(c) for general measures) to verify that $\{f_{F_n}\}$ and $\{g_{F_n}\}$ are both decreasing sequences of real-valued functions so that they converge pointwise, and

$$\lim_n f_{F_n}(x) = f_F(x) \quad \text{and} \quad \lim_n g_{F_n}(y) = g_F(y)$$

for every $x \in X$ and $y \in Y$, which define the real-valued limit functions $f_F \in \mathcal{M}(X, \mathcal{X})^+$ and $g_F \in \mathcal{M}(Y, \mathcal{Y})^+$. Proposition 2.2(d) also ensures that

$$\lim_n \pi(F_n) = \pi\Big(\bigcap\nolimits_n F_n\Big) = \pi(F).$$

Recall that $\{f_{F_n}\}$ and $\{g_{F_n}\}$ are decreasing sequences of nonnegative real-valued measurable functions, and that $\int f_{F_1} \, d\mu$ and $\int g_{F_1} \, d\nu$ are bounded by $\pi(X\times Y) = \mu(X)\,\nu(Y)$, which is finite since $\mu(X) < \infty$ and $\nu(Y) < \infty$, it follows by the Dominated Convergence Theorem (Theorem 4.7) that

$$\int_X f_F \, d\mu = \pi(E) = \int_Y g_F \, d\nu.$$

This implies that

$$F \in \mathcal{K}.$$

Hence (cf. Problem 1.15), under the finite measure assumption,

(ii) $\mathcal{K}$ is a monotone class.

According to (i) and (ii) the Monotone Class Lemma (see Problems 1.18 and 1.19) ensures that if μ and ν are finite measures, then

$$\mathcal{K} = \mathcal{X}\times\mathcal{Y}.$$

(b) If the measures μ and ν are σ-finite, then that there is a pair of *increasing* sequences $\{X_n\}$ and $\{Y_n\}$ of $\mathcal{X}$-measurable and $\mathcal{Y}$-measurable sets covering X and Y, respectively, such that $\mu(X_n) < \infty$ and $\nu(Y_n) < \infty$. For each n consider the σ-algebras $\mathcal{X}_n = \wp(X_n) \cap \mathcal{X}$ and $\mathcal{Y}_n = \wp(Y_n) \cap \mathcal{Y}$ so that μ and ν are finite measures when restricted to them. Let E be an arbitrary set in $\mathcal{X}\times\mathcal{Y}$. For each n set $E_n = E \cap (X_n \times Y_n)$ in $\mathcal{X}_n\times\mathcal{Y}_n$. Hence

$$(E_n)_x = \big(E \cap (X_n\times Y_n)\big)_x = E_x \cap (X_n\times Y_n)_x = E_x \cap Y_n \in \mathcal{Y}_n,$$

$$(E_n)^y = \big(E \cap (X_n\times Y_n)\big)^y = E^y \cap (X_n\times Y_n)^y = E^y \cap X_n \in \mathcal{X}_n.$$

Take the functions $f_{E_n}\colon X \to \overline{\mathbb{R}}$ and $g_{E_n}\colon Y \to \overline{\mathbb{R}}$ defined for each n by

$$f_{E_n}(x) = \nu\big((E_n)_x\big) = \nu\big((E_n)_x\big)\chi_{X_n},$$

$$g_{E_n}(x) = \mu\big((E_n)^y\big) = \mu\big((E_n)^y\big)\chi_{Y_n},$$

for every $x \in X$ and $y \in Y$. Let $\mu_n = \mu|_{X_n}$ and $\nu_n = \nu|_{Y_n}$ be the restrictions of μ and ν to $\mathcal{X}_n$ and $\mathcal{Y}_n$ so that $(X_n, \mathcal{X}_n, \mu_n)$ and $(Y_n, \mathcal{Y}_n, \nu_n)$ are finite measure spaces. Since the stated assertion holds for finite measures (as we saw in item (a)), it follows that $f_{E_n} \in \mathcal{M}(X, \mathcal{X})^+$, $g_{E_n} \in \mathcal{M}(Y, \mathcal{Y})^+$, and

$$\int_X f_{E_n}\, d\mu = \int_{X_n} f_{E_n}\, d\mu_n = \pi(E_n) = \int_{Y_n} g_{E_n}\, d\nu_n = \int_Y g_{E_n}\, d\nu.$$

Since $\{X_n\}$ and $\{Y_n\}$ are increasing sequences of $\mathcal{X}$-measurable and $\mathcal{Y}$-measurable sets covering X and Y, conclude that $\{E_n\}$, $\{(E_n)_x\}$, and $\{(E_n)^y\}$ are increasing sequences of $\mathcal{X}\times\mathcal{Y}$-measurable, $\mathcal{Y}$-measurable and $\mathcal{X}$-measurable sets that cover E, E_x, and E^y, respectively, so $E = \bigcup_n E_n$, $E_x = \bigcup_n (E_n)_x$, and $E^y = \bigcup_n (E_n)^y$. Then Proposition 2.2(c) ensures that $\lim_n \pi(E_n) = \pi\big(\bigcup_n E_n\big) = \pi(E)$ and

$$\lim_n f_{E_n}(x) = \lim_n \nu\big((E_n)_x\big) = \nu\Big(\bigcup\nolimits_n (E_n)_x\Big) = \nu(E_x) = f_E(x),$$

$$\lim_n g_{E_n}(y) = \lim_n \mu\big((E_n)^y\big) = \mu\Big(\bigcup\nolimits_n (E_n)^y\Big) = \mu(E^y) = g_E(y),$$

for every $x \in X$ and $y \in Y$. But $\{f_{E_n}\}$ and $\{g_{E_n}\}$ are increasing sequences (since $\{X_n\}$ and $\{Y_n\}$ are increasing) of nonnegative measurable functions (according to item (a)), so that by the Monotone Convergence Theorem (Theorem 3.4) we get $f_E \in \mathcal{M}(X, \mathcal{X})^+$, $g_E \in \mathcal{M}(Y, \mathcal{Y})^+$, and

$$\int_X f_E \, d\mu = \lim_n \int_X f_{E_n} \, d\mu = \pi(E) = \lim_n \int_Y g_{E_n} \, d\nu = \int_Y g_E \, d\nu. \qquad \square$$

9.3 Fubini and Tonelli Theorems

The following two theorems give sufficient conditions for interchanging the order of integration. The first deals with extended real-valued nonnegative functions. The second dismisses nonnegativeness but assumes integrability.

Theorem 9.8. (Tonelli Theorem). *Consider two measure spaces $(X, \mathcal{X}, \mu)$ and $(Y, \mathcal{Y}, \nu)$. If μ and ν are σ-finite and $h \in \mathcal{M}(X{\times}Y, \mathcal{X}{\times}\mathcal{Y})^+$, then the extended real-valued functions f_h and g_h defined by*

$$f_h(x) = \int_Y h_x \, d\nu \quad \text{and} \quad g_h(y) = \int_X h^y \, d\mu$$

are in $\mathcal{M}(X, \mathcal{X})^+$ and in $\mathcal{M}(Y, \mathcal{Y})^+$, respectively, and

$$\int_X f_h \, d\mu = \int_{X\times Y} h \, d\pi = \int_Y g_h \, d\nu.$$

Proof. Let $\mathcal{H}$ be the set of all functions $h \in \mathcal{M}(X{\times}Y, \mathcal{X}{\times}\mathcal{Y})^+$ such that

$$\Big\{ f_h \in \mathcal{M}(X, \mathcal{X})^+,\; g_h \in \mathcal{M}(Y, \mathcal{Y})^+, \text{ and } \textstyle\int_X f_h \, d\mu = \int_{X\times Y} h \, d\pi = \int_Y g_h \, d\nu \Big\}.$$

In other words, $\mathcal{H}$ is the subcollection of $\mathcal{M}(X{\times}Y, \mathcal{X}{\times}\mathcal{Y})^+$ for which the conclusion of the theorem holds true. The program is to prove that

$$\mathcal{H} = \mathcal{M}(X{\times}Y, \mathcal{X}{\times}\mathcal{Y})^+.$$

Equivalently, to prove that $\mathcal{M}(X{\times}Y, \mathcal{X}{\times}\mathcal{Y})^+ \subseteq \mathcal{H}$. Let χ_E be the characteristic function of E in $\mathcal{X}{\times}\mathcal{Y}$. Recall that $\chi_E \in \mathcal{M}(X{\times}Y, \mathcal{X}{\times}\mathcal{Y})^+$ (Example 1B). Note that $(\chi_E)_x = \chi_{E_x}$ and $(\chi_E)^y = \chi_{E^y}$ (Problem 9.8). Set

$$f_{\chi_E}(x) = \int_Y (\chi_E)_x \, d\nu = \int_Y \chi_{E_x} \, d\nu = \nu(E_x),$$

$$g_{\chi_E}(y) = \int_X (\chi_E)^y \, d\mu = \int_X \chi_{E^y} \, d\mu = \mu(E^y),$$

for every $x \in X$ and $y \in Y$. Since μ and ν are σ-finite measures, Lemma 9.7 ensures that $f_{\chi_E} \in \mathcal{M}(X,\mathcal{X})^+$, $g_{\chi_E} \in \mathcal{M}(Y,\mathcal{Y})^+$, and

$$\int_X f_{\chi_E}\,d\mu = \int_{X\times Y} \chi_E\,d\pi = \int_Y g_{\chi_E}\,d\nu.$$

Hence every characteristic function of sets in $\mathcal{X}\times\mathcal{Y}$ lies in $\mathcal{H}$. Therefore, by additivity and positive homogeneity in the set of positive measurable functions (Proposition 1.9), and also for the integral itself (Proposition 3.5(a,b)), we can conclude that every simple function (Definition 3.1) lies in $\mathcal{H}$. Now let h be an arbitrary function in $\mathcal{M}(X\times Y,\, \mathcal{X}\times\mathcal{Y})^+$. Problem 1.6 says that there exists an increasing sequence $\{\varphi_n\}$ of simple functions in $\mathcal{M}(X\times Y,\, \mathcal{X}\times\mathcal{Y})^+$ converging pointwise to h. For each n and for every $x \in X$ and $y \in Y$, set

$$f_{\varphi_n}(x) = \int_Y (\varphi_n)_x\,d\nu \quad\text{and}\quad g_{\varphi_n}(y) = \int_X (\varphi_n)^y\,d\mu. \qquad (*)$$

Since $\varphi_n \in \mathcal{H}$ (simple functions lie in $\mathcal{H}$), the above identities define a pair of functions f_{φ_n} and g_{φ_n} such that $f_{\varphi_n} \in \mathcal{M}(X,\mathcal{X})^+$, $g_{\varphi_n} \in \mathcal{M}(Y,\mathcal{Y})^+$, and

$$\int_X f_{\varphi_n}\,d\mu = \int_{X\times Y} \varphi_n\,d\pi = \int_Y g_{\varphi_n}\,d\nu. \qquad (**)$$

Since $\varphi_n \to h$, it follows that $(\varphi_n)_x \to h_x$ and $(\varphi_n)^y \to h^y$ for every $x \in X$ and $y \in Y$, where these convergences are pointwise. Since $\{\varphi_n\}$ is an increasing sequence of nonnegative functions, it is clear that the sequences $\{(\varphi_n)_x\}$ and $\{(\varphi_n)^y\}$ are also increasing and consist of nonnegative functions for each $x \in X$ and $y \in Y$. Furthermore, since φ_n is measurable, then so are all sections $(\varphi_n)_x$ and $(\varphi_n)^y$, for each n, by Proposition 9.6. Thus, applying the Monotone Convergence Theorem (Theorem 3.4) we get from $(*)$ that

$$\lim_n f_{\varphi_n}(x) = \lim_n \int_Y (\varphi_n)_x\,d\nu = \int_Y \lim_n\,(\varphi_n)_x\,d\nu = \int_Y h_x\,d\nu = f_h(x),$$

$$\lim_n g_{\varphi_n}(y) = \lim_n \int_X (\varphi_n)^y\,d\mu = \int_X \lim_n\,(\varphi_n)^y\,d\mu = \int_X h^y\,d\mu = g_h(y),$$

for every $x \in X$ and every $y \in Y$, so that $f_{\varphi_n} \to f_h$ and $g_{\varphi_n} \to g_h$ pointwise. Recall that $\{f_{\varphi_n}\}$ and $\{g_{\varphi_n}\}$ are increasing sequences (because $\{\varphi_n\}$ is increasing). Using the Monotone Convergence Theorem once again we get from $(**)$ that $f_h \in \mathcal{M}(X,\mathcal{X})^+$, $g_h \in \mathcal{M}(Y,\mathcal{Y})^+$, and

$$\int_X f_h\,d\mu = \int_{X\times Y} h\,d\pi = \int_Y g_h\,d\nu,$$

and hence $h \in \mathcal{H}$. Then $\mathcal{M}(X\times Y,\, \mathcal{X}\times\mathcal{Y})^+ \subseteq \mathcal{H}$. □

The conclusion of the Tonelli Theorem can be rewritten as

$$\int_X \Big(\int_Y h\,d\nu\Big)\,d\mu = \int_{X\times Y} h\,d\pi = \int_Y \Big(\int_X h\,d\mu\Big)\,d\nu.$$

This is a significant result. It says that *the order of the integrals can be interchanged*. The same applies to the next theorem, which has exactly the same conclusion but a different hypothesis, where the function h is not necessarily nonnegative but should be integrable with respect to the product measure.

Theorem 9.9. (Fubini Theorem) *Consider two measure spaces* $(X,\mathcal{X},\mu)$ *and* $(Y,\mathcal{Y},\nu)$. *If* λ *and* ν *are* σ*-finite measures and* $h \in \mathcal{L}(X{\times}Y,\ \mathcal{X}{\times}\mathcal{Y},\ \pi)$, *then there are real-valued functions* f_h *and* g_h *defined a.e. on* X *and* Y *by*

$$f_h(x) = \int_Y h_x\,d\nu \quad\text{and}\quad g_h(y) = \int_X h^y\,d\mu,$$

which are in $\mathcal{L}(X,\mathcal{X},\mu)$ *and in* $\mathcal{L}(Y,\mathcal{Y},\nu)$, *respectively, and*

$$\int_X f_h\,d\mu = \int_{X\times Y} h\,d\pi = \int_Y g_h\,d\nu.$$

Proof. Take an arbitrary h in $\mathcal{M}(X{\times}Y,\ \mathcal{X}{\times}\mathcal{Y})$, and consider its positive and negative parts h^+ and h^- in $\mathcal{M}(X{\times}Y,\ \mathcal{X}{\times}\mathcal{Y})^+$ such that $h = h^+ - h^-$ (Proposition 1.6), and also its x-section h_x in $\mathcal{M}(Y,\mathcal{Y})$ and y-section h^y in $\mathcal{M}(X,\mathcal{X})$ (Proposition 9.6). Now consider the parts of the sections $(h_x)^\pm$ in $\mathcal{M}(Y,\mathcal{Y})^+$ and $(h^y)^\pm$ in $\mathcal{M}(X,\mathcal{X})^+$, which coincide with the sections of the parts $(h^\pm)_x$ and $(h^\pm)^y$ (Problem 9.9). Since the positive and negative parts h^+ and h^- lie in $\mathcal{M}(X{\times}Y,\ \mathcal{X}{\times}\mathcal{Y})^+$, since $(h^\pm)_x = (h_x)^\pm$ and $(h^\pm)^y = (h^y)^\pm$, and since the measures μ and ν are σ-finite, the Tonelli Theorem ensures that the functions $f_{h^\pm}$ and $g_{h^\pm}$ defined on X and Y by

$$f_{h^\pm}(x) = \int_Y (h_x)^\pm\,d\nu \quad\text{and}\quad g_{h^\pm}(y) = \int_X (h^y)^\pm\,d\mu \tag{$*$}$$

are in $\mathcal{M}(X,\mathcal{X})^+$ and in $\mathcal{M}(Y,\mathcal{Y})^+$, respectively, and

$$\int_X f_{h^\pm}\,d\mu = \int_{X\times Y} h^\pm\,d\pi = \int_Y g_{h^\pm}\,d\nu. \tag{$**$}$$

In addition, if $h \in \mathcal{L}(X{\times}Y,\ \mathcal{X}{\times}\mathcal{Y},\ \pi)$, then the parts h^+ and h^- are nonnegative functions in $\mathcal{L}(X{\times}Y,\ \mathcal{X}{\times}\mathcal{Y},\ \pi)$ by Definition 4.1. Even though their sections are real-valued, the nonnegative measurable functions $f_{h^\pm}$ and $g_{h^\pm}$ are not necessarily real-valued (cf. Problem 9.11) but have finite integrals according to $(**)$. Then they are real-valued almost everywhere by Problem 3.9(b), and the differences $f_{h^+}(x) - f_{h^-}(x)$ and $g_{h^+}(y) - g_{h^-}(y)$ are defined almost everywhere with respect to μ and ν. Thus take real-valued functions f_h and g_h defined almost everywhere on X and Y, respectively, as follows.

$$f_h = f_{h^+} - f_{h^-} \quad \text{and} \quad g_h = g_{h^+} - g_{h^-}$$

and zero otherwise, which lie in $\mathcal{L}(X, \mathcal{X}, \mu)$ and $\mathcal{L}(Y, \mathcal{Y}, \nu)$. In fact (Problems 3.8 an 3.9), there is an $\mathcal{X}$-measurable set N with $\mu(N) = 0$ such that $\int_X f_{h^\pm}|_{X\backslash N}\, d\mu = \int_X f_{h^\pm}\chi_{X\backslash N}\, d\mu = \int_{X\backslash N} f_{h^\pm}\, d\mu = \int_X f_{h^\pm}\, d\mu < \infty$ for which $f_{h^\pm}|_{X\backslash N}$ is *real-valued*, and hence $f_{h^\pm}|_{X\backslash N}$ lies in $\mathcal{L}(X, \mathcal{X}, \mu)$. Thus the function f_h defined by $f_{h^+}|_{X\backslash N} - f_{h^-}|_{X\backslash N}$ on $X\backslash N$ and zero on N is real-valued and lies in $\mathcal{L}(X, \mathcal{X}, \mu)$ — see Lemma 4.5. Similarly, the function g_h is real-valued and lies in $\mathcal{L}(Y, \mathcal{Y}, \nu)$. Then, by (∗) and Definition 4.1,

$$f_h(x) = \int_Y (h_x)^+\, d\nu - \int_Y (h_x)^-\, d\nu = \int_Y h_x\, d\nu,$$

$$g_h(y) = \int_X (h^y)^+\, d\nu - \int_X (h^y)^-\, d\nu = \int_X h^y\, d\mu,$$

for μ-almost every x in X and ν-almost every y in Y. Also, by (∗∗) and applying Definition 4.1 again, and since $f_h = f_{h^+} - f_{h^-}$ on $X\backslash N$ so that $\int_X f_h\, d\mu = \int_{X\backslash N} f_{h^+}\, d\mu - \int_{X\backslash N} f_{h^-}\, d\mu$ (see Proposition 4.3), it follows that

$$\int_{X\times Y} h\, d\pi = \int_{X\times Y} h^+\, d\pi - \int_{X\times Y} h^-\, d\pi = \int_X f_{h^+}\, d\mu - \int_X f_{h^-}\, d\mu = \int_X f_h\, d\mu,$$

and similarly

$$\int_{X\times Y} h\, d\pi = \int_Y g_{h^+}\, d\mu - \int_Y g_{h^-}\, d\mu = \int_Y g_h\, d\mu. \qquad \square$$

Suppose N is an $\mathcal{X}$-measurable set such that $\mu(N) = 0$ for which $f_h(x) = \int_Y h_x(y)d\nu$ on $X\backslash N$ and zero on N (as in the preceding proof). Observe that $\int_X f_h(x)d\mu = \int_{X\backslash N} f_h(x)d\mu = \int_{X\backslash N}\big(\int_Y h_x(y)d\nu\big)d\mu = \int_X\big(\int_Y h(x,y)d\nu\big)d\mu$. Similarly, $\int_Y g_h(y)d\nu = \int_Y\big(\int_X h(x,y)d\mu\big)d\nu$. Thus, as we have commented before, the conclusion of the Fubini Theorem can also be rewritten as

$$\int_X\left(\int_Y h\, d\nu\right)d\mu = \int_{X\times Y} h\, d\pi = \int_Y\left(\int_X h\, d\mu\right)d\nu.$$

The middle integral, $\int_{X\times Y} h\, d\pi$, is the *double integral* of h. The left and right integrals, $\int_X\big(\int_Y h\, d\nu\big)d\mu$ and $\int_Y\big(\int_X h\, d\mu\big)d\nu$, are the *iterated integrals* of h.

9.4 Problems

Problem 9.1. Let $\mathfrak{R}\times\mathfrak{R}$ be the Borel algebra of subsets of $\mathbb{R}^2$ (Example 9A). If $E \in \mathfrak{R}$, then show that the *difference set* D_E is $\mathfrak{R}\times\mathfrak{R}$-measurable: that is,

$$D_E = \big\{(x,y) \in \mathbb{R}^2 \colon\, y - x \in E\big\} \;\in\; \mathfrak{R}\times\mathfrak{R}.$$

Hint:

(a) The "smart" way: $f(x,y) = y - x$ defines a continuous function f from $\mathbb{R}^2$ to $\mathbb{R}$, thus a measurable function (Section 11.1), and $D_E = f^{-1}(E)$.

(b) The "tour de force" way: Take $E \in \mathfrak{R}$ so that $E \cup \{\beta\} + \alpha \in \mathfrak{R}$ for every $\alpha, \beta \in \mathbb{R}$. If E is unbounded, then consider a countable partition of E made up of bounded measurable sets. If E is bounded, then write E as a countable union of *measurable triangles* as follows. Let $\operatorname{diam} E = \sup_{x,y\in E} |x - y| < \infty$ be the *diameter* of the bounded set $E \subseteq \mathbb{R}$. For each integer $k \in \mathbb{Z}$ set

$$E_k = (E \cup \{\sup E\}) + k \operatorname{diam} E,$$

$$F_k = E_k + \inf E = \big[(E \cup \{\sup E\}) + \inf E\big] + k \operatorname{diam} E,$$

which are subsets of $\mathbb{R}$. Now consider the following subsets of $\mathbb{R}^2$:

$$L = \big\{(x,y) \in \mathbb{R}^2 \colon\, y \le \sup E + x\big\} \quad \text{or} \quad L = \big\{(x,y) \in \mathbb{R}^2 \colon\, y < \sup E + x\big\}$$

whether $\sup E$ lies or does not lie in E,

$$U = \big\{(x,y) \in \mathbb{R}^2 \colon\, y \ge \inf E + x\big\} \quad \text{or} \quad U = \big\{(x,y) \in \mathbb{R}^2 \colon\, y > \inf E + x\big\}$$

whether $\inf E$ lies or does not lie in E, and

$$\Delta_k = (E_k \times F_k) \cap U \quad \text{and} \quad \nabla_k = (E_k \times F_{k+1}) \cap L$$

for each $k \in \mathbb{Z}$. Verify that the above *triangles* cover the "strip-shape" set D_E (as suggested in the sketch below) (Fig. 9.1):

$$D_E = \big\{(x,y) \in \mathbb{R}^2 \colon\, y - x \in E\big\} = \bigcup\nolimits_k (\Delta_k \cup \nabla_k).$$

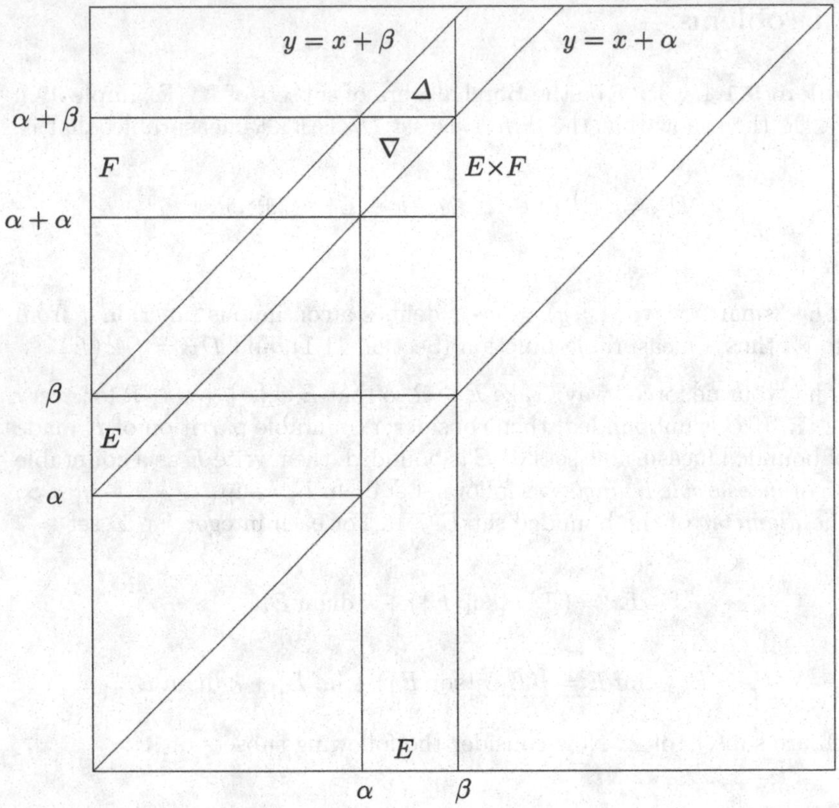

Fig. 9.1. Problem 9.1(b)

Problem 9.2. If $f : \mathbb{R} \to \mathbb{R}$ is $\mathfrak{R}$-measurable, then the function $h : \mathbb{R}^2 \to \mathbb{R}$ given by $h(x, y) = f(y - x)$ for (x, y) in $\mathbb{R}^2$ is $\mathfrak{R} \times \mathfrak{R}$-measurable. Prove it.

Hint: See Problem 9.1.

Problem 9.3. Consider the Cartesian product $(X \times Y, \mathcal{X} \times \mathcal{Y})$ of the measurable spaces $(X, \mathcal{X})$ and $(Y, \mathcal{Y})$. If $f \in \mathcal{M}(X, \mathcal{X})$ and $g \in \mathcal{M}(Y, \mathcal{Y})$ are real-valued functions, then show that the real-valued function h on $X \times Y$ defined by $h(x, y) = f(x)\, g(y)$ for every (x, y) in $X \times Y$ lies in $\mathcal{M}(X \times Y, \mathcal{X} \times \mathcal{Y})$.

Problem 9.4. Take an arbitrary subset S of the Cartesian product $X \times Y$ of two sets X and Y. Let $\{S_\alpha\}$ be an arbitrary collection of subsets of $X \times Y$, and take any x in X. Use the definition of x-section to show that

(a) $((X\times Y)\backslash S))_x = (X\times Y)_x\backslash S_x = Y\backslash S_x$,

(b) $(\bigcup_\alpha S_\alpha)_x = \bigcup_\alpha (S_\alpha)_x$.

Problem 9.5. Let $(X,\mathcal{X})$ and $(Y,\mathcal{Y})$ be measurable spaces, and take the measurable space $(X\times Y,\ \mathcal{X}\times\mathcal{Y})$ of their Cartesian product. Consider the collections of all subsets of $X\times Y$ for which all x-sections are $\mathcal{Y}$-measurable,

$$(\mathcal{X}\times\mathcal{Y})_X = \big\{E\subseteq X\times Y\colon\ E_x\in\mathcal{Y} \text{ for every } x\in X\big\},$$

and of all subsets of $X\times Y$ for which all y-sections are $\mathcal{X}$-measurable,

$$(\mathcal{X}\times\mathcal{Y})_Y = \big\{E\subseteq X\times Y\colon\ E^y\in\mathcal{X} \text{ for every } y\in Y\big\}.$$

Show that $(\mathcal{X}\times\mathcal{Y})_X$ and $(\mathcal{X}\times\mathcal{Y})_Y$ are σ-algebras of subsets of $X\times Y$ (use Problem 9.4), and that they contain all measurable rectangles (recall that sections of rectangles are either empty or sides of them), and conclude that

$$\mathcal{X}\times\mathcal{Y}\subseteq(\mathcal{X}\times\mathcal{Y})_X \quad\text{and}\quad \mathcal{X}\times\mathcal{Y}\subseteq(\mathcal{X}\times\mathcal{Y})_Y.$$

Problem 9.6. Consider the unit interval $X=[0,1]\subset\mathbb{R}$ and let $\mathcal{X}$ be the collection of all subsets of X that are either countable or are the complement of a countable set. Show that $\mathcal{X}$ is a σ-algebra of subsets of X and take the measurable space $(X\times X,\ \mathcal{X}\times\mathcal{X})$ of the Cartesian product of two copies of the measurable space $(X,\mathcal{X})$. For each $\alpha>0$ consider the *line segments*

$$I_\alpha = \big\{(x,y)\in X\times X\colon\ y=\alpha x\big\}.$$

Show that every I_α is nonmeasurable (i.e., $I_\alpha\notin\mathcal{X}\times\mathcal{X}$ for every $\alpha>0$).

Hint: Let $I_0=\{(x,y)\in X\times X\colon\ y=0\}=[0,1]\times\{0\}$ be the horizontal line segment. Take an arbitrary line segment I_α distinct from I_0 and consider the intersection of their complements, which is the *sector*

$$(X\backslash I_0)\cap(X\backslash I_\alpha) = \big\{(x,y)\in X\times X\colon\ 0<y<\alpha x\big\}.$$

Show that such a sector is not $\mathcal{X}\times\mathcal{X}$-measurable (use Proposition 9.6(a), or Problem 9.5). Also show that I_0 is $\mathcal{X}\times\mathcal{X}$-measurable. If I_α is $\mathcal{X}\times\mathcal{X}$-measurable, then verify that the intersection of their complements is measurable, which is a contradiction. Thus I_α is not $\mathcal{X}\times\mathcal{X}$-measurable.

Problem 9.7. Use Proposition 9.6(a) to prove the failure of its own converse: *There exist nonmeasurable sets for which all sections are measurable.*

Hint: See Problem 9.6.

Problem 9.8. Consider the characteristic function χ_S of a subset S of the Cartesian product $X\times Y$ of two sets X and Y. Show that for each $x \in X$ and each $y \in Y$, the sections $(\chi_S)_x$ and $(\chi_S)^y$ of the function χ_S are the characteristic functions of the sections S_x and S^y of the set S. That is,

$$(\chi_S)_x = \chi_{S_x} \quad \text{and} \quad (\chi_S)^y = \chi_{S^y}.$$

Thus conclude that if $A\times B$ is any rectangle from $X\times Y$, then

$$\chi_{A\times B}(x,y) = \chi_A(x)\chi_B(y) \quad \text{for every} \quad (x,y) \in X\times Y.$$

Problem 9.9. Let $f\colon X\times Y \colon\to \overline{\mathbb{R}}$ be an arbitrary extended real-valued function on the Cartesian Product $X\times Y$. Take its positive and negative parts, f^+ and f^-, and its x-sections and y-sections, f_x and f^y. First show that

$$(f^+)_x = (f_x)^+,\ (f^-)_x = (f_x)^- \quad \text{and} \quad (f^+)^y = (f^y)^+,\ (f^-)^y = (f^y)^-.$$

Hint: Recall that $f^+ = f\chi_{F^+}$, where $F^+ = \{(x,y) \in X\times Y\colon f(x,y) \ge 0\}$, and verify that $(f^+)_x = (f\chi_{F^+})_x = f_x(\chi_{F^+})_x = f_x\chi_{F^+} = (f_x)^+$.

Next show that the x-sections and y-sections of $f = f^+ - f^-$ are given by

$$f_x = (f^+)_x - (f^-)_x \quad \text{and} \quad f^y = (f^+)^y - (f^-)^y.$$

Problem 9.10. Consider the Borel algebra $\Re$ (or the Lebesgue algebra $\Im^*$) and take the Lebesgue measure λ on $\Re$ (or on $\Im^*$), which is referred to as the *linear Lebesgue measure* or as the *length* on $\mathbb{R}$. Consider the product of λ with itself; that is, the measure $\pi = \lambda\times\lambda$ on the σ-algebra $\Re\times\Re$ (or on $\Im^*\times\Im^*$) of subsets of the plane $\mathbb{R}^2 = \mathbb{R}\times\mathbb{R}$. This measure is referred to as the *planar Lebesgue measure* or as the *area* on $\mathbb{R}^2$. Note that $\Re\times\Re \subset \Im^*\times\Im^*$, so $\Re\times\Re$-measurable sets are $\Im^*\times\Im^*$-measurable. Thus the area of a measurable rectangle $E\times F$ in $\mathbb{R}^2$ is $\lambda(E)\lambda(F)$. Give an example of an uncountable measurable subset of $[0,1]\times[0,1]$ with area zero such that all sections of it (x and y-sections) are either empty or uncountable with length zero.

Problem 9.11. Take the Lebesgue measure space $(\mathbb{R}, \Re, \lambda)$ and the product space $(\mathbb{R}^2, \Re\times\Re, \pi)$, where $\pi = \lambda\times\lambda$ is the area on $\mathbb{R}^2$ as in Problem 9.10. Give an example of a real-valued function $f\colon \mathbb{R}^2 \to \mathbb{R}$ satisfying (1) and (2).

(1) $f = 0$ π-almost everywhere.

That is, $f = 0$ up to a measurable set of area zero, and so verify that this f is $\Re\times\Re$-measurable, and apply Proposition 4.2(a) to infer that this f is also integrable with $\int_{\mathbb{R}^2} f\, d\pi = 0$. Moreover,

(2) $\int_{\mathbb{R}} f_x\, d\lambda = \infty$ for some x-section $f_x\colon \mathbb{R} \to \mathbb{R}$ of f.

Problem 9.12. Set $E = [0,1] \subset \mathbb{R}$. Let $\mathcal{E} = \wp(E) \cap \mathfrak{R}$ be the σ-algebra of all Borel subsets of $[0,1]$. Take the Lebesgue measure λ on $\mathcal{E}$ (i.e., the restriction of the Lebesgue measure to $\mathcal{E}$ as in Problem 2.11). Let μ be the counting measure on $\mathcal{E}$ as in Problem 2.4(b), which is not σ-finite. Consider the measure spaces $(E, \mathcal{E}, \lambda)$ and $(E, \mathcal{E}, \mu)$. Apply Example 9A to verify that the *identity segment*, or the *diagonal set*, of the rectangle $[0,1]\times[0,1]$, viz., the set $I = \{(x,y) \in E\times E \colon x = y\}$, is $\mathcal{E}\times\mathcal{E}$-measurable, and show that

$$\int \mu(I_x)\, d\lambda \neq \int \lambda(I^y)\, d\mu.$$

Hence the assumption of σ-finiteness cannot be omitted from Lemma 9.7.

Problem 9.13. Consider the setup of the previous problem. Use the characteristic function $\chi_I \in \mathcal{M}(E\times E, \mathcal{E}\times\mathcal{E}, \lambda\times\mu)^+$ of the set I to show that the assumption of σ-finiteness cannot be omitted from the Tonelli Theorem.

Problem 9.14. Let $(X, \mathcal{X}, \mu)$ and $(Y, \mathcal{Y}, \nu)$ be σ-finite measure spaces. Take their product space $(X\times Y, \mathcal{X}\times\mathcal{Y}, \pi)$. If E and F are sets in $\mathcal{X}\times\mathcal{Y}$ such that $\nu(E_x) = \nu(F_x)$ for (almost) every $x \in X$ (or $\mu(E^y) = \mu(F^y)$ for (almost) every $y \in Y$), then show that $\pi(E) = \pi(F)$.

Hint: Apply the Tonelli Theorem to the characteristic functions χ_E and χ_F of the $\mathcal{X}\times\mathcal{Y}$-measurable sets E and F. Use the results of Problem 9.8.

Problem 9.15. Take a pair $f, g \in \mathcal{L}(\mathbb{R}, \mathfrak{R}, \lambda)$ of Lebesgue integrable functions. Define a function $h\colon \mathbb{R}^2 \to \mathbb{R}$ by $h(x,y) = f(x-y)$ for every (x,y) in $\mathbb{R}^2$. According to Problem 9.2, h is $\mathfrak{R}\times\mathfrak{R}$-measurable. Verify that the function $hg\colon \mathbb{R}^2 \to \mathbb{R}$ mapping each $(x,y) \in \mathbb{R}^2$ into $f(x-y)g(y) \in \mathbb{R}$ is also $\mathfrak{R}\times\mathfrak{R}$-measurable. Now apply the Fubini Theorem to show that there is a real-valued function $f * g \in \mathcal{L}(\mathbb{R}, \mathfrak{R}, \lambda)$ defined almost everywhere on $\mathbb{R}$ by

$$(f*g)(x) = \int_{\mathbb{R}} h_x\, dy = \int_{\mathbb{R}} f(x-y)g(y)\, dy.$$

This function $f*g\colon \mathbb{R} \to \mathbb{R}$ is the *convolution* of f and g. Also show that

$$\int |f*g|\, dx \leq \left(\int |f|\, dx\right)\left(\int |g|\, dy\right).$$

Problem 9.16. Let $\{a_{m,n}\}$ be a countable family of *nonnegative* real numbers, doubly indexed by $(m,n) \in \mathbb{N}\times\mathbb{N}$.

(a) Use elementary analysis to show that

$$\sum_{m=1}^{\infty}\sum_{n=1}^{\infty} a_{m,n} = \sum_{n=1}^{\infty}\sum_{m=1}^{\infty} a_{m,n}.$$

Hint: Recall that *a family of real numbers is absolutely summable if and only if it is unconditionally summable.*

Take the product space $(\mathbb{N}\times\mathbb{N},\, \wp(\mathbb{N}\times\mathbb{N}),\, \mu\times\mu)$ of two copies of the measure space $(\mathbb{N}, \wp(\mathbb{N}), \mu)$, where μ is the counting measure of Example 2B. Consider the integral of the function $\alpha_{m,n}\colon \mathbb{N}\times\mathbb{N} \to \mathbb{R}$ with respect to the product measure $\pi = \mu\times\mu$ (see Problem 3.4). That is, for every $E \in \wp(\mathbb{N}\times\mathbb{N})$,

$$\int_E \alpha_{m,n}\, d\pi = \sum\nolimits_E \alpha_{m,n}.$$

(b) If $\sum_{\mathbb{N}\times\mathbb{N}} \alpha_{m,n} < \infty$ (i.e., under the assumption of finite integral), then show that the result in (a) can be proved by using the Fubini Theorem.

Problem 9.17. Consider the setup of the previous problem, but now suppose $\{\alpha_{m,n}\}$ is an arbitrary countable family of real numbers (not necessarily nonnegative). Suppose $\alpha_{m,m} = 1$ and $\alpha_{m,m+1} = -1$ for every integer $m \in \mathbb{N}$, and $\alpha_{m,n} = 0$ otherwise. Show that

$$\sum_{m=1}^{\infty}\sum_{n=1}^{\infty} \alpha_{m,n} = 0 \quad \text{and} \quad \sum_{n=1}^{\infty}\sum_{m=1}^{\infty} \alpha_{m,n} = 1.$$

Conclude that the nonnegativeness assumption (i.e., $\alpha_{m,n} \geq 0$ for all (m, n) in $\mathbb{N}\times\mathbb{N}$) cannot the dropped in the Tonelli Theorem, and integrability (i.e., $\sum_{\mathbb{N}\times\mathbb{N}} |\alpha_{m,n}| < \infty$) cannot be omitted from the Fubini Theorem.

Hint:

$$\{\alpha_{m,n}\} = \begin{pmatrix} 1 & -1 & 0 & & \\ 0 & 1 & -1 & 0 & \\ & 0 & 1 & -1 & \\ & & 0 & 1 & \\ & & & & \ddots \end{pmatrix}.$$

Problem 9.18. Let $\big([0,1]\times[0,1],\ \wp([0,1]\times[0,1]) \cap (\Re\times\Re),\ \lambda\times\lambda\big)$ be the product space obtained by two copies of the finite Lebesgue measure space $\big([0,1],\ \wp([0,1]) \cap \Re,\ \lambda\big)$. Consider the functions $f \in \mathcal{L}\big([0,1],\ \wp([0,1]) \cap \Re,\ \lambda\big)$ and the real-valued $g \in \mathcal{M}\big([0,1],\ \wp([0,1]) \cap \Re\big)^+$ given by

$$f(x) = \begin{cases} 1, & 0 < x \leq \frac{1}{2}, \\ -1, & \frac{1}{2} < x \leq 1, \end{cases} \qquad g(y) = \begin{cases} 0, & y = 0, \\ \frac{1}{y}, & y \in (0,1]. \end{cases}$$

Let the function $h \in \mathcal{M}\big([0,1]\times[0,1],\ \wp([0,1]\times[0,1]) \cap (\Re\times\Re)\big)$ be given by

$$h(x, y) = f(x)\, g(y)$$

for every $(x, y) \in [0,1]\times[0,1]$, which is real-valued (Problem 9.3). Show that

$$\int_0^1 \left(\int_0^1 h(x,y)dx \right) dy = 0.$$

What is the value of the other iterated integral of h? Is h integrable?

Problem 9.19. Recall that an integrable function f (i.e., f in $\mathcal{L}(X,\mathcal{X},\mu)$) is a *real-valued* measurable function with a *finite integral* (i.e., $\int f^{\pm}\, d\mu < \infty$; equivalently, $\int |f|\, d\mu < \infty$ — see Lemma 4.4). Theorem 9.9 assumes that h is integrable and concludes that there are integrable functions f_h and g_h such that $\int_X f_h\, d\mu = \int_{X\times Y} h\, d\pi = \int_Y g_h\, d\nu$, where μ and ν are σ-finite measures. As integrable functions, these h, f_h, and g_h are real-valued. Show that if μ and ν are σ-finite and if h is an extended real-valued $\mathcal{X}\times\mathcal{Y}$-measurable function with a finite integral, then there are extended real-valued $\mathcal{X}$-measurable and $\mathcal{Y}$-measurable functions f_h and g_h (defined as in Theorem 9.9) with finite integrals such that $\int_X f_h\, d\mu = \int_{X\times Y} h\, d\pi = \int_Y g_h\, d\nu$.

The Fubini Theorem holds if we allow extended real-valued functions but retain the assumption of finite integrals (i.e., $\int_{X\times Y} |h|\, d\pi < \infty$).

Problem 9.20. Take a pair of measure spaces $(X,\mathcal{X},\mu)$ and $(Y,\mathcal{Y},\nu)$ and consider their product space $(X\times Y,\ \mathcal{X}\times\mathcal{Y},\ \pi)$, where $\pi = \mu\times\nu$. Let f and g be real-valued functions on X and Y, and let h be the real-valued function on $X\times Y$ defined by $h(x,y) = f(x)\, g(y)$ for every (x,y) in $X\times Y$. Show that if $f \in \mathcal{L}(X,\mathcal{X},\mu)$ and $g \in \mathcal{L}(Y,\mathcal{Y},\nu)$, then $h \in \mathcal{L}(X\times Y,\ \mathcal{X}\times\mathcal{Y},\ \pi)$ and

$$\int_{X\times Y} h\, d\pi = \left(\int_X f\, d\mu \right) \left(\int_Y g\, d\nu \right)$$

even if the measures μ and ν are not σ-finite.

Hint: Take the *kernels* of the f and g, namely, $\mathcal{N}(f) = \{x \in X\colon f(x) = 0\}$ in $\mathcal{X}$ and $\mathcal{N}(g) = \{y \in Y\colon g(y) = 0\}$ in $\mathcal{Y}$ so that $X' = X\backslash\mathcal{N}(f) \in \mathcal{X}$ and $Y' = Y\backslash\mathcal{N}(g) \in \mathcal{Y}$ are σ-finite sets with respect to μ and ν (Problem 3.9(c)). Set $\mathcal{X}' = \wp(X') \cap \mathcal{X}$ and $\mathcal{Y}' = \wp(Y') \cap \mathcal{Y}$, the sub-$\sigma$-algebras of $\mathcal{X}$ and $\mathcal{Y}$ made up of subsets of X' and Y'. Verify that μ and ν (in fact, their restrictions to $\mathcal{X}'$ and $\mathcal{Y}'$) are σ-finite measures on $\mathcal{X}'$ and $\mathcal{Y}'$. Since $h = fg$, get

$$\int_{X\times Y} h(x,y)\, d\pi = \int_{X'\times Y'} f(x)\, g(y)\, d\pi.$$

Apply the Fubini Theorem for σ-finite measures on $\mathcal{X}'$ and $\mathcal{Y}'$.

Suggested Reading

Bartle [4], Bauer [6], Berberian [7], Halmos [18], Lang [28], Royden [35].

Part II

Measures on Topological Spaces

10

Remarks on Integrals

10.1 Positive Measures

This is an introductory chapter to Part II, dealing with basic properties of integrals with respect to a positive, signed, and complex measure that will be required in the sequel. It does not yet deal with measures on topological spaces. We will not equip the underlying set X with a topology in this chapter, but we will do it from the next chapter onwards. However, to avoid trivialities, we assume that the underlying set X is nonempty. By a *positive measure* we simply mean a measure $\mu\colon \mathcal{X} \to \mathbb{R}$ on a σ-algebra $\mathcal{X}$ of subsets of a nonempty set X (so that $\mu(X) \geq 0$). (Sometimes this is used to specify a nonzero measure; that is, a measure μ such that $\mu(X) > 0$, but we allow the zero measure here.) The term *positive measure* is employed just to distinguish it from *signed measure* (also called *real measure*) and *complex measure*. In this section we summarize the basic properties of integrals with respect to a positive measure, as discussed in Chapters 3, 4, and 5. These basic properties will be extended to integrals with respect a signed measure and with respect to a complex measure in the forthcoming sections.

Remark 10.1. Consider a real-valued function $f\colon X \to \mathbb{R}$ on X. If $\mu\colon \mathcal{X} \to \mathbb{R}$ is a *finite* positive measure on a σ-algebra $\mathcal{X}$ of subsets of X, then the characteristic function $\chi_E\colon X \to \mathbb{R}$ is integrable for every measurable set $E \in \mathcal{X}$, and so is the constant function $1\colon X \to \mathbb{R}$ such that $1(x) = 1$ for every $x \in X$ (reason: $\int d\mu = \mu(X) < \infty$). If $f\colon X \to \mathbb{R}$ is an integrable function, then

C.S. Kubrusly, *Essentials of Measure Theory*,
DOI 10.1007/978-3-319-22506-7_10

$$\left|\int_E f\,d\mu\right| \le \int_E |f|\,d\mu \le \sup|f|\,\mu(E)$$

for every $E \in \mathcal{X}$ by Lemma 4.4 and Problem 5.1 (where $\sup|f| \in \overline{\mathbb{R}}$ is defined as usual: $\sup|f| = \sup_{x\in X}|f(x)|$). Therefore,

$$\sup_{|f|\le 1}\left|\int f\,d\mu\right| \le \mu(X) = \int d\mu = \int 1\,d\mu = \left|\int d\mu\right| = \left|\int 1\,d\mu\right|,$$

and so (since $\sup|1| = 1$, trivially),

$$\sup_{|f|\le 1}\left|\int f\,d\mu\right| = \max_{|f|\le 1}\left|\int f\,d\mu\right| = \left|\int 1\,d\mu\right| = \int 1\,d\mu = \mu(X),$$

where the supremum is taken over all real-valued integrable functions f on X such that $|f| \le 1$. The set $\mathcal{L} = \mathcal{L}(X, \mathcal{X}, \mu)$ of all real-valued integrable functions on X with respect to the measure μ is a linear space, and the integral $\int(\cdot)\,d\mu\colon \mathcal{L} \to \mathbb{R}$ is a linear functional (cf. Lemma 4.5). For every positive measure μ, the integral is a *positive functional* in the sense that it takes positive functions to positive numbers:

$$0 \le \int f\,d\mu \quad \text{whenever} \quad 0 \le f \in \mathcal{L}.$$

(i.e., if $f\colon X \to \mathbb{R}$ in $\mathcal{L}(X, \mathcal{X}, \mu)$ is such that $0 \le f$ μ-a.e., then $0 \le \int f\,d\mu$ — cf. Proposition 4.2(a) and Problem 4.4(b)). That $\sup_{|f|\le 1}\left|\int f\,d\mu\right| = \left|\int 1\,d\mu\right|$ (i.e., the supremum is actually attained by the function 1, which lies in $\mathcal{L}$ whenever μ is finite) is a consequence of the positivity of the linear functional $\int(\cdot)\,d\mu\colon \mathcal{L} \to \mathbb{R}$ with respect to a positive measure μ. Another consequence of the positivity of the integral with respect to a positive measure is this (see Problem 4.4(a)). If f and g are functions in $\mathcal{L}$, then

$$f \le g \quad \text{implies} \quad \int f\,d\mu \le \int g\,d\mu.$$

(In fact, since the integral is a positive functional, $0 \le \int(g - f)\,d\mu$ whenever $f \le g$ μ-a.e., and since it is linear, $\int f\,d\mu \le \int g\,d\mu$.)

Remark 10.2. Consider a complex-valued function $f\colon X \to \mathbb{C}$ on X and its Cartesian decomposition $f = f_1 + i\,f_2$, where $f_1 = \operatorname{Re} f\colon X \to \mathbb{R}$ and $f_2 = \operatorname{Im} f\colon X \to \mathbb{R}$ are real-valued functions on X (the *real* and *imaginary parts* of f). A complex-valued function f is measurable if f_1 and f_2 are measurable (Problem 1.7); and f is integrable with respect to a positive measure $\mu\colon \mathcal{X} \to \mathbb{R}$ if f_1 and f_2 are integrable with respect to μ, and its integral is

$$\int f\,d\mu = \int f_1\,d\mu + i\int f_2\,d\mu,$$

where f is integrable if and only if $|f|$ is (Problem 4.7), and for every $E \in \mathcal{X}$,

$$\left|\int_E f\,d\mu\right| \le \int_E |f|\,d\mu \le \sup|f|\,\mu(E).$$

The proof of the first inequality in the above expression requires a little care (cf. Hint to Problem 4.7). However, as a consequence, we get the same result obtained for real-valued functions, namely,

$$\sup_{|f|\le 1}\left|\int f\,d\mu\right| = \left|\int 1\,d\mu\right| = \int 1\,d\mu = \mu(X),$$

where the supremum is taken over all complex-valued integrable functions f on X such that $|f| \le 1$. The supremum is actually a maximum, attained by the real-valued function $1\colon X \to \mathbb{R}$, which is integrable whenever μ is finite, and this is a consequence of the positivity of the integral with respect to a positive measure μ (when applied to a real-valued positive function).

10.2 Real Measures

If $\lambda\colon \mathcal{X} \to \mathbb{R}$ and $\mu\colon \mathcal{X} \to \mathbb{R}$ are *finite* positive measures on a σ-algebra $\mathcal{X}$ of subsets of a set X, then consider the (finite real-valued) signed measure $\nu\colon \mathcal{X} \to \mathbb{R}$ on $\mathcal{X}$ defined by $\nu = \lambda - \mu$, and recall that $\nu = \nu^+ - \nu^-$, where $\nu^+\colon \mathcal{X} \to \mathbb{R}$ and $\nu^-\colon \mathcal{X} \to \mathbb{R}$ are finite positive measures on $\mathcal{X}$ (singular to each other), referred to as the positive and negative variation of ν, which are such that $\nu^+ \le \lambda$ and $\nu^- \le \mu$. The total variation of ν is the finite (positive) measure $|\nu|\colon \mathcal{X} \to \mathbb{R}$ defined by $|\nu| = \nu^+ + \nu^-$ (Example 7.A).

Remark 10.3. A real-valued function $f\colon X \to \mathbb{R}$ on X is integrable with respect to a signed measure $\nu = \lambda - \mu$ for arbitrary finite positive measures λ and μ, if it is integrable with respect to the positive measures λ and μ (and so if it is integrable with respect to the positive measures ν^+ and ν^-, since $\nu^+ \le \lambda$ and $\nu^- \le \mu$, so that $\int|f|\,d\nu^+ \le \int|f|\,d\lambda$ and $\int|f|\,d\nu^- \le \int|f|\,d\mu$; cf. Problem 3.3(d) and Lemma 4.4(a)). The integral of $f\colon X \to \mathbb{R}$ with respect to a signed measure ν is unambiguously defined by

$$\int f\,d\nu = \int f\,d\lambda - \int f\,d\mu = \int f\,d\nu^+ - \int f\,d\nu^-.$$

Note that $\chi_E\colon X \to \mathbb{R}$ is integrable for every $E \in \mathcal{X}$ since ν is finite, and so

$$\nu(E) = \nu^+(E) - \nu^-(E) = \int_E d\nu^+ - \int_E d\nu^- = \int_E d\nu.$$

Equivalently,

$$\nu(E) = \lambda(E) - \mu(E) = \int_E d\lambda - \int_E d\mu = \int_E d\nu.$$

This ensures that the integral with respect to a signed measure ν is equivalently defined as the difference of the integrals with respect to the positive measures ν^+ and ν^-, or with respect to the positive measures λ and μ. Similarly, the integral with respect to the positive measure $|\nu|$ is defined by

$$\int f\,d|\nu| = \int f\,d\nu^+ + \int f\,d\nu^-.$$

Again, set $\sup|f| = \sup_{x\in X}|f(x)|$. Thus (cf. Lemma 4.4), for every $E \in \mathcal{X}$,

$$\left|\int_E f\,d\nu\right| \le \int_E |f|\,d\nu^+ + \int_E |f|\,d\nu^- = \int_E |f|\,d|\nu| \le \sup|f|\,|\nu|(E),$$

so that

$$\sup_{|f|\le 1}\left|\int f\,d\nu\right| \le |\nu|(X),$$

where the supremum is taken over all real-valued integrable functions f on X such that $|f| \le 1$. Since ν^+ and ν^- are singular, there is a measurable partition $\{A^+, A^-\}$ of X such that $\nu^+(A^-) = \nu^-(A^+) = 0$, $\nu^+(A^+) = \nu^+(X)$, and $\nu^-(A^-) = \nu^-(X)$ (e.g., any Hahn decomposition of X with respect ν — see Section 7.1). Consider the function $\chi_{A^+} - \chi_{A^-}\colon X \to \mathbb{R}$ for which

$$\begin{aligned}\int(\chi_{A^+} - \chi_{A^-})\,d\nu &= \int \chi_{A^+}\,d\nu^+ + \int \chi_{A^-}\,d\nu^- = \int_{A^+} d\nu^+ + \int_{A^-} d\nu^- \\ &= \nu^+(A^+) + \nu^-(A^-) = \nu^+(X) + \nu^-(X) = |\nu|(X).\end{aligned}$$

Therefore, since $|\chi_{A^+} - \chi_{A^-}| = 1$,

$$|\nu|(X) = \int d|\nu| = \int 1\,d\nu^+ + \int 1\,d\nu^- = \int(\chi_{A^+} - \chi_{A^-})\,d\nu \le \sup_{|f|\le 1}\left|\int f\,d\nu\right|,$$

and so the supremum is actually attained by the function $\chi_{A^+} - \chi_{A^-}$, which is a consequence of the positivity of the integrals $\int(\cdot)d\nu^+$ and $\int(\cdot)d\nu^-$ with respect to the positive measures ν^+ and ν^-. Summing up:

$$\sup_{|f|\le 1}\left|\int f\,d\nu\right| = |\nu|(X).$$

Next we show another proof of the above identity. This alternate proof will be required in the sequel. Suppose f is a real-valued integrable function on X such that $\chi_F \le f \le \chi_G$, where χ_F and χ_G are characteristic functions of measurable sets $F, G \in \mathcal{X}$. If $F \subseteq G \subseteq A^+$, then $0 \le \nu(F) = \int \chi_F\,d\nu \le \int f\,d\nu \le \int \chi_G\,d\nu = \nu(G)$; on the other hand, if $F \subseteq G \subseteq A^-$, then $\nu(G) =$

$\int \chi_G \, d\nu \le \int f \, d\nu \le \int \chi_F \, d\nu = \nu(F) \le 0$. Outcome: If either $F \subseteq G \subseteq A^+$, or $F \subseteq G \subseteq A^-$, and $\chi_F \le f \le \chi_G$, then

$$|\nu(F)| \le \left| \int f \, d\nu \right| \le |\nu(G)|.$$

Note: this fails without the assumption that $F \subseteq G$ are subsets of either A^+ or of A^- (e.g., set $F = A^+$, $G = X$, $0 < -\nu(A^-) = \nu(A^+)$, and $f = 1$). However, if $\{A_i\}$ is any measurable covering of X, then consider the measurable covering $\{E_k\}$ of X defined as follows: $\{E_k\} = \{A_i^+\} \cup \{A_i^-\}$, where $A_i^+ = A_i \cap A^+$ and $A_i^- = A_i \cap A^-$ for each index i. Thus $\{E_k\}$ consists of subsets of either A^+ or A^-. So, if $F_k \subseteq G_k \subseteq E_k$ and $\chi_{F_k} \le f_k \le \chi_{G_k}$, then

$$|\nu(F_k)| \le \left| \int f_k \, d\nu \right| \le |\nu(G_k)|$$

for each k. If $\{A_i\}$ is an arbitrary *finite* measurable partition of X, then so is $\{E_k\}$. Take $G_k = E_k$ and set $f_k = \chi_{E_k}$. In this case,

$$\sup \sum\nolimits_k \left| \int \chi_{E_k} \, d\nu \right| = \sup \sum\nolimits_k |\nu(E_k)| = |\nu|(X)$$

according to Examples 2I and 7A, where the supremum is taken over all finite measurable partitions of X. Recalling that $\sum_k \chi_{E_k} = \chi_X = 1$ (and so $\left|\sum_k \chi_{E_k}\right| = 1$) and $0 \le \chi_{E_k} \le 1$, we may infer that

$$\sup \sum\nolimits_k \left| \int \chi_{E_k} \, d\nu \right| \le \sup_{|f| \le 1} \left| \int f \, d\nu \right|,$$

where the supremum on the right-hand side is taken over all real-valued integrable functions f on X such that $|f| \le 1$. Hence

$$|\nu|(X) = \sup \sum\nolimits_k \left| \int \chi_{E_k} \, d\nu \right| \le \sup_{|f| \le 1} \left| \int f \, d\nu \right| \le |\nu|(X).$$

Thus we get another proof that

$$\sup_{|f| \le 1} \left| \int f \, d\nu \right| = |\nu|(X).$$

Remark 10.4. A complex-valued function $f : X \to \mathbb{C}$ on X is integrable with respect to a signed measure $\nu : \mathcal{X} \to \mathbb{R}$ if $f_1 = \operatorname{Re} f$ and $f_2 = \operatorname{Im} f$ are integrable with respect to ν, and the integral of $f = f_1 + i f_2 : X \to \mathbb{C}$ with respect to a signed measure $\nu = \nu^+ - \nu^- : \mathcal{X} \to \mathbb{R}$ is defined by

$$\begin{aligned}
\int f\,d\nu &= \int f_1\,d\nu + i\int f_2\,d\nu \\
&= \int f_1\,d\nu^+ - \int f_1\,d\nu^- + i\left[\int f_2\,d\nu^+ - \int f_2\,d\nu^-\right] \\
&= \int f_1\,d\nu^+ + i\int f_2\,d\nu^+ - \int f_1\,d\nu^- - i\int f_2\,d\nu^- \\
&= \int f\,d\nu^+ - \int f\,d\nu^-,
\end{aligned}$$

so that, by the inequalities in Remark 10.2 and according to the definition of the integral with respect to $|\nu|$ in Remark 10.3 we get, for every $E \in \mathcal{X}$,

$$\begin{aligned}
\left|\int_E f\,d\nu\right| &\leq \left|\int_E f\,d\nu^+\right| + \left|\int_E f\,d\nu^-\right| \\
&\leq \int_E |f|\,d\nu^+ + \int_E |f|\,d\nu^- = \int_E |f|\,d|\nu| \leq \sup|f|\,|\nu|(E).
\end{aligned}$$

By the above inequality we get $\sup_{|f|\leq 1}\left|\int f\,d\nu\right| \leq |\nu|(X)$. However, as we saw in Remark 10.3, such a supremum is attained by the real-valued (as a particular case of a complex-valued) function $\chi_{A^+} - \chi_{A^-}\colon X \to \mathbb{R} \subseteq \mathbb{C}$ for any Hahn decomposition $\{A^+, A^-\}$ of X with respect to ν. Therefore,

$$\sup_{|f|\leq 1}\left|\int f\,d\nu\right| = |\nu|(X),$$

where the supremum is taken over all complex-valued integrable functions f on X such that $|f| \leq 1$.

10.3 Complex Measures

Now recall the definition of complex measure (cf. Problem 2.16): a complex measure $\eta\colon \mathcal{X} \to \mathbb{C}$ is a complex-valued set function on a σ-algebra $\mathcal{X}$ of subsets of a set X such that $\eta = \nu_1 + i\,\nu_2$, where $\nu_1 = \operatorname{Re}\eta\colon \mathcal{X} \to \mathbb{R}$ and $\nu_2 = \operatorname{Im}\eta\colon \mathcal{X} \to \mathbb{R}$, the real and imaginary parts of η, are (finite real-valued) signed measures on $\mathcal{X}$. We say that a positive measure $\mu\colon \mathcal{X} \to \mathbb{R}$ *dominates* a complex measure $\eta\colon \mathcal{X} \to \mathbb{C}$ if $|\eta(E)| \leq \mu(E)$ for every $E \in \mathcal{X}$ (warning: see Example 2G)). Set $\mu = |\nu_1| + |\nu_2|$. It is clear that $\mu\colon \mathcal{X} \to \mathbb{R}$ is a finite positive measure (since it is the sum of finite positive measures), and it is readily verified that μ dominates η. Indeed, $|\eta(E)| \leq |\nu_1(E)| + |\nu_2(E)| \leq |\nu_1|(E) + |\nu_2|(E) = (|\nu_1| + |\nu_2|)(E) < \infty$ for every $E \in \mathcal{X}$ (cf. Examples 2I and 7A). Consider the *total variation* $|\eta|\colon \mathcal{X} \to \mathbb{R}$ of $\eta\colon \mathcal{X} \to \mathbb{C}$ defined along the same line of Example 2I: for each $n \in \mathbb{N}$ let $\boldsymbol{E}(n)$ be the collection of all measurable partitions of $E \in \mathcal{X}$ containing n sets so that $\bigcup \boldsymbol{E}(n)$ is the collection of all *finite* measurable partitions of E. For each $E \in \mathcal{X}$ set

$$|\eta|(E) = \sup_{\{E_j\} \in \bigcup \boldsymbol{E}(n)} \sum_j |\eta(E_j)|.$$

Again, following the same steps of Examples 2I and 7A, we can show that the total variation $|\eta|: \mathcal{X} \to \mathbb{R}$ of the complex measure $\eta: \mathcal{X} \to \mathbb{C}$ is a finite positive measure, which is the least positive measure that dominates η (i.e., if μ dominates η, then $|\eta| \leq \mu$).

Remark 10.5. A real-valued function $f: X \to \mathbb{R}$ on X is integrable with respect to a complex measure $\eta = \nu_1 + i\,\nu_2: \mathcal{X} \to \mathbb{C}$ if it is integrable with respect to both signed measures ν_1 and ν_2, and the integral of an integrable real-valued function $f: X \to \mathbb{R}$ with respect to a complex measure $\eta = \nu_1 + i\,\nu_2$ is defined by

$$\int f\,d\eta = \int f\,d\nu_1 + i\int f\,d\nu_2.$$

Observe that since $\chi_E: X \to \mathbb{R}$ is integrable for each $E \in \mathcal{X}$ because ν_1 and ν_2 are finite, it follows that (see Remark 10.3), for every $E \in \mathcal{X}$,

$$\eta(E) = \nu_1(E) + i\,\nu_2(E) = \int_E d\nu_1 + i\int_E d\nu_2 = \int_E d\eta.$$

Also note that $\big|\int_E f\,d\eta\,\big| \leq \big|\int_E f\,d\nu_1\big| + \big|\int_E f\,d\nu_2\big| \leq \sup|f|\,(|\nu_1| + |\nu_2|)(E)$ for every $E \in \mathcal{X}$. But we can get a tighter inequality as follows. Take an arbitrary $E \in \mathcal{X}$. Let $g: X \to \mathbb{R}$ be a nonnegative measurable function, and consider the collection $\Phi_g^+(E)$ of all positive simple functions $\varphi = \sum_j \alpha_j \chi_{E_j}$, for all finite measurable partitions $\{E_j\} \in \bigcup \boldsymbol{E}(n)$, such that $0 \leq \varphi \leq g$. Note that $\big|\int_E \varphi\,d\eta\,\big| = \big|\sum_j \alpha_j\,\eta(E_j)\,\big| \leq \sum_j \alpha_j\,|\eta(E_j)| \leq \int_E \varphi\,d|\eta|$, and therefore $\big|\int_E g\,d\eta\,\big| = \sup_{\varphi\in\Phi_g(E)}\big|\int_E \varphi\,d\eta\,\big| \leq \sup_{\varphi\in\Phi_g(E)}\int_E \varphi\,d|\eta| = \int_E g\,d|\eta|$. By linearity of the integral, and recalling that $f = f^+ - f^-$ and $|f| = f^+ + f^-$, where f^+ and f^- are nonnegative and measurable, we get

$$\left|\int_E f\,d\eta\right| = \left|\int_E f^+ d\eta - \int_E f^- d\eta\right| \leq \left|\int_E f^+ d\eta\right| + \left|\int_E f^- d\eta\right|$$

$$\leq \int_E f^+ d|\eta| + \int_E f^- d|\eta| = \int_E |f|\,d|\eta| \leq \sup|f|\,|\eta|(E)$$

for every $E \in \mathcal{X}$, so that

$$\left|\int f\,d\eta\right| \leq \sup|f|\,|\eta|(X).$$

Let $\{A_1^+, A_1^-\}$ and $\{A_2^+, A_2^-\}$ be Hahn decompositions of X with respect to the signed measures ν_1 and ν_2, respectively. Consider the collection $\mathcal{A}_{1,2}^{+-} = \{A_1^+ \cap A_2^+,\, A_1^+ \cap A_2^-,\, A_1^- \cap A_2^+,\, A_1^- \cap A_2^-\}$, which is a measurable covering of the nonempty set X. Suppose f is a real-valued integrable function on X such that $\chi_F \leq f \leq \chi_G$, where χ_F and χ_G are characteristic functions of measurable subsets F, G of X. If $F \subseteq G \subseteq A$ for an arbitrary set $A \in \mathcal{A}_{1,2}^{+-}$, then $|\nu_1(F)|^2 + |\nu_2(F)|^2 \leq |\int f\,d\nu_1|^2 + |\int f\,d\nu_2|^2 \leq |\nu_1(G)|^2 + |\nu_2(G)|^2$ by Remark 10.3. Outcome: If $F \subseteq G \subseteq A \in \mathcal{A}_{1,2}^{+-}$ and $\chi_F \leq f \leq \chi_G$, then

$$|\eta(F)| \leq \left|\int f\,d\eta\right| \leq |\eta(G)|\,.$$

Consider Remark 10.3. A similar argument shows that for every measurable covering of X there is a measurable covering $\{E_k\}$ of X consisting of subsets of the four sets in $\mathcal{A}_{1,2}^{+-}$, and so if $F_k \subseteq G_k \subseteq E_k$ and $\chi_{F_k} \leq f_k \leq \chi_{G_k}$, then

$$|\eta(F_k)| \leq \left|\int f_k\,d\eta\right| \leq |\,\eta(G_k)|$$

for every k. Again, as in Remark 10.3, for every *finite* measurable partition of X there exists a *finite* measurable partition $\{E_k\}$ of X such that

$$\sup \sum\nolimits_k \left|\int \chi_{E_k}\,d\eta\right| = \sup \sum\nolimits_k |\eta(E_k)| = |\eta|(X),$$

where the supremum is taken over all finite measurable partitions of X. Thus, following the same argument in Remark 10.3, since $\sum_k \chi_{E_k} = \chi_X = 1$ (and so $|\sum_k \chi_{E_k}\,d\nu| = 1$) and $0 \leq \chi_{E_k} \leq 1$, we may infer that

$$\sup \sum\nolimits_k \left|\int \chi_{E_k}\,d\eta\right| \leq \sup_{|f|\leq 1} \left|\int f\,d\eta\right|,$$

where the supremum on the right-hand side is taken over all real-valued integrable functions f on X such that $|f| \leq 1$. So, $|\eta|(X) = \sup \sum_k |\int \chi_{E_k}\,d\eta\,| \leq \sup_{|f|\leq 1} |\int f\,d\eta\,|$. Since $\sup_{|f|\leq 1} |\int f\,d\nu| \leq |\eta|(X)$, it follows that

$$\sup_{|f|\leq 1} \left|\int f\,d\eta\right| = |\eta|(X).$$

Remark 10.6. A complex-valued function $f = f_1 + i\,f_2 \colon X \to \mathbb{C}$ on X is integrable with respect to a complex measure $\eta = \nu_1 + i\,\nu_2 \colon \mathcal{X} \to \mathbb{C}$ if the real-valued functions f_1 and f_2 (the real and imaginary parts of f) are integrable with respect to η, and the integral of an integrable function $f \colon X \to \mathbb{C}$ with respect to a complex measure $\eta \colon \mathcal{X} \to \mathbb{C}$ is given by

$$\begin{aligned}\int f\,d\eta &= \int f_1\,d\eta + i\int f_2\,d\eta\\ &= \int f_1\,d\nu_1 + i\int f_1\,d\nu_2 + i\left[\int f_2\,d\nu_1 + i\int f_2\,d\nu_2\right]\\ &= \int f_1\,d\nu_1 + i\int f_2\,d\nu_1 + i\left[\int f_1\,d\nu_2 + i\int f_2\,d\nu_2\right]\\ &= \int f\,d\nu_1 + i\int f\,d\nu_2.\end{aligned}$$

Note that $|\int_E f\,d\eta|^2 = |\int_E f_1\,d\eta|^2 + |\int_E f_2\,d\eta|^2 \leq (\sup|f_1|^2 + \sup|f_2|^2)|\eta|(E)$ for every $E \in \mathcal{X}$, by Remark 10.5. However, we can again get a tighter inequality, with $\sup|f_1|^2 + \sup|f_2|^2$ replaced by $\sup|f|^2 = \sup(|f_1|^2 + |f_2|^2)$. Indeed, write $\int f\,d\eta = \rho\, e^{i\theta}$, which implies that $e^{-i\theta}\int f\,d\eta = \rho \geq 0$. Set $g = e^{-i\theta}f\colon X \to \mathbb{C}$ and $h = \operatorname{Re} g\colon X \to \mathbb{R}$, and so $|h| \leq |g|$ and $|g| = |f|$. Then $0 \leq |\int f\,d\eta| = \rho = \int e^{-i\theta} f\,d\eta = \int g\,d\eta = \operatorname{Re}\int g\,d\eta \leq \int \operatorname{Re} g\,d\eta$. (To verify the final inequality note, by the above displayed identity, that $0 \leq \operatorname{Re}\int g\,d\eta = \int g_1\,d\nu_1 - \int g_2\,d\nu_2$, so that $\int g_2\,d\nu_2 \leq \int g_1\,d\nu_1$; thus $|\operatorname{Re}\int g\,d\eta\,| = \operatorname{Re}\int g\,d\eta \leq \int g_1\,d\nu_1$ and, since $\int \operatorname{Re} g\,d\eta = \int g_1\,d\eta = \int g_1\,d\nu_1 + i\int g_1\,d\nu_2$, we get $\operatorname{Re}\int \operatorname{Re} g\,d\eta = \int g_1\,d\nu_1$, and so $\operatorname{Re}\int g\,d\eta \leq \int \operatorname{Re} g\,d\eta$.) Moreover, since h is real-valued, we get by Remark 10.5 that $|\int f\,d\eta| \leq \int \operatorname{Re} g\,d\eta = \int h\,d\eta = |\int h\,d\eta| \leq \int |h|\,d\eta$. So, using the same argument, $|\int_E f\,d\eta| = |\int f\chi_E\,d\eta| \leq \int |h|\chi_E\,d\eta = \int_E |h|\,d\eta \leq \int_E |f|\,d\eta$. Therefore, for every $E \in \mathcal{X}$,

$$\left|\int_E f\,d\eta\right| \leq \int_E |f|\,d\eta \leq \sup|f|\,|\eta|(E),$$

so that

$$\left|\int f\,d\eta\right| \leq \sup|f|\,|\eta|(X),$$

and hence $\sup_{|f|\leq 1}|\int f\,d\nu| \leq |\eta|(X)$, where the supremum is taken over all complex-valued integrable functions f on X such that $|f| \leq 1$. However, as we saw in Remark 10.5, this inequality becomes an identity even for the particular case where the supremum is taken over the real-valued functions; and so it holds for the general case. That is,

$$\sup_{|f|\leq 1}\left|\int f\,d\eta\right| = |\eta|(X).$$

10.4 Additional Propositions

Observe that a finite positive measure (whose range lies in $[0,\infty)$) is a particular case of a signed measure (whose range lies in $\mathbb{R}$), and that a signed measure is a particular case of a complex measure (whose range lies in $\mathbb{C}$).

However, if a positive measure is not finite, then it is not a particular case of any of the above measures since its range is $[0, +\infty]$, where the symbol $+\infty$ (which is not a number; in particular, not real number) is included in it. Thus a signed measure $\nu = \nu^+ - \nu^-$ and a complex measure $\eta = \nu_1 + i\,\nu_2 = \nu_1^+ - \nu_1^- + i\,(\nu_2^+ - \nu_2^-)$ are always *finite* in the sense that $|\nu(E)| < \infty$ and $|\eta(E)| < \infty$ for every measurable set E. So they are, in particular, σ-finite in the sense that ν^+, ν^-, ν_1^+, ν_1^-, ν_2^+, ν_2^-, being all finite positive measure, are tautologically σ-finite. Let λ and μ be arbitrary measures (positive, finite positive, signed, or complex) on the same σ-algebra. Exactly as in the case of positive measures (cf. Definition 7.6), λ is *absolutely continuous* with respect to μ (notation: $\lambda \ll \mu$) if, for an arbitrary measurable set E, $\mu(E) = 0$ implies $\lambda(E) = 0$. Similarly, as in the case of positive measures (cf. Definition 7.9), λ and μ are *singular* (notation: $\lambda \perp \mu$) if $\lambda(A) = \mu(B) = 0$ for some measurable partition $\{A, B\}$ of X.

With the above extended definitions in mind, the Radon–Nikodým Theorem (Theorem 7.8) has an immediate extension to integrable functions (instead of nonnegative measurable functions), where the σ-finite positive measure λ is replaced by a (finite) signed measure ν, as follows: *Let $(X, \mathcal{X})$ be a measurable space. If ν is a signed measure and μ is a positive σ-finite measure, both on $\mathcal{X}$, and if ν is absolutely continuous with respect to μ, then there exists a unique (μ-almost everywhere unique) real-valued function f in $\mathcal{L}(X, \mathcal{X}, \mu)$ such that $\nu(E) = \int_E f\,d\mu$ for each $E \in \mathcal{X}$.* The proof follows naturally by Theorem 7.8 for $f = f^+ - f^-$ with $\nu^+(E) = \int_E f^+ d\mu$ and $\nu^-(E) = \int_E f^- d\mu$ (cf. Proposition 7.5). The original version in Theorem 7.8 is not a particular case of the above version. However, the above version a particular case of the following.

Proposition 10.A. *Let $(X, \mathcal{X})$ be a measurable space. If η is a complex measure and μ is a positive σ-finite measure, both on $\mathcal{X}$, and if η is absolutely continuous with respect to μ, then there exists a unique (μ-almost everywhere unique) complex-valued μ-integrable function f on X such that*

$$\eta(E) = \int_E f\,d\mu \quad \text{for each} \quad E \in \mathcal{X}.$$

Along the same lines, a similar approach extends the Lebesgue Decomposition Theorem (Theorem 7.10), which in fact is a corollary of the preceding Radon–Nikodým Theorem, as follows. Let $\mathbb{F}$ denote either $\mathbb{R}$ or $\mathbb{C}$ so that an $\mathbb{F}$-*valued measure* means either a finite positive measure, or a signed measure, or a complex measure.

Proposition 10.B. *Let $(X, \mathcal{X})$ be a measurable space. If μ is a positive σ-finite and η is an $\mathbb{F}$-valued, both measures on $\mathcal{X}$, then there is a unique pair of $\mathbb{F}$-valued measures η_a and η_s on $\mathcal{X}$ such that $\eta_a \ll \mu$, $\eta_s \perp \mu$, and*

$$\eta = \eta_a + \eta_s.$$

The *Fundamental Theorem of Calculus* of elementary calculus asserts that integration and differentiation are the inverse of each other as follows. Suppose, for instance, that f is a continuously differentiable real-valued function on the interval $[\alpha, \beta] \subset \mathbb{R}$. The derivative of the indefinite (Riemann) integral coincides with the function, $f(x) = \big(\int_\alpha^x f(t)\,dt\big)'$ for every $x \in [\alpha, \beta]$, and conversely the function is recovered by integrating its derivative f' (which being continuous on $[\alpha, \beta]$ is integrable), $f(x) - f(\alpha) = \int_\alpha^x f'(t)\,dt$ for every $x \in [\alpha, \beta]$. Perhaps it might come as no big surprise that such an inverse relationship between differentiation and integration may be properly stated in a measure-theoretical framework; in particular, in the context of the Lebesgue measure. As expected, the Radon–Nikodým Theorem (Theorem 7.8 — Proposition 10.A) plays a crucial role in establishing this with the help of the Lebesgue Decomposition Theorem (Theorem 7.10 — Proposition 10.B), where the reference measure μ is the Lebesgue measure.

We now proceed along the lines of the following propositions towards the Lebesgue version of the Fundamental Theorem of Calculus. In what follows, $\mu\colon \mathfrak{R} \to \overline{\mathbb{R}}$ will denote the Lebesgue measure on the Borel σ-algebra $\mathfrak{R}$ generated by the open sets of $\mathbb{R}$, equipped with its usual topology, as in Section 8.3 (see the remark that follows Problem 1.14). Recall that a *compact* subset of $\mathbb{R}$ is precisely a closed and bounded subset of $\mathbb{R}$, and also that a Borel measure on $\mathfrak{R}$ is a positive measure $\lambda\colon \mathfrak{R} \to \overline{\mathbb{R}}$ that assigns a finite value to every compact set of $\mathbb{R}$ (i.e., $\lambda(K) < \infty$ for every compact $K \subset \mathbb{R}$; cf. Problem 2.13). It is readily verified that the Lebesgue measure (which assigns its finite length to every bounded interval) is a Borel measure, and also that every Borel measure is σ-finite (cf. Problem 2.13 again). Also recall that a *finite* positive measure on $\mathfrak{R}$ (which is tautologically a Borel measure) is a particular case of a signed measure on $\mathfrak{R}$, which in turn is a particular case of a complex measure on $\mathfrak{R}$. Thus we actually have essentially two distinct cases: a positive nonfinite (possibly σ-finite) measure, or an $\mathbb{F}$-valued measure — complex, signed (i.e., real-valued), or positive finite measures. Similarly, a real-valued function on $\mathbb{R}$ can be viewed as a particular case of a complex-valued function on $\mathbb{R}$, and so we will refer to an $\mathbb{F}$*-valued function*, meaning either a real-valued or a complex-valued function. Some of the next results can be stated for measures on the σ-algebra generated by the open sets of $\mathbb{R}^n$, or for functions on $\mathbb{R}^n$. However, we will consider only measures on $\mathfrak{R}$ and functions on $\mathbb{R}$.

Let X be a nonempty open subset of $\mathbb{F}$ (particular case, $X = \mathbb{R}$). Take a function $f: X \to \mathbb{F}$, and an arbitrary point $x \in X$. Let $f'(x)$ be a number in $\mathbb{F}$ with the following property. For every $\varepsilon > 0$ there exists a $\delta > 0$ such that

$$0 < |y - x| < \delta \quad \text{implies} \quad \left| \tfrac{f(y)-f(x)}{y-x} - f'(x) \right| < \varepsilon.$$

If there exists such a number $f'(x) \in \mathbb{F}$, then it is called the *derivative of f at x*, and the function f is said to be *differentiable at x*.

Proposition 10.C. *Let η be an $\mathbb{F}$-valued measure and let μ be the Lebesgue measure, both on $\mathfrak{R}$, and consider the function $f: \mathbb{R} \to \mathbb{F}$ defined by*

$$f(x) = \eta((-\infty, x)) \quad \text{for every} \quad x \in \mathbb{R}.$$

The function f is differentiable at $x \in X$ (so that there exists $f'(x) \in \mathbb{F}$) and $f'(x) = \alpha$ if and only if for every $\varepsilon > 0$ there exists a $\delta > 0$ such that

$$\mu(I(x)) < \delta \quad \text{implies} \quad \left| \tfrac{\eta(I(x))}{\mu(I(x))} - \alpha \right| < \varepsilon$$

for every open interval $I(x)$ that contains x.

Given an arbitrary $x \in I(x) = (\alpha, \beta)$, we say that the open interval $I(x)$ *shrinks to x* (notation: $I(x) \to \{x\}$) if for every $\delta > 0$ there exists an interval $I_\delta(x) = (\alpha_\delta, \beta_\delta) \subseteq I(x) = (\alpha, \beta)$ containing x (i.e., $\alpha \le \alpha_\delta < x < \beta_\delta \le \beta$) such that $|\beta_\delta - \alpha_\delta| < \delta$. Proposition 10.C suggests the following definition. Take any point $x \in I(x) = (\alpha, \beta)$. Let $(D\eta)(x)$ be a number in $\mathbb{F}$ with the following property. For every $\varepsilon > 0$ there exists a $\delta > 0$ such that

$$|\beta_\delta - \alpha_\delta| < \delta \quad \text{implies} \quad \left| \tfrac{\eta(I_\delta(x))}{\mu(I_\delta(x))} - (D\eta)(x) \right| < \varepsilon$$

whenever the interval $I_\delta(x) = (\alpha_\delta, \beta_\delta) \subseteq I(x) = (\alpha, \beta)$ contains x. If this number $(D\eta)(x) \in \mathbb{F}$ exists, then it is called the *derivative of the measure η at a point x with respect to Lebesgue measure μ*. In other words, $(D\eta)(x)$ is the limit of $\frac{\eta(I(x))}{\mu(I(x))}$ as the open interval $I(x)$ containing x shrinks to x:

$$(D\eta)(x) = \lim_{I_x \to \{x\}} \tfrac{\eta(I(x))}{\mu(I(x))}$$

for every $x \in \mathbb{R}$ at which this limit exist. If the limit exists for every $x \in \mathbb{R}$, then the function $D\eta: \mathbb{R} \to \mathbb{F}$ defined by the preceding limit is referred to as the *derivative of the measure η with respect to Lebesgue measure μ*. Recall that η is an $\mathbb{F}$-valued measure on $\mathfrak{R}$, which may be a finite positive measure, or a signed measure, or a complex measure.

The next result says that the Radon–Nikodým derivative $\frac{d\eta}{d\mu}$ of a measure η with respect to Lebesgue measure μ coincides with the derivative $D\eta$ of the measure η with respect to Lebesgue measure μ. (Warning: all measures on $\Re$ and the reference measure must be the Lebesgue measure.)

Proposition 10.D. *Let η be an $\mathbb{F}$-valued measure and let μ be the Lebesgue measure, both on $\Re$. Suppose η is absolutely continuous with respect to μ ($\eta \ll \mu$). Let $\frac{d\eta}{d\mu}$ be the Radon–Nikodým derivative of η with respect to μ.*

Claim: $D\eta = \frac{d\eta}{d\mu}$ *(μ-almost everywhere), so that*

$$\eta(E) = \int_E (D\eta)\, d\mu \quad \text{for every} \quad E \in \Re.$$

A further form of Proposition 10.D, representing the first part of the Fundamental Theorem of Calculus extended to Lebesgue integrals, is considered in Proposition 10.E below. It requires the notion of Lebesgue points of an integrable function, which is defined as follows. If $f\colon \mathbb{R} \to \mathbb{F}$ is integrable with respect to Lebesgue measure μ, then every $x \in \mathbb{R}$ for which

$$\lim_{I_x \to \{x\}} \frac{1}{\mu(I_x)} \int_{I_x} |f - f(x)|\, d\mu = 0$$

is called a *Lebesgue point of f.*

Proposition 10.E. *Let $f\colon \mathbb{R} \to \mathbb{F}$ be μ-integrable, where μ is the Lebesgue measure on $\Re$. For each x in $\mathbb{R}$ set*

$$F(x) = \int_{-\infty}^{x} f\, d\mu$$

in $\mathbb{F}$. This defines a function $F\colon \mathbb{R} \to \mathbb{F}$, which is differentiable at every Lebesgue point x, whose derivative at x is $F'(x) = f(x)$.

A function $f\colon [\alpha, \beta] \to \mathbb{F}$ is called *absolutely continuous* on the interval $[\alpha, \beta]$ if for every $\varepsilon > 0$ there exists a $\delta > 0$ such that

$$\sum\nolimits_i (\beta_i - \alpha_i) < \delta \quad \text{implies} \quad \sum\nolimits_i |f(\beta_i) - f(\alpha_i)| < \varepsilon$$

for every *finite* disjoint collection of open intervals $(\alpha_i, \beta_i) \subseteq [\alpha, \beta]$. As a particular case, if the above holds, then it holds for every collection containing just one interval (i.e., $\beta_i - \alpha_i < \delta$ implies $|f(\beta_i) - f(\alpha_i)| < \varepsilon$ for every of open subinterval (α_i, β_i) of $[\alpha, \beta]$), leading to the standard notion of continuity. Thus absolute continuity trivially implies continuity.

The first part of the Fundamental Theorem of Calculus extended to Lebesgue integrals in Proposition 10.E asserted that the derivative of the indefinite integral coincides with the function. Conversely, the second part of the Fundamental Theorem of Calculus extended to Lebesgue integrals asserts that the function is recovered by integrating its derivative, as follows.

Proposition 10.F. *Take $f : [\alpha, \beta] \to \mathbb{F}$ on the interval $[\alpha, \beta]$ and let μ be the Lebesgue measure on $\Re$. If either f is absolutely continuous (so that it is differentiable μ-almost everywhere on $[\alpha, \beta]$ and f' is μ-integrable), or f is differentiable at every point of $[\alpha, \beta]$ and f' is μ-integrable, then*

$$f(x) - f(\alpha) = \int_\alpha^x f' \, d\mu \quad \text{for every} \quad x \in [\alpha, \beta].$$

Recall again that by an $\mathbb{F}$-valued function $F : X \to \mathbb{F}$ on a nonempty set X we mean either a complex-valued function or the particular case of a real-valued function. Similarly, by an $\mathbb{F}$-valued measure $\eta : \mathcal{X} \to \mathbb{F}$ on a σ-algebra of subsets of X we mean either a complex measure or the particular cases of a signed measure or a finite positive measure. Since the notion of integral has been extended in this chapter to $\mathbb{F}$-valued functions and $\mathbb{F}$-valued measure, the notion of the Banach space $L^1(X, \mathcal{X}, \eta)$ is naturally extended from the case of real-valued functions and positive measures, considered in Chapter 5, to $\mathbb{F}$-valued functions and measures. In particular, exactly the same expression for the norm in L^1 holds for $\mathbb{F}$-valued functions and measures due to the upper bounds obtained in Remarks 10.1 to 10.6,

$$\left| \int f \, d\eta \right| \le \int |f| \, d|\eta| = \|f\|_1 < \infty \quad \text{for every} \quad f \in L^1(X, \mathcal{X}, \eta),$$

so that the results in Propositions 5.5 to 5.9 (see also Lemmas 4.4 and 4.5) still remain in force for $\mathbb{F}$-valued functions and measures. The above inequality says that the integral, as a transformation from the normed space $(L^1(X, \mathcal{X}, \eta), \| \, \|_1)$ of integrable functions to the normed space $(\mathbb{F}, | \; |)$ — equipped with its usual norm $| \; |$ — is bounded (a contraction, actually). Since for linear transformations boundedness is equivalent to continuity, we get the next proposition (for the particular case of $\mathbb{F} = \mathbb{R}$ and $\eta = \mu$, a positive measure, see the remark that precedes Problem 5.16).

Proposition 10.G. *The integral $\int (\cdot) \, d\eta : L^1(X, \mathcal{X}, \eta) \to \mathbb{F}$, as a functional between the Banach spaces $L^1(X, \mathcal{X}, \eta)$ and $\mathbb{F}$, is linear and continuous* (i.e., *the integral is a continuous linear functional*).

Extending the notion of integral of scalar-valued (either real or complex) functions to vector-valued functions on a finite-dimensional (real or complex) normed space seems quite natural and, in fact, it is quite natural. Things become more delicate if we consider vector-valued functions on infinite-dimensional spaces.

Let Y be a normed space and let $\| \; \|$ denote the norm on Y (cf. Definition 5.1), either as a finite or an infinite-dimensional, real or complex, normed space. Let $\mathcal{Y}_\mathcal{T}$ be the σ-algebra generated by the collection $\mathcal{T}$ of all open subsets of Y (i.e., the σ-algebra generated by a topology $\mathcal{T}$ on Y — cf. Proposition 1.12 — which is called a Borel σ-algebra of subsets of Y — see the remark following Problem 1.14). Now let X be a nonempty set, and let $\mathcal{X}$ be a σ-algebra of subsets of X. Note that in accordance with the first paragraph of this chapter, no topology is being assigned to the set X.

A Y-valued function $F\colon X\to Y$ on X is said to be measurable if the inverse image of sets in $\mathcal{Y}_\mathcal{T}$ are sets in $\mathcal{X}$ (i.e., $F^{-1}(E)\in\mathcal{X}$ for every $E\in\mathcal{Y}_\mathcal{T}$ — cf. Problem 1.8). Consider the real-valued function $f\colon X\to\mathbb{R}$ defined by $f(x)=\|F(x)\|$ for every $x\in X$. When we say that $\|F(\cdot)\|$ is measurable, we mean in the sense of Definition 1.2, which coincides with the above sense (inverse image of sets in the Borel algebra $\mathfrak{R}$ lie in $\mathcal{X}$ — cf. Problem 1.8 again). Let $\mu\colon\mathcal{X}\to\overline{\mathbb{R}}$ be a positive measure on the σ-algebra $\mathcal{X}$ of subsets of the nonempty set X, and consider a measure space $(X,\mathcal{X},\mu)$. Let $\mathcal{L}(X,Y)=\mathcal{L}(X,\mathcal{X},\mu;Y)$ denote the collection of all those measurable functions $F\colon X\to Y$ for which the function $\|F(\cdot)\|\colon X\to\mathbb{R}$ is integrable; that is, for which $\|F(\cdot)\|$ is measurable and $\int\|F(\cdot)\|\,d\mu<\infty$.

Proposition 10.H. *Let $(X,\mathcal{X},\mu)$ be a measure space and let Y be a normed space.* (a) *If a Y-valued function $F\colon X\to Y$ on X is measurable, then the real-valued function $\|F(\cdot)\|\colon X\to\mathbb{R}$ on X is also measurable.* (b) *The set $\mathcal{L}(X,Y)$ is a linear space, and the function $\|\;\|\colon\mathcal{L}(X,Y)\to\mathbb{R}$ defined by*

$$\|F\| = \int\|F(\cdot)\|\,d\mu \quad \text{for every} \quad F\in\mathcal{L}(X,Y)$$

is a seminorm on the linear space $\mathcal{L}(X,Y)$.

Compare with Lemma 4.5 and Proposition 5.4. Note that we are using the same notation $\|\;\|$ for the norm $\|F(x)\|_Y$ of $F(x)$ on the linear space Y for each $x\in X$, and for the seminorm $\|F\|_{\mathcal{L}(X,Y)}$ of F on the linear space $\mathcal{L}(X,Y)$. Now suppose Y is a Banach space (i.e., a complete normed space — see the paragraph that precedes Proposition 5.4; also see Theorem 5.9).

A Banach-space-valued function $F\colon X\to Y$ on X is said to be *Bochner integrable* if it lies in $\mathcal{L}(X,Y)$ (i.e., F is measurable and $\int\|F(\cdot)\|\,d\mu<\infty$) and its range $F(X)$ is separable (i.e., $F(X)\subseteq Y$ has a countable dense subset, which means that $F(X)$ has a countable subset whose closure coincides with the closure of $F(X)$ in Y — the notion of denseness and separability will be discussed in the first section of the next chapter). Since $F(X)$ is separable whenever Y is (i.e., whenever Y has a countable dense subset; equivalently, if Y has a countable subset whose closure coincides with Y), it follows that *if Y is a separable Banach space, then $F\colon X\to Y$ is Bochner integrable if and only if $F\in\mathcal{L}(X,Y)$*. It is worth noting that $\mathbb{F}$ equipped with its usual norm (as well as $\mathbb{F}^n$ equipped with any norm) is a separable Banach space. Just as in Sections 5.1 and 5.2, consider the equivalence class

$[F]$ of all functions in $\mathcal{L}(X,Y)$ that are equal to F μ-a.e., and denote the set of these equivalence classes by $L^1(X,Y) = L^1(X,\mathcal{X},\mu;Y)$. As in Lemma 5.8, the seminorm of Proposition 10.H becomes a norm $\|\;\|_1$ on $L^1(X,Y)$,

$$\|F\|_1 = \|[F]\|_1 = \int \|F(\cdot)\|\,d\mu,$$

which does not depend on the exemplar F in $[F]$, so that $(L^1(X,Y),\|\;\|_1)$ (or simply $L^1(X,Y)$ when the norm $\|\;\|_1$ is clear in the context) is a normed space. A function $\Phi\colon X \to Y$ is a *simple function* if its range is finite, and it is an *integrable simple* function if it is simple and lies in $\mathcal{L}(X,Y)$ (i.e., $\#\Phi(X) < \infty$, Φ is measurable, and $\int\|\Phi(\cdot)\|\,d\mu < \infty$).

Proposition 10.I. *If $(X,\mathcal{X},\mu)$ is a measure space and Y is a Banach space, then $L^1(X,Y)$ is a Banach space. Moreover, if Y is a separable Banach space, then the collection of all Y-valued integrable simple functions on X is a dense linear manifold of $L^1(X,Y)$.*

The first part of the Proposition 10.I is the counterpart of Theorem 5.9; the second part (which extends Problem 5.4) allows us to define an integral for functions in $L^1(X,Y)$. Let Y be a separable Banach space and write a measurable simple function Φ as a weighted sum of characteristic functions $\chi_{E_i}\colon X \to Y$ of a finite measurable partition $\{E_i\}$ of X with a finite set of distinct elements $\{a_i\}$ from Y (the finite set $\{a_i\}$ is the range of Φ), say

$$\Phi = \sum_{i=1}^{n} a_i\,\chi_{E_i},$$

and define the *integral of Φ* in Y with respect to μ as in Definition 3.1:

$$\int \Phi\,d\mu = \sum_{i=1}^{n} a_i\,\mu(E_i),$$

which is a vector in Y whenever $\Phi \in L^1(X,Y)$. Since the collection of all Y-valued integrable simple functions on X is a dense linear manifold of $L^1(X,Y)$, every $F \in L^1(X,Y)$ is the limit (with respect to the norm on $L^1(X,Y)$) of a sequence $\{\Phi_k\}$ of integrable simple functions in $L^1(X,Y)$. That is, for every $F \in L^1(X,Y)$, there exists a sequence $\{\Phi_k\}$ such that $\|\Phi_k - F\|_1 \to 0$ as $k \to \infty$. In other words, a Bochner integrable function $F\colon X \to Y$ of X into a separable Banach space Y (i.e., $F \in L^1(X,Y)$, where Y is a separable Banach space) is, according to Proposition 10.I, the limit in $L^1(X,Y)$ of a sequence of integrable simple functions in $L^1(X,Y)$; that is,

$$F = \lim_k \Phi_k.$$

This leads to the next result, which defines the integral of F as the limit in Y of the integrals of Φ_k: for each F in $L^1(X,Y)$ there exists a unique vector $s(F)$ in Y, denoted by $s(F) = \int F\,d\mu$, such that $\|\int \Phi_k\,d\mu - s(F)\| \to 0$ as $k \to \infty$ (where $\|\,\|$ is the norm on Y), which is written as

$$\int F\,d\mu = \int \lim_k \Phi_k\,d\mu = \lim_k \int \Phi_k\,d\mu\,.$$

Such a limit in Y is ensured by the next proposition. Indeed, since *the integral transformation* $\int(\cdot)\,d\mu\colon L^1(X,Y) \to Y$ *is linear and continuous* (which means that it is linear and bounded, as in Proposition 10.J below), and since $F = \lim_k \Phi_k$ in $L^1(X,Y)$, we get $\int F\,d\mu = \int \lim_k \Phi_k\,d\mu = \lim_k \int \Phi_k\,d\mu$ (see e.g., [26, Corollary 3.8 and Theorem 4.14]).

Proposition 10.J. *If* $(X, \mathcal{X}, \mu)$ *is a measure space and* Y *is a separable Banach space, then there exists a unique bounded linear transformation* $\int(\cdot)\,d\mu\colon L^1(X,Y) \to Y$ *such that* $\int \Phi\,d\mu = \sum_{i=1}^n a_i\,\mu(E_i)$ *for every integrable simple function* $\Phi = \sum_{i=1}^n a_i \chi_{E_i} \in L^1(X,Y)$. *Moreover,*

$$\left\|\int F\,d\mu\right\| \le \int \|F(\cdot)\|\,d\mu = \|F\|_1.$$

Thus the integral $\int(\cdot)\,d\mu\colon L^1(X,Y) \to Y$ *from the Banach space* $L^1(X,Y)$ *to the separable Banach space* Y *is a continuous linear transformation.*

The value of this bounded linear transformation from $L^1(X,Y)$ to a separable Banach space Y, namely, $\int F\,d\mu$, is called the *Bochner integral* of $F\colon X \to Y$ with respect to the positive measure μ. Extensions from positive to $\mathbb{F}$-valued measures follow essentially the same path of Proposition 10.G.

For an exposition on further notions of integral, extended in a different direction from what has been done here (also referred to as generalized Riemann integral, Kurzweil–Henstock, or gauge integral), restricted to functions on the real line, that corrects some defects in the classical Riemann theory simplifying and extending the Lebesgue theory, see, for instance, [5].

Notes: The whole chapter, in particular the propositions in Section 10.4, comprise a set of basic results on integration that will be required in the sequel. The $\mathbb{F}$-valued versions of the Radon–Nikodým Theorem and of the Lebesgue Decomposition Theorem in Propositions 10.A and 10.B are natural and immediate consequences of their positive versions in Theorems 7.8 and 7.10 (e.g., also see [36, Theorem 6.10]). For the discussion on the Fundamental Theorem of Calculus extended to Lebesgue integrals, along the lines we have approached here, the reader is referred to [36, Chapter 7]. In particular, see [36, Theorems 7.1, 7.8, and 7.11] for Propositions 10.C, 10.D, and

10.E; and [36, Theorems 7.20, 7.21] for Proposition 10.F. Since the integral remains linear for $\mathbb{F}$-valued functions and measures (naturally extending the result for real-valued functions and positive measures of Lemma 4.5 and Section 5.1), it follows that Proposition 10.G is an immediate consequence of the fact that for linear transformations boundedness coincides with continuity (see e.g., [26, Theorem 4.14]). Propositions 10.H, 10.I, and 10.J lead to the notion of Bochner integral of Banach-space-valued functions (see e.g., [8, Theorems 17.8, 17.9, 17.11, 17.13, 17.14]).

Suggested Reading

Berberian [7], Brown and Pearcy [8], Cohn [10], Halmos [18], Kingman and Taylor [23], Royden [35], Rudin [36], Weir [42].

11

Borel Measure

11.1 Topological Spaces

A *topological space* was defined in Problem 1.12: a set equipped with a topology. A *topology* on a set X is a collection $\mathcal{T}$ of subsets of X satisfying the following axioms: (i) the whole set X and the empty set $\varnothing$ lie in $\mathcal{T}$, (ii) finite intersections of sets in $\mathcal{T}$ lie in $\mathcal{T}$, and (iii) arbitrary unions of sets in $\mathcal{T}$ lie in $\mathcal{T}$. The sets in $\mathcal{T}$ are called the *open* sets of X (with respect to $\mathcal{T}$).

A *metric space* is a set equipped with a metric. A *metric* in a set X is a function $d\colon X\times X \to \mathbb{R}$ such that for every $x, y, z \in X$, (i) $d(x, y) \geq 0$ and $d(x, x) = 0$, (ii) $d(x, y) = 0$ if and only if $x = y$, (iii) $d(x, y) = d(y, x)$, and (iv) $d(x, y) \leq d(x, z) + d(z, y)$ (the triangle inequality) — these are the *metric axioms*. (If $d\colon X\times X \to \mathbb{R}$ satisfies axioms (i), (iii), and (iv), but not necessarily axiom (ii), then it is called a *pseudometric*.) An *open ball* (centered at $x_0 \in X$ with radius ε) in a metric space X is the set $B_\varepsilon(x) = \{x \in X\colon d(x, x_0) < \varepsilon\}$. A set U is *open* in a metric space X if U includes a nonempty open ball centered at each one of its points. Open balls are open sets in a metric space. The collection $\mathcal{T}$ of all open sets in a metric space X satisfies the three axioms of topology; this is referred to as the *metric topology* on X, or the *topology induced* (or generated, or determined) *by a metric d*. (Whenever we refer to the topology of a metric space, it will be understood that this is the metric topology, unless otherwise stated.) Thus every metric space is a topological space, where the topology (the metric topology) is that induced by the metric. This topology $\mathcal{T}$ induced by the

C.S. Kubrusly, *Essentials of Measure Theory*,
DOI 10.1007/978-3-319-22506-7_11

metric d, and the topological space obtained by equipping X with $\mathcal{T}$, are said to be metrized by d. If X is a topological space with topology $\mathcal{T}$, and if there exists a metric d on X that metrizes $\mathcal{T}$ (i.e., if X is a metric space with respect to a metric d, and if the collection of all open sets in X with respect to d coincides with $\mathcal{T}$), then the topological space X and the topology $\mathcal{T}$ are called *metrizable*. The notion of topological space is broader than the notion of metric space. Every metric space is a topological space, but the converse fails: there are topological spaces that are not metrizable.

Definition 11.1. Let X be a topological space (i.e., a nonempty set X equipped with a topology $\mathcal{T}$, whose elements are the open subsets of X).

(a) A set $V \subseteq X$ is *closed* if its complement $X \backslash V$ is open.

(b) The *closure* A^- of $A \subseteq X$ is the smallest closed subset of X that includes A. (The intersection of all closed subsets of X that include A.)

(c) A set $A \subseteq X$ is *dense* if $A^- = X$.

(d) A topological space is *separable* if it has a countable dense subset.

(e) The *interior* A° of $A \subseteq X$ is the largest open subset of X included in A. (The union of all open subsets of X included in A.)

(f) A *covering* of a set $A \subseteq X$ is a collection of subsets of X whose union includes A. An *open covering* of A is a covering of A consisting entirely of open subsets of X.

(g) A set $K \subseteq X$ is *compact* if every open covering of K includes a finite subcovering. If X is a compact set itself, then X is a *compact space*.

(h) A set in X is *σ-compact* if it is a countable union of compact sets.

(i) A set $A \subseteq X$ is *relatively compact* (or *conditionally compact*) if its closure A^- is compact.

(j) A *base* (or a *topological base*) for X is a subcollection of $\mathcal{T}$ that covers each open subset of X (i.e., covers each set in $\mathcal{T}$). Equivalently, $\mathcal{B} \subseteq \mathcal{T}$ is a base for a topology $\mathcal{T}$ of subsets of X if, for each $U \in \mathcal{T}$ and each $x \in U$ there is a $G \in \mathcal{B}$ with $x \in G \subseteq U$.

(k) A *neighborhood* of a point x in X is any subset of X that includes an open set which contains x. An *open neighborhood* of $x \in X$ is any open subset of X that contains x.

(ℓ) A *Hausdorff* space is a topological space X such that for every pair of distinct points x and y in X there exist neighborhoods N_x and N_y of x and y, respectively, such that $N_x \cap N_y = \varnothing$.

(m) A *locally compact space* is a topological space X such that every point of X has a compact neighborhood.

The concepts of closed and open are dual of each other in the sense that a set is open if and only if its complement is closed, and vice versa. The three axioms of topology say that $\varnothing$ and X are open, intersection of a finite collection of open sets is open, and union of an arbitrary collection of open sets is open. It is readily verified that the dual result for closed sets reads as follows: $\varnothing$ and X are closed, union of a finite collection of closed sets is closed, and intersection of an arbitrary collection of closed sets is closed. If a set is both open and closed, then its is said to be a *clopen* set. The standard properties stated in the next lemma will be required in the sequel.

Lemma 11.2. *Let A, B, G, U, V, K be subsets of a topological space X.*

(a) *A closed subset of a compact set is compact.*

(b) *If $A \subseteq B$ and B^- is compact, then A^- is compact.*

(c) *A nonempty subset of a topological space is open if and only if it is (or includes) a neighborhood of each one of its points.*

(d) *Every metric space is Hausdorff.*

(e) *Every singleton in a Hausdorff space is closed.*

(f) *If K is compact, X is Hausdorff, and $x \in X \backslash K$, then there is an open set U with $K \subseteq U$ and a neighborhood N_x of x such that $U \cap N_x = \varnothing$.*

(g) *Every compact subset of a Hausdorff space is closed.*

(h) *If X is Hausdorff, V is closed, and K is compact, then $V \cap K$ is compact.*

(i) *If intersection of an infinite (not necessarily countable) collection of compact subsets of a Hausdorff space is empty, then there is a finite subcollection whose intersection is also empty.*

Proof. Let X be topological space.

(a) *If $V \subseteq K$, where K is compact and V is closed, then V is compact.* In fact, suppose V is a closed subset of a compact set $K \subseteq X$. Take an arbitrary covering of V, say $\mathcal{U}$, consisting of open subsets of X. So $\mathcal{U} \cup \{X \backslash V\}$ is an open covering of K. Since K is compact, this covering includes a finite subcovering, say $\mathcal{U}'$, so that $\mathcal{U}' \backslash \{X \backslash V\} \subseteq \mathcal{U}$ is a finite subcovering of V. Therefore, every open covering of V has a finite subcovering.

(b) This is a direct consequence of (a), since $A \subseteq B$ implies $A^- \subseteq B^-$.

(c) If $A \subseteq X$ is open and nonempty, then A trivially is (and so A trivially includes) an open neighborhood of each one of its points. Conversely, take $A \subseteq X$. If A includes (or if A is) a neighborhood N_a of every $a \in A$, then there is an open set $U_a \subseteq N_a \subseteq A$ such that $a \in U_a$ for every $a \in A$. Since $A = \bigcup_{a\in A} U_a$, it follows that A is open (cf. axiom (iii) of a topological space).

(d) It is readily verified (by the triangle inequality) that for every pair of distinct points x and y in a metric space X there are nonempty open balls $B_\varepsilon(x)$ and $B_\rho(y)$ centered at x and y (which clearly are neighborhoods of x and y), respectively, such that $B_\varepsilon(x) \cap B_\rho(y) = \varnothing$.

(e) Take an arbitrary $x \in X$, consider the singleton $\{x\} \subseteq X$ and its complement $X\backslash\{x\} \subseteq X$, and take an arbitrary $y \in X\backslash\{x\}$. If X is Hausdorff, then there exists a neighborhood N_y of y such that $N_y \subseteq X\backslash\{x\}$. Hence there exists an open $U_y \subseteq X\backslash\{x\}$ such that $y \in U_y$. Thus $X\backslash\{x\}$ is itself a neighborhood of each one of its points, and so $X\backslash\{x\}$ is open by item (c).

(f) Let K be a compact proper subset of X. If $K = \varnothing$, then the result is trivially verified for $U = \varnothing$. Thus suppose $K \neq \varnothing$ and take an arbitrary point x in $X\backslash K$. Since x is distinct from every point in K, it follows that for every $y \in K$ there exists an open neighborhood K_y of y and an open neighborhood X_y of x such that $K_y \cap X_y = \varnothing$ (reason: X is a Hausdorff space). But $K \subseteq \bigcup_{y\in K} K_y$ so that $\{K_y\}_{y\in K}$ is a covering of K consisting of nonempty open subsets of X. If K is compact, then there is a finite subset of K, say $\{y_i\}_{i=1}^n$, such that $K \subseteq U = \bigcup_{i=1}^n K_{y_i}$, which is open (each K_{y_i} is open). Set $N_x = \bigcap_{i=1}^n X_{y_i}$, which is a neighborhood of x (in fact an open neighborhood of x since it is a finite intersection of open neighborhoods X_{y_i} of x). Since $K_{y_i} \cap X_{y_i} = \varnothing$, it follows that $K_{y_i} \cap N_x = \varnothing$, for each i, and therefore $(\bigcup_{i=1}^n K_{y_i}) \cap N_x = \varnothing$. That is, $U \cap N_x = \varnothing$.

(g) Let K be a compact subset of a Hausdorff space X. If either $K = \varnothing$ or $K = X$, then K is trivially closed in X. Thus suppose $\varnothing \neq K \neq X$. According to (f), for every $x \in X\backslash K$ there exists an open set U with $K \subseteq U$ and a (open) neighborhood N_x of x such that $U \cap N_x = \varnothing$. Hence $N_x \subseteq X\backslash K$. Thus $X\backslash K$ includes a (open) neighborhood of each one of its points, which means by (c) that $X\backslash K$ is open, and so K is closed.

(h) *If X is Hausdorff, then $V \cap K$ is compact whenever K is compact and V is closed.* Indeed, let K be a compact subset and V a closed subset of a Hausdorff space. By item (g), K is closed. Thus $V \cap K$ is a closed subset of the compact K, and therefore compact itself according to (a).

(i) *If an infinite collection $\{K_\gamma\}$ of compacts sets in a Hausdorff space X is such that $\bigcap_\gamma K_\gamma = \varnothing$, then there exists a finite subcollection $\{K_i\}$ of $\{K_\gamma\}$ such that $\bigcap_i K_i = \varnothing$.* Indeed, set $U_\gamma = X\backslash K_\gamma$, which is open if X is Hausdorff according to (g). Take an arbitrary K from the collection $\{K_\gamma\}$. If $\bigcap_\gamma K_\gamma = \varnothing$, then there is no point of K that belongs to every K_γ. Hence

$\{U_\gamma\}$ is an open covering of K. Since K is compact, there is a finite subcollection $\{U_i\}$ of $\{U_\gamma\}$ such that $K \subseteq \bigcup_i U_i$. Thus $X\backslash\bigcup_i U_i \subseteq X\backslash K$, so that $\bigcap_i K_i = \bigcap_i X\backslash U_i = X\backslash\bigcup_i U_i \subseteq X\backslash K$. Then $K \cap \bigcap_i K_i = \varnothing$. □

Remarks on Boundedness: (a) METRIC SPACES. A set B in a metric space (X, d) is *bounded* if $\sup_{x,y\in B} d(x, y) < \infty$ (i.e., if it has a finite diameter); and B is *totally bounded* if for every $\varepsilon > 0$ there exists a finite ε-net B_ε for B (a subset B_ε of B is an *ε-net* for B if for every point x of B there exists a point y in B_ε such that $d(x, y) < \varepsilon$). It is readily verified that total boundedness implies boundedness, but the converse fails (the unit ball in ℓ^p equipped with the usual metric is bounded but not totally bounded). The Compactness Theorem says that *in a metric space, a set is compact if and only if it is complete and totally bounded* (see, e.g., [26, Corollary 3.81] — a set in a metric space is complete if every Cauchy sequence in it converges to a point in it — Section 5.1). Since metric spaces are Hausdorff, compact sets in a metric space are closed (Lemma 11.2(d,g)). Therefore, *compact sets in a metric space are closed and bounded.* The Heine–Borel Theorem states the converse for the metric space $\mathbb{F}^n$ (where $\mathbb{F}$ denotes either $\mathbb{R}$ or $\mathbb{C}$) equipped with their usual metric — see, e.g., [26, Theorem 3.83 and Corollary 4.32]): *in $\mathbb{F}^n$, compact means closed and bounded.* Since $\mathbb{F}^n$ is a complete metric space, it follows by the Compactness Theorem that what the Heine–Borel Theorem says is that in $\mathbb{F}^n$ boundedness implies total boundedness (i.e., in $\mathbb{F}^n$, boundedness coincides with total boundedness). Note that the metric (thus Hausdorff) space $\mathbb{F}^n$ is locally compact (but the metric space ℓ^p is not). In fact, $\mathbb{F}^n$ *is a prototype of a locally compact Hausdorff space.*

(b) LOCALLY COMPACT SPACES. In light of these facts, a set in a locally compact space is said to be *bounded* (or *topologically bounded*) if it is included in a compact set. So, also in this case, *every compact set is bounded* (i.e., topologically bounded); actually, every relatively compact set is bounded. Thus, in this case, a closed and bounded set is a closed set included in a compact set, and hence (by Lemma 11.2(a)) *a closed and bounded set is compact.* The converse holds in a locally compact Hausdorff space, where compact sets are closed (Lemma 11.2(g)): *in a locally compact Hausdorff space, compact means closed and bounded* (where bounded means relatively compact — Lemma 11.2(b)). A set in a locally compact space is *σ-bounded* if it is included in a σ-compact set. Finally, note that *in a locally compact complete metric space, topologically bounded means totally bounded.* Indeed, B totally bounded $\Longrightarrow B^-$ closed and totally bounded $\Longleftrightarrow B$ compact (reason: in a complete metric space a set is complete if and only if it closed) $\Longrightarrow B$ topologically bounded $\Longrightarrow B^-$ closed and topologically bounded $\Longleftrightarrow B^-$ compact $\Longrightarrow B^-$ totally bounded $\Longrightarrow B$ totally bounded.

Remarks on Denseness: It is readily verified (cf. Proposition 11.B) that *if a topological space has a countable base, then it is separable.* The converse is not true in general. Example: if X is uncountable with topology consisting of the empty set and the complements of the finite sets, then X is separable but has no countable base — cf. [21, p. 49]. However, the converse holds in a metric space: *a metric space is separable if and only if it has a countable base* (see, e.g., [26, Theorem 3.35]).

Theorem 11.3. *If $K \subseteq U$, where K is compact and U is open in a locally compact Hausdorff space X, then there exists an open set G with compact closure such that*

$$K \subseteq G \subseteq G^{-} \subseteq U.$$

In particular, for each $x \in U$ there exists an open neighborhood N of x and a compact C, such that $N \subseteq C \subseteq U$.

Proof. Since X is locally compact, every point of X, in particular, every point x of $K \subseteq X$ has a compact neighborhood, and therefore an open neighborhood B_x with a compact closure. Thus, $K \subseteq \bigcup_x B_x$. Since K is compact, this open covering has a finite subcovering $\{B_i\}$ such that $K \subseteq \bigcup_i B_i$. Set $B = \bigcup_i B_i$, so that B is open and $K \subseteq B$, and hence $B^{-} = \bigcup_i B_i^{-}$ (recall: the closure of a finite union of sets coincides with the union of their closures). Since each B_i^{-} is compact, and since a finite union of compact sets is clearly compact, it follows that B^{-} is compact. If the open set U is such that $U = X$, then the result is verified with $G = B$ (i.e., $K \subseteq B \subseteq B^{-} \subseteq X$). Thus suppose $U \subset X$, and take an arbitrary y in the closed set $V = X \backslash U$. Since $K \subseteq U$, it follows that $y \in X \backslash K$. Thus, since X is Hausdorff, Lemma 11.2(f) ensures that there exists an open set U_y' with $K \subseteq U_y'$ and $y \notin U_y'$. Set $K_y = V \cap B^{-} \cap U_y'^{-}$, which is compact by Lemma 11.2 (h) — B^{-} is compact and X is Hausdorff. Since $\{K_y\}_{y \in X \backslash U}$ is a collection of compact sets such that $\bigcap_y K_y = \varnothing$ (because $\bigcap_y U_y' = \varnothing$ — if there exists $z \in X \backslash U$ such that $z \in \bigcap_y U_y'$, then $z \in U_z'$, which is a contradiction), it follows by Lemma 11.2 (i) that there exists a finite subcollection, say, $\{K_i\}$ such that $\bigcap_i K_i = \varnothing$. Hence $(X \backslash U) \cap B^{-} \cap \bigcap_i U_i'^{-} = \varnothing$ so that $B^{-} \cap \bigcap_i U_i'^{-} \subseteq U$. Recall that $K \subseteq B \cap \bigcap_i U_i'$. Thus, by setting $G = B \cap \bigcap_i U_i'$, which is open (finite intersection of open sets), we get $G^{-} \subseteq B^{-} \cap \bigcap_i U_i'^{-}$, and so

$$K \subseteq G \subseteq G^{-} \subseteq U.$$

In particular, since a singleton is clearly compact, the above inclusion ensures that $\{x\} \subseteq N \subseteq C \subseteq U$ for every $x \in U$, where $\{x\} = K$ is compact, $N = G$ is an open neighborhood of x, and $C = G^{-}$ is compact. □

Remarks on Borel σ-algebras: Let X be a topological space with topology $\mathcal{T}$. Let $\mathcal{X}_{\mathcal{T}}$ be the σ-algebra generated by the topology $\mathcal{T}$. That is, $\mathcal{X}_{\mathcal{T}}$ is the smallest σ-algebra of subsets of X that includes $\mathcal{T}$. The elements of $\mathcal{X}_{\mathcal{T}}$ (i.e., the $\mathcal{X}_{\mathcal{T}}$-measurable sets) are referred to as the *Borel sets* of X — see the remark that follows Problem 1.14. A *Borel σ-algebra* of subsets of X is any σ-algebra $\mathcal{A}$ of subsets of X that includes $\mathcal{X}_{\mathcal{T}}$. Observe that all open and all closed subsets of X are Borel sets (i.e., they belong to the Borel σ-algebra $\mathcal{X}_{\mathcal{T}}$). A subset of X is a G_δ (read: *G-delta*) if it is a countable intersection of open subsets of X, and an F_σ (read: *F-sigma*) if it is a countable union of closed subsets of X. These are also Borel sets. In a Hausdorff space, all compact sets are Borel sets, since they are closed by Lemma 11.2(g).

A function F between topological spaces X and Y is *continuous* if the inverse image of open (closed) sets is open (closed), and *measurable* if the inverse image of open (closed) sets is measurable (with respect to a σ-algebra $\mathcal{X}$ of subsets of X). In other words, if X and Y are topological spaces with topologies $\mathcal{T}_X$ and $\mathcal{T}_Y$, and if $F\colon X\to Y$ is a function of X into Y, then

$$F\colon X\to Y \text{ is continuous} \iff F^{-1}(U)\in\mathcal{T}_X \text{ for every } U\in\mathcal{T}_Y$$

and, if $\mathcal{X}$ is an arbitrary σ-algebra of subsets of X, then

$$F\colon X\to Y \text{ is } \mathcal{X}\text{-measurable} \iff F^{-1}(U)\in\mathcal{X} \text{ for every } U\in\mathcal{T}_Y.$$

Theorem 11.4(b) says that *a measurable function is precisely a measurable transformation* (in the sense of Problem 1.8) if $\mathcal{Y}=\mathcal{Y}_{\mathcal{T}}$ (i.e., if $\mathcal{Y}$ is the Borel σ-algebra $\mathcal{Y}_{\mathcal{T}}$ of subsets of Y generated by the topology $\mathcal{T}_Y$ on Y):

$$F^{-1}(U)\in\mathcal{X} \text{ for every } U\in\mathcal{T}_Y \iff F^{-1}(E)\in\mathcal{X} \text{ for every } E\in\mathcal{Y}_{\mathcal{T}}.$$

In particular, if $\mathcal{A}\supseteq\mathcal{X}_{\mathcal{T}}$ is a Borel σ-algebra of subsets of X (where $\mathcal{X}_{\mathcal{T}}$ is the σ-algebra generated by a topology $\mathcal{T}_X$ on X), and if $\mathcal{T}_Y$ is a topology on Y, then a function $F\colon X\to Y$ is *Borel measurable* if it is $\mathcal{A}$-measurable (i.e., if $F^{-1}(U)\in\mathcal{A}$ for every $U\in\mathcal{T}_Y$). So the implication below is clear.

Continuous functions are Borel measurable.

(Since $\mathcal{T}_X\subseteq\mathcal{X}_{\mathcal{T}}$, $F^{-1}(U)\in\mathcal{T}_X$ for every $U\in\mathcal{T}_Y$ implies $F^{-1}(U)\in\mathcal{X}_{\mathcal{T}}\subseteq\mathcal{A}$ for every $U\in\mathcal{T}_Y$.) A *Borel function* is simply a Borel measurable function.

Theorem 11.4. *Let $\mathcal{X}$ be a σ-algebra of subsets of a set X, let Y and Z be topological spaces, and consider the functions $F\colon X\to Y$ and $G\colon Y\to Z$.*

(a) *The collection $\mathcal{Y}=\{E\subseteq Y\colon F^{-1}(E)\in\mathcal{X}\}$ is a σ-algebra.*

(b) *If F is $\mathcal{X}$-measurable, then $\mathcal{Y}_{\mathcal{T}}\subseteq\mathcal{Y}$ (i.e., $\mathcal{Y}$ is a Borel σ-algebra). In other words, if F is $\mathcal{X}$-measurable, then $F^{-1}(E)\in\mathcal{X}$ for every $E\in\mathcal{Y}_{\mathcal{T}}$.*

(c) *If* $Y = \mathbb{R}$ *(equipped with the usual topology), then* F *is* $\mathcal{X}$*-measurable if and only if* $F^{-1}((\alpha,\infty)) \in \mathcal{X}$ *for every* $\alpha \in \mathbb{R}$.

(d) *If* G *is a Borel function and* F *is* $\mathcal{X}$*-measurable, then the composition* $H = G \circ F \colon X \to Z$ *is* $\mathcal{X}$*-measurable.*

(e) *If* G *is continuous and* F *is* $\mathcal{X}$*-measurable, then the composition* $H = G \circ F \colon X \to Z$ *is* $\mathcal{X}$*-measurable.*

Proof. Let $\mathcal{X}$ be a σ-algebra of subsets of X.

(a) Observe that $F^{-1}(\varnothing) = \varnothing$ and $F^{-1}(Y) = X$. Moreover, $F^{-1}(Y \backslash E) = X \backslash F^{-1}(E)$ for every $E \in \mathcal{X}$ and $F^{-1}\big(\bigcup_n E_n\big) = \bigcup_n F^{-1}(E_n)$ for every countable collection $\{E_n\}$ of $\mathcal{X}$-measurable sets. Thus $\mathcal{Y}$ is a σ-algebra according to Definition 1.1.

(b) If F is $\mathcal{X}$-measurable (inverse image of open sets are measurable), then the σ-algebra $\mathcal{Y}$ of item (a) includes all open sets of Y. That is, it includes the topology $\mathcal{T}_Y$ on Y: $\mathcal{T}_Y \subseteq \mathcal{Y}$. Thus, since $\mathcal{Y}$ is a σ-algebra that includes $\mathcal{T}_Y$, and $\mathcal{Y}_\mathcal{T}$ is the smallest σ-algebra that includes $\mathcal{T}_Y$ (i.e., the σ-algebra generated by the topology $\mathcal{T}_Y$), it follows that $\mathcal{Y}_\mathcal{T} \subseteq \mathcal{Y}$.

(c) Let $Y = \mathbb{R}$ equipped with the usual topology, take $F \colon X \to Y$, and set $\mathcal{Y} = \{E \subseteq Y \colon F^{-1}(E) \in \mathcal{X}\}$. Take an arbitrary $\alpha \in \mathbb{R}$. Suppose $F^{-1}(\alpha,\infty)$ lies in $\mathcal{X}$, so that (α,∞) lies in $\mathcal{Y}$. Let $\{\alpha_n\}$ be a real-valued sequence converging to α and such that $\alpha_n < \alpha$ for each n. Since (i) each $(\alpha_n,\infty) \in \mathcal{Y}$, (ii) $(-\infty,\alpha) = \bigcup_n(-\infty,\alpha_n] = \bigcup_n Y \backslash (\alpha_n,\infty)$, and (iii) $\mathcal{Y}$ is a σ-algebra according to (a), it follows that $(-\infty,\alpha) \in \mathcal{Y}$. So $(\alpha,\beta) = (-\infty,\beta) \cap (\alpha,\infty) \in \mathcal{Y}$ for every $\beta \in \mathbb{R}$, which means that every interval of the real line is $\mathcal{Y}$-measurable. Since every open set of $\mathbb{R}$ is a countable union of open intervals — every separable metric space has a countable topological base of open balls (see, e.g., [26, Corollary 3.16 and Theorem 3.35]), and since $\mathcal{Y}$ is a σ-algebra, it follows that every open set in $\mathbb{R}$ lies in $\mathcal{Y}$, and therefore, according to the definition of $\mathcal{Y}$, $F^{-1}(U) \in \mathcal{X}$ for every open set $U \subseteq \mathbb{R}$, which means that F is $\mathcal{X}$-measurable. Outcome: if $F^{-1}(\alpha,\infty) \in \mathcal{X}$, then F is $\mathcal{X}$-measurable. The converse is trivial since (α,∞) is open in Y: if F is $\mathcal{X}$-measurable, then $F^{-1}(\alpha,\infty) \in \mathcal{X}$.

(d) Take $U \in \mathcal{T}_Z$; an arbitrary open subset U of Z. If $G \colon Y \to Z$ is a Borel function, then $G^{-1}(U) \in \mathcal{Y}_\mathcal{T}$, where $\mathcal{Y}_\mathcal{T}$ is the Borel σ-algebra generated by the topology $\mathcal{T}_Y$. Thus, if F is $\mathcal{X}$-measurable, then $F^{-1}(G^{-1}(U)) \in \mathcal{X}$ by (b). If $H = G \circ F \colon X \to Z$, then $H^{-1}(C) = F^{-1}(G^{-1}(C))$ for every $C \subseteq Z$. (Indeed, $x \in H^{-1}(C) \Leftrightarrow H(x) \in C \Leftrightarrow G(F(x)) \in C \Leftrightarrow F(x) \in G^{-1}(C) \Leftrightarrow x \in F^{-1}(G^{-1}(C))$.) Hence $H^{-1}(U) \in \mathcal{X}$, and so H is $\mathcal{X}$-measurable.

(e) This is a particular case of (d), since every continuous function is a Borel function (i.e., continuous functions are Borel measurable). □

11.2 Regular Measures

Set $X = \mathbb{R}$, the real line equipped with its usual (metric) topology, and take $\mathcal{X}_{\mathcal{T}} = \Re$, the Borel algebra (i.e., the σ-algebra generated by the open intervals, which coincides with the σ-algebra generated by the topology of $\mathbb{R}$ — the σ-algebra of subsets of $\mathbb{R}$ generated by the open sets of $\mathbb{R}$; see the remark following Problem 1.14). When dealing with a measure $\mu\colon \Re \to \overline{\mathbb{R}}$ on the Borel algebra $\Re$ we always assume the usual topology of $\mathbb{R}$. If $\mu(K) < \infty$ for every compact (i.e., closed and bounded) subset K of $\mathbb{R}$, then μ is called a Borel measure (note: $K \in \Re$ since closed subsets of $\mathbb{R}$ lie in $\Re$; see Problem 2.13). We generalize the notion of Borel measure from the concrete space $\mathbb{R}$ to general topological spaces and, in particular, to Hausdorff spaces.

Remarks on Borel Measures: Let $\mathcal{X}_{\mathcal{T}}$ be the Borel σ-algebra of subsets of a topological space X generated by a topology $\mathcal{T}$ on X. Every closed subset of X is a Borel set. Suppose X is Hausdorff. Thus every compact set is a Borel set (Lemma 11.2(g)). Take any Borel σ-algebra $\mathcal{A}$ of subsets of X (i.e., any σ-algebra $\mathcal{A}$ that includes $\mathcal{X}_{\mathcal{T}}$; in particular, $\mathcal{X}_{\mathcal{T}}$ itself, or its completion). A *Borel measure* is a (positive) measure $\mu\colon \mathcal{A} \to \overline{\mathbb{R}}$ on $\mathcal{A}$ such that $\mu(K) < \infty$ for every compact set $K \in \mathcal{A}$. Every finite measure on a Borel σ-algebra $\mathcal{A}$ is a Borel measure. If X is σ-compact, then every Borel measure is σ-finite.

Let $\mathcal{A}$ be a Borel σ-algebra of subsets of a Hausdorff space X (where compact sets are Borel sets, so that if U is an open subset of X and K is a compact subset of X, then U and K lie in $\mathcal{A}$). An arbitrary set E in $\mathcal{A}$ is *outer regular* with respect to a measure $\mu\colon \mathcal{A} \to \overline{\mathbb{R}}$ on $\mathcal{A}$ if

$$\mu(E) = \inf\big\{\mu(U)\colon E \subseteq U,\ U \text{ open in } X\big\},$$

and *inner regular* with respect to a measure $\mu\colon \mathcal{A} \to \overline{\mathbb{R}}$ on $\mathcal{A}$ if

$$\mu(E) = \sup\big\{\mu(K)\colon K \subseteq E,\ K \text{ compact in } X\big\}.$$

A set E in $\mathcal{A}$ is *regular* with respect to μ if it is both outer and inner regular with respect to μ. A measure μ on a Borel σ-algebra $\mathcal{A}$ is *regular* (*outer regular, inner regular*) if every set E in $\mathcal{A}$ is regular (outer regular, inner regular) with respect to it. A set E in $\mathcal{A}$ is *quasiregular* with respect to μ if it is outer regular, and, if it is open, then it is inner regular; that is,

$$\mu(E) = \sup\big\{\mu(K)\colon K \subseteq E,\ K \text{ compact in } X\big\} \quad \text{if } E \text{ is open in } X.$$

The difference between regular and quasiregular is that a nonopen quasiregular set may not be inner regular. A measure μ on a Borel σ-algebra $\mathcal{A}$ is *quasiregular* if every E set in $\mathcal{A}$ is quasiregular with respect to it. Clearly, regularity implies quasiregularity: a regular measure is quasiregular.

Lemma 11.5. *Take the Borel σ-algebra $\mathcal{X}_\mathcal{T}$ of subsets of a Hausdorff space X. All sets below are Borel sets, and measures are Borel measures on $\mathcal{X}_\mathcal{T}$.*

(a) *A set of infinite measure is outer regular.*

(b) *A set of measure zero is inner regular.*

(c) *An open set is outer regular.*

(d) *A compact set is inner regular.*

(e) *A countable union of open sets is outer regular.*

(f) *A finite intersection of compact sets is inner regular.*

(g) *A finite intersection of open sets is outer regular.*

The analogous result for inner regular sets reads: *A finite union of compact sets is inner regular*. However, more is true as stated in (i).

(h) *A countable intersection of open sets of finite measure is outer regular.*

(i) *A countable union of compact sets is inner regular.*

(j) *A finite disjoint union of inner regular sets of finite measure is inner regular.*

The analogous result for outer regular sets reads: *A finite disjoint union of outer regular sets of finite measure is outer regular*. However, more is true.

(k) *A countable union of outer regular sets is outer regular.*

(ℓ) *A countable intersection of inner regular sets of finite measure is inner regular.*

(m) *An increasing countable union of inner regular sets is inner regular.*

(n) *A decreasing countable intersection of outer regular sets of finite measure is outer regular.*

Proof. Consider the definitions of outer and inner regular Borel sets with respect to a Borel measure μ on $\mathcal{X}_\mathcal{T}$.

(a) This is immediate by the definition of outer regular sets: X is an open Borel set, and $\mu(E) = \infty$ implies $\mu(X) = \infty$ for every Borel set $E \subseteq X$.

(b) This is the dual of item (a) since $\varnothing$ is a compact Borel set with $\mu(\varnothing) = 0$, and $\varnothing \subseteq E$ for every Borel set E.

(c) This is trivial by the definition of outer regular sets (set $E = U$).

(d) This is the dual of item (c). Also trivial by the definition of inner regular sets (set $E = K$).

(e) A countable union of Borel sets is a Borel set, an arbitrary union of open sets is an open set, and open sets are outer regular by (c).

(f) This is the dual of item (e). A countable intersection of Borel sets is a Borel set, a finite intersection of compact sets is a compact set, and compact sets are inner regular by (d).

(g) A countable intersection of Borel sets is a Borel set, a finite intersection of open sets is an open set, and open sets are outer regular by (c).

(h) Let $\{E_k\}_{k=1}^{\infty}$ be a sequence of open sets with $\mu(E_k) < \infty$. Take a sequence $\{U_n\}_{n=1}^{\infty}$ of open sets with $U_n = \bigcap_{i=1}^{n} E_i$. Set $E = \bigcap_{k=1}^{\infty} E_k = \bigcap_{n=1}^{\infty} U_n$, which is a Borel set. Since $\{U_n\}_{n=1}^{\infty}$ is a decreasing sequence with $\mu(U_n) \leq \min_{1\leq i\leq n} \mu(E_i) < \infty$ for each n, it follows by Proposition 2.2(d) that

$$\mu(E) = \lim_n \mu(U_n),$$

where $\{U_n\}_{n=1}^{\infty}$ is a decreasing sequence of open sets such that $E \subseteq U_n$. This implies that E is outer regular.

(i) This is the dual of item (h). Let $\{E_k\}_{k=1}^{\infty}$ be a sequence of compact sets. Take a sequence $\{K_n\}_{n=1}^{\infty}$ of compact sets with $K_n = \bigcup_{i=1}^{n} E_i$. Set $E = \bigcup_{k=1}^{\infty} E_k = \bigcup_{n=1}^{\infty} K_n$, which is a Borel set. Since $\{K_n\}_{n=1}^{\infty}$ is an increasing sequence, it follows by Proposition 2.2(c) that

$$\mu(E) = \lim_n \mu(K_n),$$

where $\{K_n\}_{n=1}^{\infty}$ is an increasing sequence of compact sets such that $K_n \subseteq E$. This implies that E is inner regular.

(j) Take an arbitrary $\varepsilon > 0$. If $\{E_i\}_{i=1}^{n}$ is a finite collection of disjoint inner regular Borel sets of finite measure, then there is a compact set $K_i \subseteq E_i$ such that $\mu(E_i) \leq \mu(K_i) + \frac{\varepsilon}{n}$ for each i. Set $K = \bigcup_{i=1}^{n} K_i$, which is compact (finite union of compact sets), and $E = \bigcup_{i=1}^{n} E_i$. These are Borel sets such that $K \subseteq E$. Since $\{E_i\}_{i=1}^{n}$ is a disjoint collection,

$$\mu(E) = \sum_{i=1}^{n} \mu(E_i) \leq \sum_{i=1}^{n} \mu(K_i) + \varepsilon = \mu(K) + \varepsilon,$$

which implies that E is inner regular.

(k) Again take an arbitrary $\varepsilon > 0$. If $\{E_k\}_{k=1}^{\infty}$ is a sequence of outer regular Borel sets, then there exists an open set U_k such that $E_k \subseteq U_k$ and $\mu(U_k) \leq \mu(E_k) + \frac{\varepsilon}{2^k}$ for each k. Set $U = \bigcup_{k=1}^{\infty} U_k$, which is open, and

$E = \bigcup_{k=1}^{\infty} E_k$. These are Borel sets (countable union of Borel sets) such that $E \subseteq U$. If $\mu(E) = \infty$, then E is outer regular by (c). If $\mu(E) < \infty$, then

$$\begin{aligned}\mu(U) - \mu(E) = \mu(U \backslash E) &\leq \mu\Big(\bigcup_{k=1}^{\infty} U_k \backslash E_k\Big) \\ &\leq \sum_{k=1}^{\infty} \mu(U_k \backslash E_k) = \sum_{k=1}^{\infty} \big(\mu(U_k) - \mu(E_k)\big) \leq \sum_{k=1}^{\infty} \tfrac{\varepsilon}{2^k} = \varepsilon.\end{aligned}$$

Thus $\mu(U) \leq \mu(E) + \varepsilon$, which implies that E is outer regular.

(ℓ) This is the dual of item (k). Take an arbitrary $\varepsilon > 0$. If $\{E_k\}_{k=1}^{\infty}$ is a sequence of inner regular Borel sets such that $\mu(E_k) < \infty$ for every k, then there exists a compact set K_k such that $K_k \subseteq E_k$ and $\mu(E_k) \leq \mu(K_k) + \frac{\varepsilon}{2^k}$ for each k. Set $K = \bigcap_{k=1}^{\infty} K_k$, which is compact, and $E = \bigcap_{k=1}^{\infty} E_k$. These are Borel sets (countable intersection of Borel sets) such that $K \subseteq E$. Moreover, since $E \backslash K = \bigcap_{k=1}^{\infty} E_k \backslash \bigcap_{k=1}^{\infty} K_k) \subseteq \bigcup_{k=1}^{\infty}(E_k \backslash K_k)$, we get

$$\begin{aligned}\mu(E) - \mu(K) = \mu(E \backslash K) &\leq \mu\Big(\bigcup_{k=1}^{\infty} E_k \backslash K_k\Big) \\ &\leq \sum_{k=1}^{\infty} \mu(E_k \backslash K_k) = \sum_{k=1}^{\infty} \big(\mu(E_k) - \mu(K_k)\big) \leq \sum_{i=1}^{\infty} \tfrac{\varepsilon}{2^k} = \varepsilon.\end{aligned}$$

Thus $\mu(E) \leq \mu(K) + \varepsilon$, which implies that E is inner regular.

(m) Suppose $\{E_k\}_{k=1}^{\infty}$ is a increasing sequence of inner regular Borel sets. Set $E = \bigcup_{k=1}^{\infty} E_k$, which is a Borel set. If $\mu(E) = 0$, then E is inner regular by item (b). Thus suppose $\mu(E) > 0$. Take any real number δ such that $0 < \delta < \mu(E)$. Since $\{E_k\}_{k=1}^{\infty}$ is increasing, it follows that $\delta < \sup_k \mu(E_k) = \lim_k(E_k) = \mu(\bigcup_{k=1}^{\infty} E_k) = \mu(E)$ — cf. Proposition 2.2(c). Since each E_k is inner regular (and recalling again that $\{E_k\}_{k=1}^{\infty}$ is increasing), there exists a compact set K such that $K \subseteq \sup_k E_k = \bigcup_{k=1}^{\infty} E_k = E$ and $\delta < \mu(K) \leq \sup_k \mu(E_k) = \mu(E)$. Summing up: for every $0 < \delta < \mu(E)$ there exists a compact set K such that $K \subseteq E$ and

$$\delta < \mu(K) \leq \mu(E).$$

This implies that E is inner regular.

(n) This is the dual of item (m). Suppose $\{E_k\}_{k=1}^{\infty}$ is a decreasing sequence of outer regular Borel sets with $\mu(E_k) < \infty$. Set $E = \bigcap_{k=1}^{\infty} E_k$, which is a Borel set with $\mu(E) < \infty$. Take any $\delta > 0$ such that $\mu(E) < \delta$. Since $\{E_k\}_{k=1}^{\infty}$ is decreasing, $\mu(E) = \mu(\bigcap_{k=1}^{\infty} E_k) = \lim_k(E_k) = \inf_k \mu(E_k) < \delta$ — cf. Proposition 2.2(d). Since each E_k is outer regular (and since $\{E_k\}_{k=1}^{\infty}$ is decreasing), there is an open set U such that $E = \bigcap_{k=1}^{\infty} E_k = \inf_k E_k \subseteq U$

and $\mu(E) = \inf_k \mu(E_k) \le \mu(U) < \delta$. Summing up: for every $\mu(E) < \delta$ there is an open set U such that $E \subseteq U$ and

$$\mu(E) \le \mu(U) < \delta.$$

This implies that E is outer regular. □

Theorem 11.6. *Let μ be a Borel measure on the Borel σ-algebra $\mathcal{X}_{\mathcal{T}}$ of subsets of a locally compact Hausdorff space X. The following assertions are pairwise equivalent.*

(a) *Every compact set is outer regular.*

(b) *Every bounded open set is inner regular.*

(c) *Every bounded set is regular.*

(d) *Every σ-bounded set is regular.*

If X is σ-compact, then the above equivalent assertions are also equivalent to the following assertion.

(e) *The Borel measure $\mu\colon \mathcal{X}_{\mathcal{T}} \to \overline{\mathbb{R}}$ is regular.*

Proof. Sets are subsets of a locally compact Hausdorff space (where compact means closed and bounded, and bounded means relatively compact), and μ is a Borel measure, thus finite at every Borel set included in a compact set.

(a) ⇒(b) Take an arbitrary $\varepsilon > 0$. Let U be an arbitrary bounded open set. Thus $U \subseteq K$ for some compact set K. Note that $K\backslash U$ is compact (intersection of a compact with a closed set in a Hausdorff space — Lemma 11.2(h)). If every compact set is outer regular, then $K\backslash U$ is outer regular, so that there exists an open set G such that $K\backslash U \subseteq G$ and $\mu(G) \le \mu(K\backslash U) + \varepsilon$. Note that $K\backslash G$ is compact and that $K\backslash G \subseteq K\backslash(K\backslash U) = U$. Thus

$$\begin{aligned}\mu(U) - \mu(K\backslash G) &= \mu(U\backslash(K\backslash G)) = \mu(U \cap G) = \mu(G \cap (K\backslash[K\backslash U])\\ &\le \mu(G\backslash[K\backslash U]) = \mu(G) - \mu(K\backslash U) \le \varepsilon.\end{aligned}$$

Hence $\mu(U) \le \mu(K\backslash G) + \varepsilon$, which implies that U is inner regular.

(b) ⇒(a) Dually, again take an arbitrary $\varepsilon > 0$. Let K be an arbitrary compact set. Let U be a bounded open set such that $K \subseteq U$. Note that $U\backslash K$ is a bounded open set (intersection of a bounded open set with an open set). If every bounded open set is inner regular, then $U\backslash K$ is inner regular, so that there exists a compact set C such that $C \subseteq U\backslash K$ and $\mu(U\backslash K) \le \mu(C) + \varepsilon$. Note that $U\backslash C$ is open and that $K = U\backslash(U\backslash K) \subseteq U\backslash C$. Thus

$$\begin{aligned}\mu(U\backslash C) - \mu(K) &= \mu((U\backslash C)\backslash K) = \mu((U\backslash K)\backslash C)\\ &\le \mu(U\backslash K) - \mu(C) \le \varepsilon.\end{aligned}$$

Hence $\mu(U\backslash C) \leq \mu(K) + \varepsilon$, which implies that K is outer regular.

(d) $\Rightarrow$(c) $\Rightarrow$(a,b) Trivial.

(c) $\Rightarrow$(d) Let $E = \bigcup_k E_k$ be an arbitrary σ-bounded Borel set, where each E_k is a Borel bounded set. If (c) holds, then each E_k is, in particular, outer regular. Thus Lemma 11.5(k) ensures that E is outer regular. Now set $F_n = \bigcup_{i=1}^n E_i$, so that $\{F_n\}_{n=1}^\infty$ is an increasing sequence of bounded Borel sets, which are, in particular, inner regular according to (c). Since $E = \bigcup_n F_n$, it follows by Lemma 11.5(k) that E is inner regular. Therefore, E is regular.

(a,b) $\Rightarrow$(c) Take an arbitrary $\varepsilon > 0$. Let E be any bounded Borel set, so that $E^\circ \subseteq E \subseteq E^-$, where the interior E^o is open and bounded, and the closure E^- is compact. If (a,b) holds, then E° and E^- are regular (cf. Lemma 11.5(c,d)). Therefore, since $\mu(E^-) < \infty$, Proposition 11.F(a) ensures that there exist compact sets C and K and open sets G and U such that $C \subseteq E^\circ \subseteq G$ and $K \subseteq E^- \subseteq U$, for which $\mu(G\backslash C) = \mu(G) - \mu(C) < \varepsilon$ and $\mu(U\backslash K) = \mu(U) - \mu(K) < \varepsilon$. In fact, we may take K and G such that $\mu(K) \leq \mu(G)$ (e.g., set $G = E^\circ$ and $K = E^-$). Thus $C \subseteq E \subseteq K$ and

$$\begin{aligned}\mu(U) - \mu(C) &= \mu(U) - \mu(K) + \mu(K) - \mu(C)\\ &\leq \mu(U) - \mu(K) + \mu(G) - \mu(C) \leq 2\varepsilon,\end{aligned}$$

which means by Proposition 11.F(a) that E is regular.

(d) $\Leftrightarrow$(e) Assertion (e) implies (d) trivially and, if X is σ-compact, then every Borel set is σ-bounded, so that (d) implies (e). □

11.3 Construction of Borel Measures

An *outer measure* (or a *plain outer measure*) is an extended real-valued set function $\mu^*\colon \wp(X) \to \overline{\mathbb{R}}$ on the power set $\wp(X)$ of a given set X such that

(a) $\mu^*(\varnothing) = 0$,

(b) $\mu^*(S) \geq 0$ for every $S \in \wp(X)$,

(c) $\mu^*(S_1) \leq \mu^*(S_2)$ whenever $S_1 \subseteq S_2 \subseteq X$,

(d) $\mu^*\big(\bigcup_n S_n\big) \leq \sum_n \mu^*(S_n)$ for countable families $\{S_n\}$ of sets in $\wp(X)$.

Note that the *outer measure generated by a measure on an algebra*, as in Definition 8.2, is a (plain) outer measure according to Proposition 8.3. By analogy with Section 8.2, given an outer measure $\mu^*\colon \wp(X) \to \overline{\mathbb{R}}$, we say that a set $E \in \wp(X)$ is μ^*-*measurable* if

$$\mu^*(S) = \mu^*(S \cap E) + \mu^*(S\backslash E)$$

for every $S \in \wp(X)$. Observe that $S\backslash E = S \cap (X\backslash E) = S\backslash(S \cap E)$ so that $S = (S \cap E) \cup (S\backslash E)$ and $(S \cap E) \cap (S\backslash E) = \varnothing$. Since μ^* is subadditive (i.e., since (d) holds for the sets $S \cap E$ and $S\backslash E$), it follows that $E \in \wp(X)$ is μ^*-measurable if and only if

$$\mu^*(S \cap E) + \mu^*(S\backslash E) \le \mu^*(S)$$

for every $S \in \wp(X)$. A *quasiregular outer measure* (or a *topologically regular outer measure*) on the power set $\wp(X)$ of a locally compact Hausdorff space X is an outer measure μ^* such that

(i) $\mu^*(S) = \inf\{\mu^*(U) \colon S \subseteq U,\ U \in \wp(X) \text{ open}\}$ for every $S \in \wp(X)$,

(ii) $\mu^*(G \cup U) = \mu^*(G) + \mu^*(U)$ if G and U are disjoint open in $\wp(X)$,

(iii) $\mu^*(U) = \sup\{\mu^*(K) \colon K \subseteq U,\ K \in \wp(X) \text{ compact}\}$ for U open in $\wp(X)$.

Property (iii) justifies the terminology *quasiregular*. It is readily verified that (iii) is equivalent to the following property: for each open set $U \in \wp(X)$,

(iii′) $\mu^*(U) = \sup\{\mu^*(G) \colon G^- \subseteq U,\ G \in \wp(X) \text{ open with } G^- \text{ compact}\}$.

The next theorem is the counterpart of Theorem 8.4, building a Borel measure from an outer measure (instead of building a measure on a σ-algebra from an outer measure generated by a measure on an algebra).

Theorem 11.7. *If* $\mu^* \colon \wp(X) \to \overline{\mathbb{R}}$ *is a quasiregular outer measure on the power set of a locally compact Hausdorff space* X*, then the collection* $\mathcal{A}^*$ *of all* μ^**-measurable subsets of* X*,*

$$\mathcal{A}^* = \{E \in \wp(X) \colon E \text{ is } \mu^*\text{-measurable}\},$$

is a Borel σ*-algebra* (i.e., *it includes the* σ*-algebra* $\mathcal{X}_\mathcal{T}$)*. The restriction*

$$\lambda^* = \mu^*|_{\mathcal{A}^*} \colon \mathcal{A}^* \to \overline{\mathbb{R}}$$

of μ^* *to* $\mathcal{A}^*$ *is a quasiregular complete measure. Moreover, If* $\mu^*(S) < \infty$ *for every bounded set* $S \subseteq X$*, then* $\lambda^* \colon \mathcal{A}^* \supseteq \mathcal{X}_\mathcal{T} \to \overline{\mathbb{R}}$ *is a Borel measure. Furthermore, the restriction of* μ^* *to the Borel* σ*-algebra* $\mathcal{X}_\mathcal{T}$*,*

$$\lambda = \lambda^*|_{\mathcal{X}_\mathcal{T}} = \mu^*|_{\mathcal{X}_\mathcal{T}} \colon \mathcal{X}_\mathcal{T} \to \overline{\mathbb{R}},$$

is a quasiregular measure on $\mathcal{X}_\mathcal{T}$*, which is a Borel measure if* $\mu^*(S) < \infty$ *for every bounded set* $S \subseteq X$*.* (However λ is not necessarily complete.)

Proof. Let $\mu^* \colon \wp(X) \to \overline{\mathbb{R}}$ be a quasiregular outer measure on $\wp(X)$, where X is a locally compact Hausdorff space.

(a) That the class $\mathcal{A}^* = \{E \in \wp(X) : E \text{ is } \mu^*\text{-measurable}\}$ of all μ^*-measurable sets form a σ-algebra has already been verified in the proof of Theorem 8.4.

(b) Thus we proceed to show that every Borel set in the Borel σ-algebra $\mathcal{X}_\mathcal{T}$ is μ^*-measurable (i.e., $\mathcal{X}_\mathcal{T} \subseteq \mathcal{A}^*$). Let X be a locally compact Hausdorff space. It can be verified by property (i) of quasiregular outer measures that a set $E \in \wp(X)$ is μ^*-measurable if and only if, for every open set $U \in \wp(X)$,

$$\mu^*(U \cap E) + \mu^*(U \backslash E) \leq \mu^*(U).$$

Since this holds trivially if $\mu^*(U) = \infty$, it follows that *a set E in $\wp(X)$ is μ^*-measurable if and only if the above inequality holds for every open set $U \in \wp(X)$ with $\mu^*(U) < \infty$*. Since $\mathcal{X}_\mathcal{T}$ is a σ-algebra generated by $\mathcal{T}$, and since $\mathcal{A}^*$ is a σ-algebra, in order to show that $\mathcal{X}_\mathcal{T} \subseteq \mathcal{A}^*$ it is enough to verify that the above inequality holds for every open set E in $\mathcal{T} \subseteq \mathcal{X}_\mathcal{T}$ (instead of for every set $E \in \mathcal{X}_\mathcal{T}$). That is, it suffices to show that each open set E is μ^*-measurable. Thus take arbitrary open sets E and U in $\mathcal{T} \subseteq \mathcal{X}_\mathcal{T} \subseteq \wp(X)$ with $\mu^*(U) < \infty$, and take an arbitrary $\varepsilon > 0$. Property (iii′) of quasiregular outer measures says that for each open set U',

$$\mu^*(U') = \sup\{\mu^*(G) : G^- \subseteq U',\ G \text{ open with } G^- \text{ compact}\}.$$

Since $U \cap E$ is an open set such that $\mu^*(U \cap E) < \infty$, there exists an open set G for which $G \subseteq U \cap E$ and $\mu^*(U \cap E) < \mu^*(G) + \varepsilon$. Note that $U \backslash E = U \cap (X \backslash E) \subseteq U \backslash G$ and $(U \backslash G) \cap G = \varnothing$ (and hence $\mu^*(G) + \mu^*(U \backslash G) = \mu^*(G \cup (U \backslash G))$ by property (ii) of quasiregular outer measures). Therefore,

$$\begin{aligned} \mu^*(U \cap E) + \mu^*(U \backslash E) &< \mu^*(G) + \mu^*(U \backslash G) + \varepsilon = \mu^*(G \cup (U \backslash G)) + \varepsilon \\ &= \mu^*(U \cup G) + \varepsilon = \mu^*(U) + \varepsilon. \end{aligned}$$

Thus $\mu^*(U \cap E) + \mu^*(U \backslash E) \leq \mu^*(U)$, and so E is μ^*-measurable.

(c) That the restriction $\lambda^* = \mu^*|_{\mathcal{A}^*} : \mathcal{A}^* \to \overline{\mathbb{R}}$ of μ^* to $\mathcal{A}^*$ is a complete measure has been verified in the proofs of Theorem 8.4 and Proposition 8.5.

(d) Now we show that $\lambda^* = \mu^*|_{\mathcal{A}^*} : \mathcal{A}^* \supseteq \mathcal{X}_\mathcal{T} \to \overline{\mathbb{R}}$ is a quasiregular measure. Take an arbitrary set $E \in \mathcal{A}^*$. Since every open set U lies in $\mathcal{X}_\mathcal{T} \subseteq \mathcal{A}^*$, it follows by property (i) of quasiregular outer measures that

$$\begin{aligned} \lambda^*(E) = \mu^*(E) &= \inf\{\mu^*(U) : E \subseteq U,\ U \text{ open}\} \\ &= \inf\{\lambda^*(U) : E \subseteq U,\ U \text{ open}\}. \end{aligned}$$

Moreover, since every compact set K lies in $\mathcal{X}_\mathcal{T} \subseteq \mathcal{A}^*$, it follows by property (iii) of quasiregular outer measures that

$$\lambda^*(E) = \mu^*(E) = \sup\{\mu^*(K)\colon K \subseteq E,\ K \text{ compact}\}$$
$$= \sup\{\lambda^*(K)\colon K \subseteq E,\ K \text{ compact}\}$$

whenever E is open (and so E lies in $\mathcal{X}_{\mathcal{T}} \subseteq \mathcal{A}^*$).

(e) That $\lambda^*\colon \mathcal{A}^* \supseteq \mathcal{X}_{\mathcal{T}} \to \overline{\mathbb{R}}$ is a Borel measure goes as follows. If $\mu^*(S) < \infty$ for every bounded subset S of X (i.e., for every $S \in \wp(X)$ with compact closure), and since every compact set lies in $\mathcal{X}_{\mathcal{T}}$ and is closed (because X is Hausdorff), we get $\lambda^*(K) = \mu^*(K) < \infty$ for every compact set K.

(f) Finally, observe that all properties of the measure λ^* except for completeness are readily transferred to the measure λ. □

A *content on a topology* (or simply a *content*) is an extended real-valued set function $\mu^{\#}\colon \mathcal{T} \to \overline{\mathbb{R}}$ on the topology $\mathcal{T}$ of a topological space X such that for arbitrary open sets U, G, and U_n in $\mathcal{T}$,

(1) $\mu^{\#}(\varnothing) = 0$,

(2) $\mu^{\#}(U) \geq 0$,

(3) $\mu^{\#}(U) < \infty$ whenever U^- is compact,

(4) $\mu^{\#}(G) \leq \mu^{\#}(U)$ whenever $G \subseteq U$,

(5) $\mu^{\#}(G \cup U) = \mu^{\#}(G) + \mu^{\#}(U)$ whenever $G \cap U = \varnothing$,

(6) $\mu^{\#}\big(\bigcup_n U_n\big) \leq \sum_n \mu^{\#}(U_n)$ for countable families $\{U_n\}$.

It is an *inner content* if, in addition,

(7) $\mu^{\#}(U) = \sup\{\mu^{\#}(G)\colon G^- \subseteq U,\ G^- \text{compact}\}$.

Lemma 11.8. *Let $\mu^{\#}\colon \mathcal{T} \to \overline{\mathbb{R}}$ be an inner content. If X is a locally compact Hausdorff space, then the set function $\mu^*\colon \wp(X) \to \overline{\mathbb{R}}$ given by*

$$\mu^*(S) = \inf\{\mu^{\#}(U)\colon S \subseteq U,\ U \text{ open}\} \quad \text{for every} \quad S \in \wp(X)$$

is a quasiregular outer measure such that $\mu^*(S) < \infty$ for every bounded S.

Proof. Properties (a), (b), (c), and (d) in the definition of an outer measure μ^* follow at once by properties (1) — null measure of the empty set, (2) — nonnegativeness, (4) — monotonicity, and (6) — countable subadditivity in the definition of the content $\mu^{\#}$, respectively.

Thus μ^* is an outer measure.

Property (4) in the definition of the content $\mu^{\#}$ ensures that

$$\mu^*(U) = \mu^{\#}(U) \quad \text{for every open set } U.$$

Thus property (i) in the definition of a quasiregular outer measure holds by the definition of μ^*. Moreover, the above identity also shows that property (7) in the definition of an inner content ensures that property (iii′) in the definition of a quasiregular outer measure holds. Observe that property (5) in the definition of an inner content trivially implies property (ii) in the definition of a quasiregular outer measure.

Thus the outer measure μ^* is quasiregular.

Finally, property (3) in the definition of the content $\mu^\#$ and property (c) in the definition of an outer measure μ^* imply that

$$\mu^*(S) < \infty \text{ for every bounded set } S \in \wp(X)$$

(i.e., for every $S \in \wp(X)$ with compact closure). □

The combination of Lemma 11.8 and Theorem 11.7 concludes the program of constructing a Borel measure out of an inner content: an inner content $\mu^\#$ generates a quasiregular outer measure μ^* which is finite at bounded sets, and this in turn generates a quasiregular Borel measure λ.

11.4 Additional Propositions

A collection $\mathcal{C}$ of sets has the *finite intersection property* if every finite subcollection of $\mathcal{C}$ has a nonempty intersection.

Proposition 11.A. *A space is compact if and only if every family of closed sets that has the finite intersection property has a nonempty intersection.*

A *Lindelöf* space is a topological space X such that every open covering of X includes a countable subcovering.

Proposition 11.B. *If a topological space has a countable base, then it is a Lindelöf space and separable. Every σ-compact space is a Lindelöf space. Every locally compact space with a countable basis is σ-compact. A locally compact Hausdorff space is Lindelöf if and only if it is σ-compact.*

Proposition 11.C. *Take the Borel σ-algebra $\mathcal{X}_{\mathcal{T}}$ of subsets of a locally compact Hausdorff space. If $(E \cap K) \in \mathcal{X}_{\mathcal{T}}$ for every compact K, then $E \in \mathcal{X}_{\mathcal{T}}$.*

Proposition 11.D. *Let K_1 and K_2 be distinct compact sets in a Borel σ-algebra of subsets of a locally compact Hausdorff space.*

(a) *If every compact set is outer regular, then so is $K_1 \backslash K_2$.*

(b) *If every bounded open set is inner regular, then so is $K_1 \backslash K_2$.*

Proposition 11.E. *Let μ be a finite Borel measure on a Borel σ-algebra of subsets of a locally compact Hausdorff space. If every Borel set is inner regular, then μ is regular.*

Proposition 11.F. *Let μ be a Borel measure on a Borel σ-algebra of subsets of a locally compact Hausdorff space.*

(a) *A Borel set E of finite measure is regular if and only if for every $\varepsilon > 0$ there exists a compact set K and an open set U such that $K \subseteq E \subseteq U$ and $\mu(U \backslash K) < \varepsilon$.*

(b) *If E and F are regular Borel sets of finite measure, then $E \backslash F$ is also regular of finite measure.*

Proposition 11.G. *Let μ be a Borel measure on a Borel σ-algebra of subsets of a locally compact Hausdorff space. If every open set is σ-compact, then μ is regular.*

Proposition 11.H. *Lebesgue is a regular Borel measure — either viewed as a measure on the Borel algebra $\mathfrak{R}$ generated by the usual topology of $\mathbb{R}$, or as a measure on its completion $\mathfrak{I}^*$, the Lebesgue algebra.* (See Section 8.3 — compare with Problems 8.10 and 8.11.)

Recall from Problem 2.13 the definition of support of a measure on the Borel algebra $\mathfrak{R}$ generated by the open sets from $\mathbb{R}$, and extend it to the σ-algebra $\mathcal{X}_{\mathcal{T}}$ of Borel sets of a locally compact Hausdorff space X. The *support* $[\mu]$ of a measure $\mu\colon \mathcal{X}_{\mathcal{T}} \to \overline{\mathbb{R}}$ is the (unique) set $[\mu] = X \backslash U$, where U is the union of all open sets of measure zero, so that $[\mu]$ is a closed set in $\mathcal{X}_{\mathcal{T}}$ such that $X \backslash [\mu]$ is the largest open set of measure zero:

$$X \backslash [\mu] = \sup\big\{U \in \mathcal{X}_{\mathcal{T}}\colon\ U \text{ is open in } X \text{ and } \mu(U) = 0\big\}.$$

Proposition 11.I. *If μ is a regular Borel probability measure* (i.e., $\mu(X) = 1$) *on a compact Hausdorff space X, then its support $[\mu]$ is the smallest compact set such that $\mu([\mu]) = 1$ and $\mu(K) < 1$ for every compact set $K \subset [\mu]$.*

If X is a locally compact Hausdorff space, then let $\mathcal{X}_G$ denote the σ-algebra of subsets of X generated by the compact G_δ's in X. The sets in $\mathcal{X}_G$ are referred to as *Baire sets*. Since every G_δ in X is a Borel set, it follows that $\mathcal{X}_G \subseteq \mathcal{X}_{\mathcal{T}}$: every Baire set is a Borel set. The notions of outer, inner, regular, and quasiregular extend naturally to Baire sets. A measure $\mu\colon \mathcal{X}_G \to \overline{\mathbb{R}}$ which is finite for every compact set is called a *Baire measure*.

Proposition 11.J. *If $E \in \mathcal{X}_G$, then E or $X \backslash E$ is σ-bounded. Every Baire set is σ-bounded if and only if X is σ-compact.*

Proposition 11.K. *A σ-bounded Baire set is the countable disjoint union of bounded Baire sets.*

Proposition 11.L. *Compact Baire sets are G_δ; open Baire sets are F_σ.*

Proposition 11.M. *Let μ be a measure on a σ-algebra including $\mathcal{X}_G$. If μ is inner regular or quasiregular, then for each measurable set E with finite measure there is a Baire set B such that $\mu(E \Delta B) = 0$.* (See Problem 6.5.)

Proposition 11.N. *If X is a separable locally compact metric space, then $\mathcal{X}_G = \mathcal{X}_\mathcal{T}$* (i.e., *every Borel set is a Baire set*).

Proposition 11.O. *If X is a σ-compact locally compact Hausdorff space, then every Baire set is regular* (i.e., *every Baire measure is regular*).

Let $\mu\colon \mathcal{X}_G \to \overline{\mathbb{R}}$ be a Baire measure on the σ-algebra $\mathcal{X}_G$ of all Baire sets in a locally compact Hausdorff space X, and let $[\mu] \in \mathcal{X}_G$ be the support of μ, now with respect to $\mathcal{X}_G$; that is, the set $[\mu] = X \backslash U$, where U is the union of all open Baire sets of measure zero.

Proposition 11.P. *Consider the support $[\mu]$ of a Baire measure μ.*

(a) *If U is an open Baire set such that $U \cap [\mu] \neq \varnothing$, then $\mu(U) > 0$.*

(b) *If K is a compact Baire set such that $K \cap [\mu] = \varnothing$, then $\mu(K) = 0$.*

(c) *If E is a σ-bounded Baire set such that $E \cap [\mu] = \varnothing$, then $\mu(E) = 0$.*

A measure on a Borel σ-algebra of subsets of a locally compact Hausdorff space X is a *locally finite measure* if every $x \in X$ has an open neighborhood of finite measure (i.e., a measure $\mu : \mathcal{A} \to \overline{\mathbb{R}}$, where $\mathcal{A}$ is a Borel σ-algebra of subsets of a locally compact Hausdorff space X, is *locally finite* if for every $x \in X$ there exists an open set $U \subseteq X$ such that $x \in U$ and $\mu(U) < \infty$). A measure μ on a Borel σ-algebra of subsets of a locally compact Hausdorff space is a *Radon measure* if it is both locally finite and inner regular.

Proposition 11.Q. *Every locally finite measure is a Borel measure.*

A *neighborhood base* of a point x in a topological space (or a *local base* at x) is a family of neighborhoods of x such that every neighborhood of x includes a member of the family (e.g., the family of all open neighborhoods of x is trivially a neighborhood base of x). Note that *if a topological space has a countable base, then it is a topological space for which every point has a countable neighborhood base* (see Definition 1.1 and Proposition 11.B).

Proposition 11.R. *Consider a Borel σ-algebra of subsets of a locally compact Hausdorff space X. Suppose every point in X has a countable neighborhood base (in particular, if X has a countable base; more particularly, if*

X is separable). In this case, every inner regular Borel measure is locally finite, and so is a Radon measure. Equivalently, under the above hypothesis, a Radon measure is precisely an inner regular Borel measure.

Consider a metric space (X,d). The *diameter* $\operatorname{diam}(S)$ of a nonempty subset S of X is defined by $\operatorname{diam}(S)=\sup_{x,y\in S} d(x,y)$. Thus a nonempty set S is bounded in a metric space (X,d) if and only if $\operatorname{diam}(S)<\infty$. By convention the empty set $\varnothing$ is bounded and $\operatorname{diam}(\varnothing)=0$. Take an arbitrary $\varepsilon>0$. An ε-covering of a set $S\subseteq X$ is a covering of S made up of subsets of X of diameter not greater than ε.

Proposition 11.S. *Let X be a locally compact metric space. For every pair of real numbers $p\geq 0$ and $\varepsilon>0$ consider the nonnegative extended real*

$$\mu^*_{\varepsilon,p}(S)=\inf\sum\nolimits_i \operatorname{diam}(S_i)^p \quad \text{for every} \quad S\in\wp(X),$$

where the infimum is taken over all finite open ε-coverings $\{S_i\}$ of S. This defines an outer measure $\mu^*_{\varepsilon,p}\colon\wp(X)\to\overline{\mathbb{R}}$. Set $\mu^*_p(S)=\lim_{\varepsilon\to 0}\mu^*_{\varepsilon,p}(S)$ for every $S\in\wp(X)$. The limit exists in $\overline{\mathbb{R}}$ and coincides with $\sup_{\varepsilon>0}\mu^*_{\varepsilon,p}(S)$. This defines another outer measure $\mu^*_p\colon\wp(X)\to\overline{\mathbb{R}}$. Following the setup of Theorem 11.7, *let $\mathcal{A}^*_p$ be the Borel σ-algebra of all μ^*_p-measurable sets (so that $\mathcal{A}^*_p$ includes the Borel σ-algebra $\mathcal{X}_\mathcal{T}$). The restriction λ^*_p of μ^*_p to $\mathcal{A}^*_p$, and the restriction λ_p of μ^*_p to $\mathcal{X}_\mathcal{T}$, viz.,*

$$\lambda^*_p=\mu^*_p|_{\mathcal{A}^*_p}\colon\mathcal{A}^*_p\to\overline{\mathbb{R}},$$

$$\lambda_p=\lambda^*_p|_{\mathcal{X}_\mathcal{T}}=\mu^*_p|_{\mathcal{X}_\mathcal{T}}\colon\mathcal{X}_\mathcal{T}\to\overline{\mathbb{R}},$$

are measures on $\mathcal{A}^*_p$ and on $\mathcal{X}_\mathcal{T}$. The outer measure μ^*_p and the measures λ^*_p and λ_p, are called p-dimensional Hausdorff measures. *Both λ^*_p and λ_p are Borel measures if $\mu^*_p(S)<\infty$ for every bounded set $S\in\wp(X)$, and λ^*_p is complete. For each set $S\in\wp(X)$ there is a unique real number* $\dim_H(S)=\inf\{p\geq 0\colon\mu^*_p(S)=0\}=\sup\{p>0\colon\mu^*_p(S)=\infty\}$, *called the* Hausdorff dimension *of S, such that $\mu^*_p(S)=\infty$ if $p<\dim_H(S)$, ${\mu^*}_{\dim_H(S)}(S)\in[0,\infty]$, and $\mu^*_p(S)=0$ if $\dim_H(S)<p$. Examples: If $X=\mathbb{R}$, then μ^*_1 coincides with the outer measure ℓ^* generated by the length function ℓ (and so λ_1 coincides with the Lebesgue measure); if C is the Cantor set, then $\dim_H(C)=\log 2/\log 3$ and ${\mu^*}_{\dim_H(C)}(C)=1$; if U is an open set in $X=\mathbb{R}^n$, then $\dim_H(U)=n$.*

If $X=\mathbb{R}^n$ (or $X=\mathbb{C}^n$) equipped with the usual (Euclidean) topology (or any equivalent topology), then X is a σ-compact locally compact metric (thus Hausdorff) space with a countable base (thus separable), and so it satisfies all the assumptions in every of the preceding propositions.

Notes: For the general topology summarized in Section 11.1 see, e.g., [12], [21]. This final section contains standard results on measures on topological spaces whose proofs can be found in many texts. For instance, Proposition 11.A can be found in [21, Theorem 5.1] and Proposition 11.B can be found in [21, Theorems 1.14, 1.15, Problem 5.Y(b)], [35, Problem 8.16, Theorem 9.21], and [6, Example IV.29-2(c)]. For Proposition 11.C see, e.g., [35, Lemma 13.9], and for Proposition 11.D see, e.g., [18, Theorem 52.A]. Propositions 11.E and 11.F are related to regular Borel measures (see, e.g., [35, Proposition 13.10] and [8, Problem 10,F]). Proposition 11.G (see, e.g., [36, Theorem 2.18]) is a first step towards Proposition 11.H, which says that Lebesgue measure is a prototype of a regular Borel measure (see, e.g., [35, Proposition 3.15]). Proposition 11.I deals with support of probability measures; see, e.g., [36, Exercise 2.9]. Proposition 11.J, 11.K, 11.L, and 11.M introduce Baire sets and measures (see, e.g., [35, Lemmas 13.6, 13.7, Problem 13.12(e,f), Proposition 13.15]). Proposition 11.N shows when Baire and Borel coincide, and Proposition 11.O gives a necessary condition for regularity (see, e.g., [35, Problem 13.1, Corollary 13.12]). Proposition 11.P deals with supports of Baire measures (see, e.g., [35, Problem 13.24]). Propositions 11.Q and 11.R on locally finite and Radon measures establish the connection between Borel and Radon measure. For the Hausdorff measures of Proposition 11.S see, e.g., [14, Chapter 1], [15, Chapter 3], and also [34].

Suggested Reading

Bauer [6], Brown and Pearcy [8], Halmos [18], Royden [35], Rudin [36].

12

Representation Theorems

12.1 Continuous Functions and Compact Support

Let $\mathbb{F}$ denote either the real field $\mathbb{R}$ or the complex field $\mathbb{C}$. A *scalar-valued* function $f\colon X \to \mathbb{F}$ on a nonempty set X is called *real-valued* if $\mathbb{F} = \mathbb{R}$, and *complex-valued* if $\mathbb{F} = \mathbb{C}$. If it is immaterial whether a scalar-valued function is real-valued or complex-valued, then we refer to it as $\mathbb{F}$-valued. The *kernel* $\mathcal{N}(f)$ and *range* $\mathcal{R}(f)$ of a function $f\colon X \to \mathbb{F}$ are the sets

$$\mathcal{N}(f) = f^{-1}(\{0\}) = \{x \in X\colon\ f(x) = 0\},$$
$$\mathcal{R}(f) = f(X) = \{\gamma \in \mathbb{F}\colon\ \gamma = f(x) \text{ for some } x \in X\}.$$

The *support* $[f]$ of an $\mathbb{F}$-valued function $f\colon X \to \mathbb{F}$ on a topological space X is the closure of the complement of its kernel:

$$[f] = (X \backslash \mathcal{N}(f))^{-} = \{x \in X\colon\ f(x) \neq 0\}^{-}.$$

From now on suppose X is a topological space and consider the usual (metric) topology of $\mathbb{F}$. Let $C_c(X)$ denote the collection of all $\mathbb{F}$-valued continuous functions on X with *compact support*; that is,

$$C_c(X) = \{f\colon X \to \mathbb{F}\colon\ f \text{ is continuous and } [f] \text{ is compact in } X\}.$$

If it is necessary to make it clear whether a collection of scalar-valued continuous functions on X consists of real-valued functions only, or whether it may contain complex-valued functions as well, then we may use the notation

$$C_c(X,\mathbb{R}) = \{f\colon X \to \mathbb{R}\colon\ f \text{ is continuous and } [f] \text{ is compact in } X\},$$

C.S. Kubrusly, *Essentials of Measure Theory*,
DOI 10.1007/978-3-319-22506-7_12

or

$$C_c(X,\mathbb{C}) = \{f\colon X \to \mathbb{C}\colon\ f \text{ is continuous and } [f] \text{ is compact in } X\}.$$

It is worth noticing that if X is a locally compact Hausdorff space, then the class of Baire sets coincides with the smallest σ-algebra of subsets of X upon which every function in $C_c(X,\mathbb{R})$ is measurable (see e.g., [35, p. 331]), which in turn is included in the Borel σ-algebra $\mathcal{X}_{\mathcal{T}}$ (cf. Section 11.4).

Lemma 12.1. *The range of a function $f \in C_c(X)$ is a compact subset of $\mathbb{F}$ (either the complex plane $\mathbb{C}$ or the real line $\mathbb{R}$ equipped with usual topology).*

Proof. Let K be the support of $f \in C_c(X)$. Since f is continuous and K is closed, it follows that $f(X) = f(K)$. We show that $f(K)$ is compact in $\mathbb{F}$.

Claim. Let $F\colon X \to Y$ be a continuous mapping of a topological space X into a topological space Y. If K is compact X, then $F(K)$ is compact in Y (i.e., continuous image of a compact set is compact).

Proof. Let $\mathcal{U}$ be a covering of $F(K)$ (i.e., $F(A) \subseteq \bigcup_{U\in\mathcal{U}} U$) consisting of open subsets U of Y. If F is continuous, then $F^{-1}(U)$ is an open subset of X for every $U \in \mathcal{U}$. Set $\mathcal{F}^{-1}(\mathcal{U}) = \{F^{-1}(U)\colon U \in \mathcal{U}\}$, a collection of open subsets of X. Since $K \subseteq F^{-1}(F(K)) \subseteq F^{-1}\big(\bigcup_{U\in\mathcal{U}} U\big) = \bigcup_{U\in\mathcal{U}} F^{-1}(U)$, it follows that $\mathcal{F}^{-1}(\mathcal{U})$ is a covering of K made up of open subsets of X. If K is compact, then there exists a finite subcollection of $\mathcal{F}^{-1}(\mathcal{U})$ covering K; that is, there exists $\{U_i\}_{i=1}^n \subseteq \mathcal{U}$ such that $K \subseteq \bigcup_{i=1}^n F^{-1}(U_i) \subseteq X$. So $F(K) \subseteq F\big(\bigcup_{i=1}^n F^{-1}(U_i)\big) \subseteq \bigcup_{i=1}^n U_i \subseteq Y$, and hence $F(K)$ is compact. □

Consider the collection $C_c(X)$ of all continuous $\mathbb{F}$-valued functions f on X with compact support $[f]$. Let K be a compact subset, and U an open subset, of the topological space X (with topology $\mathcal{T}$). For each K and U set

$$C_c(X)_K = \{f \in C_c(X)\colon\ f(X) \subseteq [0,1] \text{ and } f(K) = 1\},$$

$$C_c(X)^U = \{f \in C_c(X)\colon\ f(X) \subseteq [0,1] \text{ and } [f] \subseteq U\}.$$

Observe that when dealing with $C_c(X)_K$ or $C_c(X)^U$ we work only with functions in $C_c(X,\mathbb{R})$, since $f(X) \subseteq [0,1]\mathbb{R}$. A real-valued (or extended real-valued) function $f\colon X \to \mathbb{R}$ (or $f\colon X \to \overline{\mathbb{R}}$) on a topological space X is said to be *lower semicontinuous* if the inverse image of the open interval (α,∞) under f is an open subset of X for every real number α,

$$f^{-1}((\alpha,\infty)) = \{x \in X\colon\ f(x) > \alpha\} \in \mathcal{T} \quad \text{for every} \quad \alpha \in \mathbb{R}$$

(see Definition 1.2), and *upper semicontinuous* if the inverse image of the open interval $(-\infty,\alpha)$ under f is an open subset of X for every real α,

$$f^{-1}((-\infty,\alpha)) = \{x \in X\colon\ f(x) < \alpha\} \in \mathcal{T} \quad \text{for every} \quad \alpha \in \mathbb{R}.$$

Equipping $\mathbb{R}$ with its usual topology, it is readily verified that f *is continuous if and only if it is both lower and upper semicontinuous.* Examples of functions that are either lower or upper semicontinuous (but possibly not both): a characteristic function $\chi_U \colon X \to \mathbb{R}$ of an open set $U \subseteq X$ is lower semicontinuous (compare with Example 1.B) and, dually, a characteristic function $\chi_V \colon X \to \mathbb{R}$ of a closed set $V \subseteq X$ is upper semicontinuous (both of which are not continuous if $X = \mathbb{R}$ is equipped with its usual topology). Consider the definition of $\sup_\gamma f_\gamma$ and of $\inf_\gamma f_\gamma$ for any family $\{f_\gamma\}$ of functions $f_\gamma \colon X \to \mathbb{R}$ (or $f_\gamma \colon X \to \overline{\mathbb{R}}$) as in Section 1.4. It is also readily verified that $\sup_\gamma f_\gamma \colon \mathcal{X} \to \overline{\mathbb{R}}$ *is lower semicontinuous whenever each* f_γ *is*, and $\inf_\gamma f_\gamma \colon \mathcal{X} \to \overline{\mathbb{R}}$ *is upper semicontinuous whenever each* f_γ *is.*

Thus we can interpret continuity of functions in $C_c(X)_K$ and in $C_c(X)^U$ as functions being both lower and upper semicontinuous. *Urysohn Lemma* (stated below) is the result that says that $C_c(X)_K \cap C_c(X)^U \neq \varnothing$ if $K \subseteq U$.

Lemma 12.2. *If* U *and* K *are open and compact sets in a locally compact Hausdorff space* X *such that* $K \subseteq U$, *then there exists* $f \in C_c(X)_K \cap C_c(X)^U$ (i.e., *then there exists a function* $f \in C_c(X)$ *such that* $\chi_K \leq f \leq \chi_U$).

Proof. Let U and K be open and compact subsets of X such that $K \subseteq U$.

Claim. There exists a countable family of open subsets of X with compact closure, say $\{G_q\}_{q \in \mathbb{Q} \cap [0,1]}$ indexed by the rationals in $[0,1]$, such that

$$K \subseteq G_1^-, \quad G_0 \subseteq U, \quad \text{and} \quad G_q^- \subseteq G_p \;\; \text{whenever} \;\; p < q.$$

Proof. Let $\{q_k\}_{k=1}^\infty$ be an enumeration of the countable set $\mathbb{Q} \cap (0,1)[0,1]\mathbb{R}$ of all rational numbers in the open interval $(0,1)$. Set $q_0 = 0$ and $q_\infty = 1$. According to Theorem 11.3, there exist open subsets G_0 and G_1 of X with compact closures such that

$$K \subseteq G_1 \subseteq G_1^- \subseteq G_0 \subseteq G_0^- \subseteq U.$$

Take any $q \in \mathbb{Q} \cap (0,1)$. Using Theorem 11.3 again, there is an open set with compact closure G_q associated with q such that

$$G_1^- \subseteq G_q \subseteq G_q^- \subseteq G_0 \, .$$

Take any $q' \in \mathbb{Q} \cap (0,1)$ such that $q' \neq q$. Use Theorem 11.3 and get an open set with compact closure $G_{q'}$ such that if $q' < q$, then $G_q \subseteq G_{q'}$ and

$$G_1^- \subseteq G_q \subseteq G_q^- \subseteq G_{q'} \subseteq G_{q'}^- \subseteq G_0 \, ;$$

if $q < q'$, then $G_{q'} \subseteq G_q$ and

$$G_1^- \subseteq G_{q'} \subseteq G_{q'}^- \subseteq G_q \subseteq G_q^- \subseteq G_0 \,.$$

Proceeding along this line, if we already have a finite set $\{q_i\}_{i=1}^n$ of rational numbers from $\mathbb{Q} \cap (0,1)$, and an associated set of open sets with compact closure $\{G_{q_i}\}_{i=1}^n$ such that $G_{q_{i+1}} \subseteq G_{q_i}$ for $q_i < q_{i+1}$ and

$$G_{q_{i+1}} \subseteq G_{q_{i+1}}^- \subseteq G_{q_i} \subseteq G_{q_i}^-,$$

then take any $q_{n+1} \in \mathbb{Q} \cap (0,1)$ such that $q_{n+1} \neq q_i$ for every integer $i \in [1,n]$. Using Theorem 11.3 once again we get an open set with compact closure $G_{q_{n+1}}$ such that

$$G_{q_j}^- \subseteq G_{q_{n+1}} \subseteq G_{q_{n+1}}^- \subseteq G_{q_\ell},$$

where q_ℓ is the largest number from $\{q_i\}_{i=1}^n$ which is smaller than q_{n+1}, and q_j is the smallest number from $\{q_i\}_{i=1}^n$ which is greater than q_{n+1}. Continuing this way we obtain, for each $q \in \{q_k\}_{k=1}^\infty \cup \{q_0, q_\infty\} = \mathbb{Q} \cap [0,1]$, an open set with compact closure G_q, such that $K \subseteq G_1$, $G_0^- \subseteq U$, and $G_q^- \subseteq G_p$ whenever $p < q$. This concludes the proof of the claimed statement. □

For each $q \in \{q_k\}_{k=1}^\infty \cup \{q_0, q_\infty\} = \mathbb{Q} \cap [0,1]$ take the characteristic functions of G_q and $X \backslash G_q^-$, and consider the following functions of X into $\mathbb{R}$:

$$f_q = q\,\chi_{G_q}, \qquad g_q = q\,\chi_{X\backslash G_q^-} + \chi_{G_q^-},$$
$$f = \sup_k f_{q_k}, \qquad g = \inf_k g_{q_k}.$$

As we saw before, characteristic functions of open sets are lower semicontinuous, and so are the sup of lower semicontinuous functions; dually, characteristic functions of closed sets are upper semicontinuous, and so are the inf of upper semicontinuous functions. Therefore, it is readily verified that each f_q and f are lower semicontinuous functions, while each g_q and g are upper semicontinuous functions. Now we show that

$$f \leq g \quad \text{and} \quad g \leq f, \quad \text{and hence} \quad f = g.$$

Indeed, if $g_p(x) < f_q(x)$, then $q < p$, $x \in G_p$ and $x \notin G_q^-$ (i.e., $x \in (G_p \backslash G_q^-)$). But $q < p$ implies $G_p^- \subseteq G_q$ so that $G_p \backslash G_q^- = \varnothing$. So $f_q < g_p$ for every p and q in $\mathbb{Q} \cap [0,1]$, which implies that $f \leq g$. On the other hand, if $f(x) < g(x)$ for some $x \in X$, then there are $p, q \in \mathbb{Q} \cap [0,1]$ such that $f(x) < p < q < g(x)$. But $f(x) < p$ implies that $x \notin G_p$, and $q < g(x)$ implies that $x \in G_q^-$, so that $x \in G_q^- \backslash G_p$. Again, $p < q$ implies $G_q^- \subseteq G_p$, so that $G_q^- \backslash G_p = \varnothing$. Then we get $g \leq f$. Hence $f = g$. Therefore, since f is lower semicontinuous, g is upper semicontinuous, and $f = g$, it follows that

$$f \colon X \to \mathbb{R} \text{ is continuous}.$$

Clearly, f_q has a compact support $G_q^- \subseteq [0,1]$ and $f_q(X) \subseteq [0,1]$ for every q in $\mathbb{Q} \cap [0,1]$, and so (i) f has compact support and (ii) $f(X) \subseteq [0,1]$. Since $K \subseteq [f_q]$ for all $q \in \mathbb{Q} \cap [0,1]$ and $\sup(\mathbb{Q} \cap [0,1]) = 1$, we get (iii) $f(K) = 1$. Since $\bigcup_q G_q \subseteq G_0 \subseteq U$, it is also follows that (iv) $[f] \subseteq U$. Outcome:

$$f \in C_c(X)_K \cap C_c(X)^U. \qquad \square$$

Corollary 12.3. *Let U_i and K be subsets of a locally compact Hausdorff space. If $\{U_i\}_{i=1}^n$ is a open covering of a compact set K, then there exists $f_i \in C_c(X)^{U_i}$ for each $i \in [1,n]$ such that $\sum_{i=1}^n f_i(x) = 1$ for every $x \in K$.*

Proof. Since $K \subseteq U = \bigcup_{i=1}^n U_i$, where K is compact and U is open, there exists an open set G with compact closure (cf. Theorem 11.3) such that

$$K \subseteq G \subseteq G^- \subseteq U.$$

Take an arbitrary integer $i \in [1,n]$. Set $G_i = G \cap U_i \subseteq U_i$, which is an open set (intersection of two open sets) with compact closure (since $(G \cap U_i)^- \subseteq G_i^-$, which is compact — see Lemma 11.2). Let K_i be the finite union of all sets G_j^- such that $G_j \subseteq U_i$. Thus K_i is compact (since each G_j^- is compact and K_i is a finite union of them) and $K_i \subseteq U_i$, where U_i is open. Therefore, by Lemma 12.2, there exists a function $g_i: X \to \mathbb{R}$ such that

$$g_i \in C_c(X)_{K_i} \cap C_c(X)^{U_i}.$$

Now set

$$f_1 = g_1 \quad \text{and} \quad f_i = \prod_{j=1}^{i-1}(1 - g_j)g_i \quad \text{if} \quad i \geq 2.$$

Since $g_i \in C_c(X)^{U_i}$, it follows by the definition of $C_c(X)^{U_i}$ that

$$f_i \in C_c(X)^{U_i}.$$

It is readily verified by induction that

$$\sum_{i=1}^n f_i = 1 - \prod_{i=1}^n (1 - g_i).$$

Next note that $K \subseteq \bigcup_{i=1}^n K_i$. Indeed, since $\bigcup_{i=1}^n G_i^- = \bigcup_{i=1}^n K_i$, it follows that $K \subseteq G \subseteq \bigcup_{i=1}^n (G \cap U_i) = \bigcup_{i=1}^n G_i \subseteq \bigcup_{i=1}^n G_i^- = \bigcup_{i=1}^n K_i$. If $g_i(x) < 1$ for every $i \in [1,n]$ for some $x \in K \subset \bigcup_{j=1}^n K_j$, then $g_i \notin C_c(X)_{K_i}$, which is a contradiction. Thus for every $x \in K$ there exists an integer $i' \in [1,n]$ such that $g_{i'}(x) = 1$. Hence $1 - g_{i'}(x) = 0$, and so $\prod_{i=1}^n (1 - g_i(x)) = 0$, for every $x \in K$. Therefore,

$$\sum_{i=1}^n f_i(x) = 1 \quad \text{for every} \quad x \in K. \qquad \square$$

The preceding consequence of the Urysohn Lemma (i.e., Corollary 12.3), together with the Urysohn Lemma itself (i.e., Lemma 12.2), will play an important role for proving the first version of the Riesz Representation Theorem, namely, Theorem 12.5 in Section 12.3.

12.2 Bounded Linear Functionals

Let S be a nonempty set and consider the set $\mathbb{F}^S$ of all functions $f\colon S \to \mathbb{F}$ of S into $\mathbb{F}$ (i.e., of all $\mathbb{F}$-valued function on S). It is readily verified that $\mathbb{F}^S$ is a *linear space* (over the field $\mathbb{F}$) where addition and scalar multiplication in $\mathbb{F}^S$ are defined pointwise (thus the linear structure of $\mathbb{F}^S$ is inherited by that in $\mathbb{F}$). Let $L(S) \subseteq \mathbb{F}^S$ be any linear manifold of the linear space $\mathbb{F}^S$. (That is, $L(S) \subseteq \mathbb{F}^S$ is such that $f + g$ and αf — defined as $(f+g)(s) = f(s) + g(s)$ and $(\alpha f)(s) = \alpha f(s)$ for every $s \in S$ and every $\alpha \in \mathbb{F}$ — lie in $L(S)$ whenever $f, g \in L(S)$ and $\alpha \in \mathbb{F}$.) So $L(S)$ is itself a linear space over $\mathbb{F}$; if $\mathbb{F} = \mathbb{C}$ or $\mathbb{F} = \mathbb{R}$, then $L(S)$ is referred to as a *complex linear space* or a *real linear space* and, when necessary, it will be denoted by $L(S, \mathbb{C})$ or $L(S, \mathbb{R})$, respectively — see the remarks that follow Lemma 4.5. Take a *linear functional* $\Phi\colon L(S) \to \mathbb{F}$ on the linear space $L(S)$, which means that Φ is *additive* and *homogeneous* (i.e., $\Phi(f+g) = \Phi(f) + \Phi(g)$ and $\Phi(\alpha f) = \alpha\,\Phi(f)$ for every pair of functions $f, g \in L(S)$ and every scalar $\alpha \in \mathbb{F}$). A functional Φ on $L(S)$ is said to be *positive* if $\Phi(f) \geq 0$ for every $f \geq 0$ (i.e., for every $f \in L(S)$ with nonnegative range $f(S)$); in other words, if

$$\Phi(f) \geq 0 \text{ for every } f \in L(S) \text{ such that } f(s) \geq 0 \text{ for every } s \in S.$$

Let $\Phi\colon L(S) \to \mathbb{F}$ be a positive linear functional on the linear space $L(S)$. If $f, g \in L(S)$ are such that $f \leq g$, then $0 \leq \Phi(g - f) = \Phi(g) - \Phi(f)$. So

$$\Phi(f) \leq \Phi(g) \quad \text{whenever} \quad f \leq g.$$

Moreover, if $f\colon S \to \mathbb{R}$ is a real-valued function in $L(S)$, then (considering the decomposition $f = f^+ - f^-$ of Section 1.2 where $f^+ \geq 0$ and $f^- \geq 0$)

$$\begin{aligned} |\Phi(f)| &= |\Phi(f^+ - f^-)| = |\Phi(f^+) - \Phi(f^-)| \\ &\leq |\Phi(f^+)| + |\Phi(f^-)| = \Phi(f^+) + \Phi(f^-) = \Phi(f^+ + f^-) = \Phi(|f|). \end{aligned}$$

Also, if the function $1\colon S \to \mathbb{F}$ (i.e., $1(s) = 1$ for all $s \in S$) lies in $L(S)$, and if $f\colon S \to \mathbb{F}$ is such that $|f| \leq 1$, then $0 \leq \Phi(1 - |f|) = \Phi(1) - \Phi(|f|)$. Thus

$$\Phi(|f|) \leq \Phi(1) \quad \text{whenever} \quad |f| \leq 1.$$

Summing up: *If $L(S)$ is a linear space of $\mathbb{F}$-valued functions on a nonempty set S, if $\Phi\colon L(S) \to \mathbb{F}$ is a positive linear functional, if $f\colon S \to \mathbb{R}$ is a real-valued function in $L(S)$, and if the constant function* 1 *lies in $L(S)$, then*

$$|\Phi(f)| \leq \Phi(|f|) \leq \Phi(1) \quad \text{whenever} \quad |f| \leq 1.$$

An important particular case: A *bounded function* $f: S \to \mathbb{F}$ is a function such that its range $f(S)$ is a bounded subset of the metric space $\mathbb{F}$ (equipped with the usual metric — i.e., a function $f: S \to \mathbb{F}$ such that $\sup_{s\in S}|f(s)| < \infty$ — cf. Remarks on Boundedness in Section 11.1). Now consider the collection $B(S)$ of all $\mathbb{F}$-valued bounded functions on S,

$$B(S) = \{f: S \to \mathbb{F}: f \text{ is bounded}\},$$

which is a linear space (sum and scalar multiplication of bounded functions are again bounded functions); in fact, it is a normed space with *sup-norm*

$$\|f\|_\infty = \sup_{s\in S}|f(s)|.$$

Note that the constant function 1 lies in $B(S)$. If a function f lies in $B(S)$, then $|f| \le 1$ means $|f(s)| \le 1$ for every $s \in S$, which is equivalent to saying that $\sup_{s\in S}|f(s)| \le 1$. Therefore,

$$\|f\|_\infty \le 1 \quad \text{if and only if} \quad |f| \le 1.$$

If $L(S) \subseteq B(S)$ is a linear manifold of $B(S)$, then $L(S)$ inherits the norm of $B(S)$ and $(L(S), \|\cdot\|_\infty)$ is a normed space. A *bounded linear functional* $\varPhi: L(S) \to \mathbb{F}$ on the normed space $L(S)$ is a linear functional such that $\sup_{f\neq 0}\frac{|\varPhi(f)|}{\|f\|_\infty} < \infty$. (It is worth noticing that, if $\varPhi \neq 0$, then the image of $\varPhi$, viz., $\varPhi(L(S))$, is not bounded in $\mathbb{F}$ — indeed, $\varPhi(L(S)) = \mathbb{F}$: nonzero *linear* functionals are surjective.) The *induced uniform norm* of a bounded linear functional (in fact, the induced uniform norm on the normed space of all $\mathbb{F}$-valued bounded linear functionals on $L(S)$ — the *dual* of $L(S)$) is

$$\|\varPhi\| = \sup_{f\neq 0}\frac{|\varPhi(f)|}{\|f\|_\infty} = \sup_{\|f\|_\infty\le 1}|\varPhi(f)| = \sup_{|f|\le 1}|\varPhi(f)|.$$

Hence, if the bounded linear functional $\varPhi$ is positive and if $1 \in L(S)$, then $\varPhi(1) \le \|\varPhi\| = \sup_{|f|\le 1}|\varPhi(f)|$, and also $\sup_{|f|\le 1}\varPhi(|f|) \le \varPhi(1)$. Moreover, if $f \in L(S)$ is real-valued, then $|\varPhi(f)| \le \varPhi(|f|)$. Therefore, if $\mathbb{F} = \mathbb{R}$, then

$$\varPhi(1) \le \|\varPhi\| = \sup_{|f|\le 1}|\varPhi(f)| \le \sup_{|f|\le 1}\varPhi(|f|) \le \varPhi(1).$$

Summing up: *If $L(S)$ is a real linear space of real-valued bounded functions on a nonempty set S, if $\varPhi$ is a real-valued positive bounded linear functional on $L(S)$, and if $1 \in L(S)$, then*

$$\|\varPhi\| = \varPhi(1).$$

Again, the complex linear space of all *complex-valued bounded functions* on S, and the real linear space of all *real-valued bounded functions* on S will, when necessary, be denoted respectively by

$$B(S,\mathbb{C}) = \{f\colon S \to \mathbb{C}\colon\ f \text{ is bounded}\},$$

and

$$B(S,\mathbb{R}) = \{f\colon S \to \mathbb{R}\colon\ f \text{ is bounded}\}.$$

A *linear lattice* is a linear space equipped with a partial ordering such that every pair of elements has an infimum and a supremum in it. That is,

$$f \wedge g = \inf\{f,g\} \quad \text{and} \quad f \vee g = \sup\{f,g\}$$

lie in the linear space for every f and g in the linear space. Set $\mathbb{F} = \mathbb{R}$ and let $L(S,\mathbb{R})$ be a real linear space of real-valued functions on S. For f, g in $L(S,\mathbb{R})$ let $f \wedge g$ and $f \vee g$ be defined by $(f \wedge g)(s) = \min\{f(s), g(s)\}$ and $(f \vee g)(s) = \max\{f(s), g(s)\}$ for each $s \in S$. If $L(S,\mathbb{R})$ has (in addition to the linear properties) the property that $f \wedge g$ and $f \vee g$ lie in $L(S,\mathbb{R})$ whenever f, g lie in $L(S,\mathbb{R})$, then $L(S,\mathbb{R})$ is a *linear lattice* of real-valued functions on S. Example: the linear space $B(S,\mathbb{R})$ of all *real-valued bounded functions* on S equipped with the partial ordering $\leq$ (defined by $f \leq g$ if $f(s) \leq g(s)$ for every $s \in S$), which induces the above binary operations $\wedge$ and $\vee$, is a linear lattice that contains the constant function $1(s) = 1$ for all $s \in S$.

Now we equip S with a topology. Let $S = X$ be a topological space and consider the set $C(X)$ of all $\mathbb{F}$-valued continuous functions on X:

$$C(X) = \{f\colon X \to \mathbb{F}\colon\ f \text{ is continuous}\}.$$

Since sum and scalar multiples of continuous functions are again continuous functions, $C(X)$ is a linear manifold of the linear space $\mathbb{F}^X$ of all $\mathbb{F}$-valued functions on X, and so $C(X)$ is itself a linear space. Clearly, $C_c(X) \subseteq C(X)$. Since the support of a sum of two functions in $C_c(X)$ is the union of their support (because finite union of compact sets is compact), we can infer that $C_c(X)$ is a linear manifold of $C(X)$, and hence $C_c(X)$ is itself a linear space. If X is a compact space, then closed subsets of X are compact sets (Lemma 11.2(a)) so that $C_c(X) = C(X)$ by the definition of $C_c(X)$; that is,

$$C_c(X) = C(X) \quad \text{whenever } X \text{ is a compact topological space.}$$

By Lemma 12.1 (recalling that in $\mathbb{F}$ compact means closed and bounded),

$$C_c(X) \subseteq B(X),$$

and so the linear space $C_c(X)$, and also $C(X)$ if X is compact, is again a normed space equipped with the sup-norm $\|\cdot\|_\infty$.

Once again, the complex linear space of all *complex-valued continuous functions* on X, and the real linear space of all *real-valued continuous functions* on X will, when necessary, be denoted respectively by

$$C(X,\mathbb{C}) = \{f\colon X \to \mathbb{C}\colon f \text{ is continuous}\},$$

and

$$C(X,\mathbb{R}) = \{f\colon X \to \mathbb{R}\colon f \text{ is continuous}\}.$$

Note that $C(X,\mathbb{R})$ has the additional property that if f,g lie in $C(X,\mathbb{R})$, then so does the functions $f \wedge g$ and $f \vee g$. This means that $C(X,\mathbb{R})$ is a linear lattice of real-valued functions on X. Moreover, the constant function $1(x)=1$ for all $x \in X$ lies in $C(X)$. Also, if X is compact, then $C(X) \subseteq B(X)$ ($\mathbb{F}$-valued continuous functions on a compact space are bounded — cf. Claim in the proof of Lemma 12.1 plus Remarks on Boundedness in Section 11.1).

Lemma 12.4. *Let $L(S,\mathbb{R})$ be a linear lattice of real-valued bounded functions on a nonempty set S. If $\Phi\colon L(S,\mathbb{R}) \to \mathbb{R}$ is a real-valued bounded linear functional on $L(S,\mathbb{R})$, then there are two positive real-valued bounded linear functionals $\Phi^+\colon L(S,\mathbb{R}) \to \mathbb{R}$ and $\Phi^-\colon L(S,\mathbb{R}) \to \mathbb{R}$ on $L(S,\mathbb{R})$ such that*

$$\Phi = \Phi^+ - \Phi^-.$$

If $L(S,\mathbb{R})$ contains the function 1, then $\|\Phi\| = \Phi^+(1) + \Phi^-(1)$.

Proof. Let $\Phi\colon L(S,\mathbb{R}) \to \mathbb{R}$ be a bounded linear functional. Set

$$\Psi^+(f) = \sup_{0 \le \varphi \le f} \Phi(\varphi) \quad \text{for each} \quad 0 \le f \in L(S,\mathbb{R}).$$

Thus $\Psi^+(0) = 0$ (since $\Phi(0) = 0$ by linearity of Φ) and hence $\Psi^+(f) \ge 0$ for every $0 \le f \in L(S,\mathbb{R})$, and $\sup_{0 \le f} \Psi^+(f) < \infty$ (since Φ is bounded). Note that since Φ is linear, for every $0 \le \alpha \in \mathbb{R}$ and every $0 \le f \in L(S,\mathbb{R})$,

$$\text{(i)} \qquad \Psi^+(\alpha f) = \alpha \Psi^+(f).$$

Now take an arbitrary pair of functions $0 \le f,g$ in $L(S,\mathbb{R})$. Let φ, ψ be functions in $L(S,\mathbb{R})$ such that $0 \le \varphi \le f$ and $0 \le \psi \le g$. Thus, by the definition of Ψ^+, since Φ is additive, $\Phi(\varphi) + \Phi(\psi) = \Phi(\varphi + \psi) \le \Psi^+(f+g)$, and so (definition of Ψ^+ again — taking the sup for all such φ and ψ)

$$\Psi^+(f) + \Psi^+(g) \le \Psi^+(f+g).$$

Conversely, take any function $h \in L(S,\mathbb{R})$ such that $0 \le h \le f+g$. Since $0 \le h \wedge f = \inf\{h,f\} = f$, and since $0 \le h \wedge f \le h \le f+g$, it follows that $0 \le h - (h \wedge f) \le g$. (Indeed, $(h - (h \wedge f))(s) = 0$ if $h(s) \le f(s)$ and $h - (h \wedge f)(s) = (h-f)(s) \le g(s)$ if $f(s) \le h(s)$.) Again, by the definition of Ψ^+, since Φ is linear, $\Phi(h) = \Phi(h \wedge f) + \Phi(h - (h \wedge f)) \le \Psi^+(f) + \Psi^+(g)$ and so (definition of Ψ^+ again — taking the sup for all such h)

$$\Psi^+(f+g) \le \Psi^+(f) + \Psi^+(g).$$

Therefore, for every $0 \le f, g \in L(S,\mathbb{R})$,

$$\text{(ii)} \qquad \Psi^+(f+g) = \Psi^+(f) + \Psi^+(g).$$

Since every real-valued function $f \in L(S,\mathbb{R})$ can be written as $f = f^+ - f^-$, where $f^+ = f \vee 0$ and $f^- = -f \wedge 0$ (positive and negative parts of f) are nonnegative functions in $L(S,\mathbb{R})$, the values $\Psi^+(f^+)$ and $\Psi^+(f^-)$ are well defined. Thus consider the functional $\Phi^+ \colon L(S,\mathbb{R}) \to \mathbb{R}$ defined on the whole $L(S,\mathbb{R})$ as follows. For every $f \in L(S,\mathbb{R})$,

$$\Phi^+(f) = \Psi^+(f^+) - \Psi^+(f^-).$$

First we show that Φ^+ is homogeneous. Take any $f \in L(S,\mathbb{R})$. Since $\alpha \ge 0$ implies $(\alpha f)^+ = \alpha f^+$ and $(\alpha f)^- = \alpha f^-$, we get by (i) that $\Psi^+((\alpha f)^+) = \alpha \Psi^+(f^+)$ and $\Psi^+((\alpha f)^-) = \alpha \Psi^+(f^-)$, and hence $\Phi^+(\alpha f) = \alpha \Phi^+(f)$ whenever $\alpha \ge 0$. Next observe that since $f^- = (-f)^+$ and $f^+ = (-f)^-$,

$$\Phi^+(-f) = \Psi^+((-f)^+) - \Psi^+((-f)^-) = \Psi^+(f^-) - \Psi^+(f^+) = -\Phi^+(f).$$

So $\Phi^+(\alpha f) = \Phi^+(-|\alpha| f) = -\Phi^+(|\alpha| f) = -|\alpha|\Phi^+(f) = \alpha\Phi^+(f)$ if $\alpha \le 0$. Therefore, for every $\alpha \in \mathbb{R}$ and every $f \in L(S,\mathbb{R})$,

$$\text{(i}'\text{)} \qquad \Phi^+(\alpha f) = \alpha \Phi^+(f).$$

In order to show additivity, take any f in $L(S,\mathbb{R})$, and let φ, ψ in $L(S,\mathbb{R})$ be such that $0 \le \varphi$, $0 \le \psi$, $0 \le f + \varphi$ and $0 \le f + \psi$. Note that by (ii),

$$\Psi^+(f+\varphi) + \Psi^+(\psi) = \Psi^+(f+\varphi+\psi) = \Psi^+(f+\psi) + \Psi^+(\varphi),$$

and so

$$\Psi^+(f+\varphi) - \Psi^+(\varphi) = \Psi^+(f+\psi) - \Psi^+(\psi).$$

In particular, with $\varphi = f^-$ (so that $f + \varphi = f^+$), we get

$$\Phi^+(f) = \Psi^+(f^+) - \Psi^+(f^-) = \Psi^+(f+\psi) - \Psi^+(\psi).$$

Summing up: *For every* $f \in L(S,\mathbb{R})$,

$$\Phi^+(f) = \Psi^+(f+\psi) - \Psi^+(\psi)$$

whenever $0 \le \psi$ *is such that* $0 \le f + \psi$. Thus take any $g \in L(S,\mathbb{R})$, and let ψ be such that, in addition, $0 \le g + \psi$. Then

$$\Phi^+(g) = \Psi^+(g+\psi) - \Psi^+(\psi).$$

Therefore, by using (i) and (ii),

$$\begin{aligned}\Phi^+(f)+\Phi^+(g) &= \Psi^+(f+\psi)+\Psi^+(g+\psi)-2\,\Psi^+(\psi)\\ &= \Psi^+(f+g+2\,\psi)-\Psi^+(2\,\psi).\end{aligned}$$

But $0 \leq f+\psi$ and $0 \leq g+\psi$ imply that $0 \leq f+g+2\,\psi$. Hence

$$\Phi^+(f+g) = \Psi^+(f+g+2\,\psi)-\Psi^+(2\,\psi).$$

Outcome: For every $f, g \in L(S,\mathbb{R})$,

(ii′) $$\Phi^+(f+g) = \Phi^+(f)+\Phi^+(g).$$

Properties (i′) and (ii′) mean that the functional $\Phi^+\colon L(S,\mathbb{R}) \to \mathbb{R}$ is linear. It is also bounded since $\sup_{f\geq 0}\Psi^+(f) < \infty$. Indeed,

$$\sup_{f\in L(S,\mathbb{R})} \Phi^+(f) = \sup_{f\in L(S,\mathbb{R})} \big(\Psi^+(f^+)-\Psi^+(f^-)\big) \leq 2\sup_{f\geq 0}\Psi^+(f) < \infty.$$

Moreover, observe that Φ^+ is positive as well:

$$\Phi^+(f) = \Psi^+(f) \geq 0 \quad \text{for every} \quad f \geq 0.$$

Next consider the functional $\Phi^-\colon L(S,\mathbb{R}) \to \mathbb{R}$ defined on $L(S,\mathbb{R})$ by

$$\Phi^- = \Phi^+ - \Phi,$$

which is trivially linear and bounded (since Φ and Φ^+ are). Since $\Phi(f) \leq \Psi^+(f)$ for every $f \geq 0$ by the definition of Ψ^+, it follows that Φ^- is positive:

$$\Phi^-(f) = \Phi^+(f)-\Phi(f) = \Psi^+(f)-\Phi(f) \geq 0 \quad \text{for every} \quad f \geq 0.$$

Finally, suppose $L(S,\mathbb{R})$ contains the function 1. Since $L(S,\mathbb{R})$ is a real linear space of real-valued bounded functions, since Φ^+ and Φ^- are real-valued positive bounded linear functionals on $L(S,\mathbb{R})$, and since $1 \in L(S,\mathbb{R})$,

$$\|\Phi^+\| = \Phi^+(1) \quad \text{and} \quad \|\Phi^-\| = \Phi^-(1),$$

so that

$$\|\Phi\| = \|\Phi^+ - \Phi^-\| \leq \|\Phi^+\| + \|\Phi^-\| = \Phi^+(1)+\Phi^-(1).$$

On the other hand, take any $\varphi \in L(S,\mathbb{R})$ such that $0 \leq \varphi \leq 1$, equivalently, such that $|2\,\varphi - 1| \leq 1$. First recall by the definition of Φ^- that

$$\Phi^+(1)+\Phi^-(1) = 2\,\Phi^+(1)+\Phi(1).$$

Next observe that according to the definitions of Ψ^+ and Φ^+,

$$\Phi^+(1) = \Psi^+(1) = \sup_{0\leq\varphi\leq 1}\Phi(\varphi).$$

Moreover, since $|2\varphi - 1| \le 1$, and since Φ is linear,

$$2\Phi(\varphi) - \Phi(1) = \Phi(2\varphi - 1) \le \sup_{|f|\le 1} \|\Phi(f)\| \le \|\Phi\|.$$

Hence,

$$\Phi^+(1) + \Phi^-(1) = \sup_{0\le\varphi\le 1} 2\Phi(\varphi) + \Phi(1) \le \|\Phi\|.$$

Therefore, $\Phi^+(1) + \Phi^-(1) = \|\Phi\|$. □

12.3 The Riesz Representation Theorem

Recall that *continuous functions are Borel measurable*. So *every* $f \in C_c(X)$ *is* $\mathcal{X}_\mathcal{T}$*-measurable*. Actually, $f \in C_c(X)$ *is integrable with respect to a Borel measure* (and so is $f \in C(X) = C_c(X)$ if X is compact), since these functions are bounded ($C_c(X) \subseteq B(X)$), and since Borel measures are finite on compact sets (cf. Lemma 4.4(b) and Problem 3.3(a,b)).

This section contains the central theme of the chapter. The next theorem, Theorem 12.5, is one of the forms of the celebrated *Riesz Representation Theorem*, which might be called *Riesz–Markov–Radon–Banach–Kakutani–Halmos Theorem*, since many versions of it have traveled a long way from Riesz's original 1909 paper to the general form on locally compact Hausdorff space developed in Halmos's book [18]. For a short account of such a long story see [35, p. 354]. We will see further versions up to Section 12.4.

Theorem 12.5. *Let X be a locally compact Hausdorff space. If Φ is a positive linear functional on $C_c(X)$, then there exists a unique quasiregular Borel measure μ on the Borel σ-algebra $\mathcal{X}_\mathcal{T}$ of subsets of X such that*

$$\Phi(f) = \int f\,d\mu \quad \text{for every} \quad f \in C_c(X).$$

Proof. Let X be a locally compact Hausdorff space. Take the linear space of all $\mathbb{F}$-valued continuous functions on X with a compact support,

$$C_c(X) = \big\{f\colon X \to \mathbb{F}\colon\ f \text{ is continuous and } [f] \text{ is compact in } X\big\},$$

and let $\Phi\colon C_c(X) \to \mathbb{F}$ be a positive linear functional,

$$\Phi(f) \ge 0 \text{ for every } f \in C_c(X) \text{ such that } f(x) \ge 0 \text{ for every } x \in X,$$

and consider the set function $\mu^\#\colon \mathcal{T} \to \overline{\mathbb{R}}$ on the topology of X defined by

$$\mu^\#(U) = \sup_{f \in C_c(X)^U} \Phi(f),$$

where $C_c(X)^U$ is a collection of *positive* functions for every open set $U \in \mathcal{T}$:

$$C_c(X)^U = \{f \in C_c(X)\colon\ f(X) \subseteq [0,1] \text{ and } [f] \subseteq U\} \subseteq C_c(X, \mathbb{R}).$$

First we show that $\mu^{\#}$ satisfies the properties (1) to (7) in the definition of an inner content (Section 11.3). Let G and U be arbitrary sets in $\mathcal{T}$. Clearly,

$$\text{(1)} \qquad \mu^{\#}(\varnothing) = 0, \qquad \text{(2)} \qquad \mu^{\#}(U) \geq 0,$$

since Φ positive, and it is readily verified that

$$\text{(3)} \qquad \mu^{\#}(U) < \infty \quad \text{whenever} \quad U^- \text{ is compact},$$

$$\text{(4)} \qquad \mu^{\#}(G) \leq \mu^{\#}(U) \quad \text{whenever} \quad G \subseteq U.$$

Before verifying property (5) we need to show property (6). Let $\{U_k\}$ be a countable family of open sets and set $U = \bigcup_k U_k$, which is open. Take an arbitrary $f \in C_c(X)^U$. Thus $\{U_k\}$ is an open covering of the compact set $[f]$, and hence there exists a finite open subcovering $\{U_i\}_{i=1}^n \subseteq \{U_k\}$ of $[f]$. In this case, Corollary 12.3 ensures that there exists $f_i \in C_c(X)^{U_i}$ for each $i \in [1, n]$ such that $\sum_{i=1}^n f_i(x) = 1$ for every $x \in [f]$. Thus

$$f = \sum_{i=1}^{n} f_i f \quad \text{where each } f_i f \text{ lies in } C_c(X)^{U_i}.$$

Hence, by the definition of $\mu^{\#}$, and since Φ is linear,

$$\Phi(f) = \sum_{i=1}^{n} \Phi(f_i f) \leq \sum_{i=1}^{n} \mu^{\#}(U_i) \leq \sum_{i=1}^{\infty} \mu^{\#}(U_i),$$

and so

$$\mu^{\#}(U) = \sup_{f \in C_c(X)^U} \Phi(f) \leq \sum_{i=1}^{\infty} \mu^{\#}(U_i).$$

Therefore,

$$\text{(6)} \qquad \mu^{\#}\big(\textstyle\bigcup_k U_n\big) \leq \sum_k \mu^{\#}(U_k) \quad \text{for every countable family } \{U_k\} \subseteq \mathcal{T}.$$

Now suppose that $G \cap U = \varnothing$. If $f \in C_c(X)^U$ and if $g \in C_c(X)^G$, then $f + g \in C_c(X)^{U \cup G}$. Thus (cf. definition of $\mu^{\#}$),

$$\Phi(f) + \Phi(g) = \Phi(f + g) \leq \mu^{\#}(U \cup G)$$

for every $f \in C_c(X)^U$ and every $g \in C_c(X)^G$. This implies that

$$\begin{aligned} \mu^{\#}(U) + \mu^{\#}(G) &= \sup_{f \in C_c(X)^U} \Phi(f) + \sup_{g \in C_c(X)^G} \Phi(g) \\ &\leq \sup_{f \in C_c(X)^U,\ g \in C_c(X)^G} \big(\Phi(f) + \Phi(g)\big) \leq \mu^{\#}(U \cup G). \end{aligned}$$

Since we have already verified that $\mu^{\#}(U \cup G) \leq \mu^{\#}(U) + \mu^{\#}(G)$ (independently of the disjointness assumption), it follows that

$$(5) \qquad \mu^{\#}(G \cup U) = \mu^{\#}(G) + \mu^{\#}(U) \quad \text{whenever} \quad G \cap U = \varnothing.$$

Thus $\mu^{\#}$ is a content, which in fact is an inner content, since

$$\sup_{G \in \mathcal{G}(U)} \mu^{\#}(G) = \sup_{G \in \mathcal{G}(U)} \sup_{f \in C_c(X)^G} \Phi(f) = \sup_{f \in C_c(X)^U} \Phi(f) = \mu^{\#}(U),$$

with $\mathcal{G}(U) = \{G \in \mathcal{T}\colon G^- \subseteq U,\ G \text{ compact}\}$ for each $U \in \mathcal{T}$, so that

$$(7) \qquad \mu^{\#}(U) = \sup\big\{\mu^{\#}(G)\colon\ G^- \subseteq U,\ G^- \text{ compact}\big\}.$$

Therefore, by Lemma 11.8, the set function $\mu^*\colon \wp(X) \to \overline{\mathbb{R}}$ defined by

$$\mu^*(S) = \inf\big\{\mu^{\#}(U)\colon\ S \subseteq U\big\} \quad \text{for every} \quad S \in \wp(X)$$

is a quasiregular outer measure such that $\mu^*(S) < \infty$ if S is bounded. So

$$\mu = \mu^*|_{\mathcal{X}_\mathcal{T}}\colon \mathcal{X}_\mathcal{T} \to \overline{\mathbb{R}}$$

is a quasiregular Borel measure on the Borel σ-algebra $\mathcal{X}_\mathcal{T}$ generated by the topology $\mathcal{T}$ according to Theorem 11.7. Next we show that

$$\Phi(f) = \int f\,d\mu \quad \text{for every} \quad f \in C_c(X).$$

Any complex-valued function f in $C_c(X)$ can be written as $f = f_1 + i\,f_2$, where $f_1 = \operatorname{Re} f$ and $f_2 = \operatorname{Im} f$ (the real and imaginary parts of f) are real-valued functions in $C_c(X)$, and any real-valued function f in $C_c(X)$ can be written as $f = f^+ - f^-$, where f^+ and f^- (positive and negative parts of f) are nonnegative functions in $C_c(X)$. In fact, such a decomposition is possible since every function in $C_c(X)$ is bounded (reason: $f(X)$ is compact by Lemma 12.1 and so it is bounded — cf. Remarks on Boundedness in Section 11.1). Therefore, since Φ and $\int(\cdot)\,d\mu$ are linear functionals, it is enough to consider nonnegative functions. That is, if the above identity holds for every nonnegative function f in $C_c(X)$, then it holds for every $\mathbb{F}$-valued function f in $C_c(X)$. Thus take an arbitrary nonnegative function f in $C_c(X)$. Since Φ and $\int(\cdot)\,d\mu$ are linear functionals and $f(X)$ is bounded, we may assume without loss of generality that $f \leq 1$. Thus suppose $f(X) \subseteq [0,1]$. Take an arbitrary bounded open set $U \in \mathcal{T}$ such that $[f] \subseteq U$. In other words, take an arbitrary bounded open set U such that

$$f \in C_c(X)^U.$$

Take an arbitrary positive integer n and set, for each integer $k \geq 0$,

$$U_0 = U, \qquad U_k = \{x \in X\colon \tfrac{k-1}{n} < f(x)\} = f^{-1}((\tfrac{k-1}{n}, \infty)) \quad \text{for } k \geq 1,$$

which are open (inverse image of open sets under a continuous function). Note that $U_1 = X \backslash f^{-1}(\{0\}) = X \backslash \mathcal{N}(f)$ so that $U_1^- = [f]$, $U_{k+1}^- \subseteq U_k$, and $U_k = \varnothing$ for every $k \geq n+1$. Set, for each integer $k \in [1, n]$,

$$f_k(x) = \begin{cases} 1, & x \in U_{k+1} \quad \text{(i.e., if } \frac{k}{n} < f(x)), \\ nf(x) - (k-1), & x \in U_k \backslash U_{k+1} \quad \text{(i.e., if } \frac{k-1}{n} < f(x) \leq \frac{k}{n}), \\ 0, & x \in X \backslash U_k \quad \text{(i.e., if } f(x) \leq \frac{k-1}{n}), \end{cases}$$

which defines a collection of n real-valued (since f is real-valued) functions f_k on X with compact support. In fact, $[f_k] \subseteq U_k^- \subseteq U_{k-1}$. Each f_k is continuous (reason: $nf(x) - (k-1) = 0$ if $f(x) = \frac{k-1}{n}$, $nf(x) - (k-1) = 1$ if $f(x) = \frac{k}{n}$, and f is continuous). Thus $f_k(X) \subseteq [0, 1]$, and so $f_k \in C_c(X)^{G_k}$ for each $k \in [1, n]$ and each open set G_k such that $[f_k] \subseteq G_k$. In particular,

$$f_k \in C_c(X)^{U_{k-1}}.$$

Thus, according to the definitions of the inner content $\mu^{\#}\colon \mathcal{T} \to \overline{\mathbb{R}}$ and of the quasiregular outer measure $\mu^*\colon \wp(X) \to \overline{\mathbb{R}}$ induced by it, and recalling that the quasiregular Borel measure $\mu\colon \mathcal{X}_{\mathcal{T}} \to \overline{\mathbb{R}}$ generated by them is such that $\mu|_{\mathcal{T}} = \mu^*|_{\mathcal{T}} = \mu^{\#}$, we get, for each integer $k \in [1, n]$,

$$\mu^{\#}(U_{k+1}) = \mu^*(U_{k+1}) = \mu(U_{k+1}) \leq \int f_k \, d\mu \leq \mu(U_k) = \mu^*(U_k) = \mu^{\#}(U_k).$$

Since $f_k = 1$ on $U_{k+1} \subseteq U_k$, and $[f_k] \subseteq U_k^- \subseteq U_{k=1}$, it also follows that

$$\mu^{\#}(U_{k+1}) = \sup_{g \in C_c(X)^{U_{k+1}}} \Phi(g) \leq \Phi(f_k) \leq \sup_{g \in C_c(X)^{U_{k-1}}} \Phi(g) = \mu^{\#}(U_{k-1}).$$

Hence,

$$\int f_{k+1} \, d\mu \leq \Phi(f_k) \qquad \text{and} \qquad \Phi(f_{k+2}) \leq \int f_k \, d\mu$$

for every integer $k \in [1, n-1]$ if $n > 1$, and for every integer $k \in [1, n-2]$ if $n > 2$, respectively.

Claim. $\qquad nf(x) = \sum_{k=1}^{n} f_k(x) \quad$ for every $\quad x \in X$.

Proof. If $n = 1$, then $f = f_1$. Thus suppose $n > 1$. Take an arbitrary $x \in X$. If $\frac{i-1}{n} < f(x) \leq \frac{i}{n}$ for some integer i in $(1, n)$, then

$$f_k(x) = \begin{cases} 1, & \text{if } 1 \le k \le i-1, \\ 0, & \text{if } i+1 \le k \le n, \end{cases}$$

for every integer $k \neq i$ in $[1,n]$. Hence,

$$\sum_{k=1}^{n} f_k(x) = \sum_{k=1}^{i-1} f_k(x) + f_i(x) + \sum_{k=i+1}^{n} f_k(x) = (i-1) + nf(x) - (i-1) = nf(n). \quad \square$$

Now suppose $n > 1$ so that $\int f_{k+1}\,d\mu \le \Phi(f_k)$ for each integer $k \in [1, n-1]$. Since Φ and $\int(\cdot)\,d\mu$ are linear functionals, we get

$$\begin{aligned} n\int f\,d\mu - \int f_1\,d\mu &= \int (nf - f_1)\,d\mu = \int\Big(\sum_{i=1}^{n} f_k\,d\mu - f_1\Big)\,d\mu \\ &= \int \sum_{i=1}^{n-1} f_{k+1}\,d\mu = \sum_{i=1}^{n-1}\int f_{k+1}\,d\mu \le \sum_{i=1}^{n-1}\Phi(f_k) = \Phi\Big(\sum_{i=1}^{n-1} f_k\Big) \\ &= \Phi\Big(\sum_{i=1}^{n} f_k - f_n\Big) = \Phi(nf - f_n) = n\Phi(f) - \Phi(f_n). \end{aligned}$$

Recall that $\int f_1\,d\mu \le \mu(U_0)$, with $U_0 = U$, where U is bounded, and so $\mu(U) < \infty$, and also that $\Phi(f_n) \ge 0$ (because Φ is positive). Then

$$\int f\,d\mu \le \Phi(f) + \tfrac{1}{n}\int f_1\,d\mu - \tfrac{1}{n}\Phi(f_n) \le \Phi(f) + \tfrac{1}{n}\mu(U_0).$$

Since the preceding inequality holds for every integer $n > 1$, it follows that

$$\int f\,d\mu \le \Phi(f).$$

Next suppose $n > 2$ so that $\Phi(f_{k+2}) \le \int f_k\,d\mu$ for each integer $k \in [1, n-2]$. Again, since Φ and $\int(\cdot)\,d\mu$ are linear functionals, we get

$$\begin{aligned} n\,\Phi(f) - \Phi(f_1) - \Phi(f_2) &= \Phi(nf - f_1 - f_2) = \Phi\Big(\sum_{i=1}^{n} f_k - f_1 - f_2\Big) \\ &= \sum_{i=1}^{n-2}\Phi(f_{k-2}) \le \sum_{i=1}^{n-2}\int f_k\,d\mu = \int\Big(\sum_{i=1}^{n} f_k - f_{n-1} - f_n\Big)\,d\mu \\ &= \int (nf_k - f_{n-1} - f_n)\,d\mu = n\int f\,d\mu - \int f_{n-1}\,d\mu - \int f_n\,d\mu. \end{aligned}$$

Recall that $\Phi(f_1) + \Phi(f_2) \le \mu(U_0) + \mu(U_1) \le 2\,\mu(U) < \infty$, and also that each $\int f_k\,d\mu \ge 0$. Then

$$\Phi(f) \le \int f\,d\mu + \tfrac{1}{n}\Big(\Phi(f_1) + \Phi(f_2) - \int f_{n-1}\,d\mu - \int f_n\,d\mu\Big) \le \int f\,d\mu + \tfrac{2}{n}\mu(U).$$

Since the preceding inequality holds for every integer $n > 2$, it follows that

$$\Phi(f) \leq \int f\,d\mu.$$

Outcome:

$$\int f\,d\mu = \Phi(f).$$

Finally, we show that the quasiregular measure $\mu\colon \mathcal{X}_{\mathcal{T}} \to \overline{\mathbb{R}}$ on the Borel σ-algebra $\mathcal{X}_{\mathcal{T}}$ of subsets of the locally compact Hausdorff space X generated by the topology $\mathcal{T}$ on X, which satisfies the above identity, is unique. Suppose μ and μ' on $\mathcal{X}_{\mathcal{T}}$ are quasiregular measures that satisfy the above identity. Observe by the definition of quasiregular measures on $\mathcal{X}_{\mathcal{T}}$ that they are determined by their values on open sets, which in turn are determined by their values on compact sets. So, in order to show that $\mu' = \mu$ on $\mathcal{X}_{\mathcal{T}}$, it suffices to verify that $\mu'(K) = \mu(K)$ for every compact set K. Thus take an arbitrary compact set $K \subseteq X$ and an arbitrary $\varepsilon > 0$. Since μ' is outer regular, there exists an open set U such that $K \subseteq U$ and $\mu'(U) \leq \mu'(K) + \varepsilon$. By Lemma 12.2, let f be any function in $C_c(X)_K \cap C_c(X)^U$. Therefore, since $f(X) \subseteq [0,1]$, $f(K) = 1$, and $[f] \subseteq U$, it follows that

$$\mu(K) \leq \int f\,d\mu = \Phi(f) = \int f\,d\mu' \leq \mu'(U) \leq \mu'(K) + \varepsilon,$$

and hence $\mu(K) \leq \mu'(K)$. Swapping μ with μ' we get $\mu(K)' \leq \mu(K)$. Thus $\mu(K)' = \mu(K)$, so that $\mu' = \mu$. □

A major application of Theorem 12.5 occurs in the proof of the Spectral Theorem for normal operators (see, e.g., [27, Proof of Theorem 3.11]). Comparing with the previous version, the next one exchange positiveness with boundedness for the linear functional Φ, and the positive Borel measure is replaced with a finite signed measure. The price for it is that the domain on which such a functional acts is the real linear space $C(X, \mathbb{R})$ for a compact X, rather than the linear space $C_c(X)$ on a locally compact X.

Theorem 12.6. *Let X be a compact Hausdorff space. If Φ is a real-valued bounded linear functional on the real linear space $C(X, \mathbb{R})$, then there is a unique signed measure ν on the Borel σ-algebra $\mathcal{X}_{\mathcal{T}}$ of subsets of X such that*

$$\Phi(f) = \int f\,d\nu \quad \text{for every} \quad f \in C(X, \mathbb{R}).$$

Moreover, $\|\Phi\| = |\nu|(X)$.

Proof. Let X be a compact Hausdorff space. Recall that the real linear space

$$C(X, \mathbb{R}) = \{f\colon X \to \mathbb{R}\colon\ f \text{ is continuous}\}$$

of all real-valued continuous functions on X is a linear lattice, and let $\Phi\colon C(X,\mathbb{R}) \to \mathbb{R}$ be a real-valued bounded linear functional on $C(X,\mathbb{R})$. Lemma 12.4 ensures that there exist two positive real-valued bounded linear functionals $\Phi^+\colon C(X,\mathbb{R}) \to \mathbb{R}$ and $\Phi^-\colon C(X,\mathbb{R}) \to \mathbb{R}$ on $C(X,\mathbb{R})$ such that

$$\Phi = \Phi^+ - \Phi^-.$$

Take an arbitrary $f \in C(X,\mathbb{R})$. Since X is compact, $C(X) = C_c(X)$, and so Theorem 12.5 says that there is a Borel measure λ, and a Borel measure μ, both unique on $\mathcal{X}_\mathcal{T}$ (which are finite because X is compact), for which

$$\Phi^+(f) = \int f\,d\lambda \quad \text{and} \quad \Phi^-(f) = \int f\,d\mu.$$

Thus, the (finite) real-valued signed measure $\nu = \lambda - \mu$ on $\mathcal{X}_\mathcal{T}$ is such that

$$\Phi(f) = \Phi^+(f) - \Phi^-(f) = \int f\,d\lambda - \int f\,d\mu = \int f\,d\nu,$$

which proves the existence. To verify uniqueness suppose ν and ν' are (finite) real-valued signed measures on $\mathcal{X}_\mathcal{T}$ such that for an arbitrary $f \in C(X,\mathbb{R})$,

$$\int f\,d\nu = \Phi(f) = \int f\,d\nu'.$$

Set $\tilde{\nu} = \nu - \nu'$. This is a (finite) real-valued signed measure on $\mathcal{X}_\mathcal{T}$ for which

$$\int f\,d\tilde{\nu} = \int f\,d\tilde{\nu}^+ - \int f\,d\tilde{\nu}^- = 0,$$

where $\tilde{\nu}^+$ and $\tilde{\nu}^-$ are finite positive measures on $\mathcal{X}_\mathcal{T}$ associated with the decomposition $\tilde{\nu} = \tilde{\nu}^+ - \tilde{\nu}^-$: the positive and negative variations of $\tilde{\nu}$. Now consider the (real-valued) positive linear functional Ψ on $C(X,\mathbb{R})$ such that

$$\Psi(f) = \int f\,d\tilde{\nu}^+ = \int f\,d\tilde{\nu}^-.$$

By uniqueness of the Borel measure in Theorem 12.5 we get $\tilde{\nu}^+ = \tilde{\nu}^-$, so that $\tilde{\nu} = 0$, and hence $\nu = \nu'$, which proves uniqueness. Finally, recall that $C(X) \subseteq B(X)$ whenever X is compact, and so $\sup|f| < \infty$, and also that

$$|\Phi(f)| = \left|\int f\,d\nu\right| \le \int |f|\,d|\nu| \le \sup|f|\,|\nu|(X)$$

for every $f \in C(X,\mathbb{R})$ (as we saw in Remark 10.3). Hence

$$\|\Phi\| = \sup_{|f|\le 1} |\Phi(f)| \le |\nu|(X).$$

On the other hand, according to Example 7A and Theorem 7.4,

$$|\nu|(X) = \nu^+(X) + \nu^-(X) \le \lambda(X) + \mu(X) = \Phi^+(1) + \Phi^-(1) = \|\Phi\|$$

by Lemma 12.4 since $C(X,\mathbb{R})$ contains the function 1. Therefore,

$$\|\Phi\| = |\nu|(X). \qquad \square$$

The next result extends Theorem 12.6 to bounded linear functionals on $C(X,\mathbb{C})$, where finite signed measures are replaced with complex measures.

Theorem 12.7. *Let X be a compact Hausdorff space. If Φ is a complex-valued bounded linear functional on $C(X,\mathbb{C})$, then there exists a unique complex measure η on a Borel σ-algebra $\mathcal{X}_{\mathcal{T}}$ of subsets of X such that*

$$\Phi(f) = \int f\,d\eta \quad \text{for every} \quad f \in C(X,\mathbb{C}).$$

Moreover, $\|\Phi\| = |\eta|(X)$.

Proof. Let X be a compact Hausdorff space, let

$$C(X,\mathbb{C}) = \{f\colon X \to \mathbb{C}\colon\ f \text{ is continuous}\}$$

be the complex linear space of all complex-valued continuous functions on X, and let $\Phi\colon C(X,\mathbb{C}) \to \mathbb{C}$ be a complex-valued bounded linear functional on $C(X,\mathbb{C})$. A function f in $C(X,\mathbb{C})$ can be written as $f = f_1 + i\,f_2$, where f_1 and f_2 (real and imaginary parts of f) are real-valued continuous functions in the real linear space $C(X,\mathbb{R}) = \{\varphi\colon X \to \mathbb{R}\colon\ \varphi \text{ is continuous}\}$. Moreover, write the complex numbers $\Phi(f_1)$ and $\Phi(f_2)$ in its Cartesian representation, $\Phi(f_1) = \operatorname{Re}\Phi(f_1) + i\operatorname{Im}\Phi(f_1)$ and $\Phi(f_2) = \operatorname{Re}\Phi(f_2) + i\operatorname{Im}\Phi(f_2)$, where $\operatorname{Re}\Phi(f_i)$ and $\operatorname{Im}\Phi(f_i)$ are real numbers. Thus, for each complex-valued functional $\Phi\colon C(X,\mathbb{C}) \to \mathbb{C}$ on the complex linear space $C(X,\mathbb{C})$, we can associate two real-valued functionals, say, $\Phi_1\colon C(X,\mathbb{R}) \to \mathbb{R}$ and $\Phi_2\colon C(X,\mathbb{R}) \to \mathbb{R}$ on the real linear space $C(X,\mathbb{R})$, defined as follows.

$$\Phi_1(\varphi) = \operatorname{Re}\Phi(\varphi) \quad \text{and} \quad \Phi_2(\varphi) = \operatorname{Im}\Phi(\varphi),$$

so that

$$\Phi(\varphi) = \Phi_1(\varphi) + i\,\Phi_2(\varphi),$$

for every $\varphi \in C(X,\mathbb{R})$. These real-valued functionals Φ_1 and Φ_2 on $C(X,\mathbb{R})$ are linear and bounded. Indeed, since Φ is linear, it follows that for every $\alpha \in \mathbb{R}$ and every $\varphi, \psi \in C(X,\mathbb{R})$,

$$\begin{aligned}
\alpha\,\Phi_1(\varphi) + i\,\alpha\,\Phi_2(\varphi) &= \alpha\,\Phi(\varphi) = \Phi(\alpha\,\varphi) = \Phi_1(\alpha\,\varphi) + i\,\Phi_2(\alpha\,\varphi),\\
\Phi_1(\varphi+\psi) + i\,\Phi_2(\varphi+\psi) &= \Phi(\varphi+\psi) = \Phi(\varphi) + \Phi(\psi)\\
&= \big(\Phi_1(\varphi) + \Phi_1(\psi)\big) + i\big(\Phi_2(\varphi) + \Phi_2(\psi)\big),
\end{aligned}$$

and so Φ_1 and Φ_2 are linear, which are also bounded since Φ is bounded:

$$\begin{aligned}\max\{\|\Phi_1\|^2, \|\Phi_2\|^2\} &= \max\Big\{\sup_{|\varphi|\leq 1}|\Phi_1(\varphi)|^2, \sup_{|\psi|\leq 1}|\Phi_2(\psi)|^2\Big\}\\ &\leq \sup_{|\varphi|\leq 1}\big(|\Phi_1(\varphi)|^2 + |\Phi_2(\varphi)|^2\big)\\ &= \sup_{|\varphi|\leq 1}|\Phi_1(\varphi) + i\,\Phi_2(\varphi)|^2\\ &\leq \sup_{|\varphi|\leq 1}|\Phi(\varphi)|^2 \leq \sup_{|f|\leq 1}|\Phi(f)|^2 = \|\Phi\|^2.\end{aligned}$$

Therefore, according to Theorem 12.6, there exist unique finite signed measures ν_1 and ν_2 on $\mathcal{X}_\mathcal{T}$ such that

$$\Phi_1(f_1) = \int f_1\,d\nu_1, \qquad \Phi_1(f_2) = \int f_2\,d\nu_1,$$

$$\Phi_2(f_1) = \int f_1\,d\nu_2, \qquad \Phi_2(f_2) = \int f_2\,d\nu_2.$$

Observe that by the linearity of Φ we get for each $f \in C(X,\mathbb{C})$,

$$\begin{aligned}\Phi(f) = \Phi(f_1 + i\,f_2) &= \Phi(f_1) + i\,\Phi(f_2)\\ &= \Phi_1(f_1) + i\,\Phi_2(f_1) + i\,(\Phi_1(f_2) + i\,\Phi_2(f_2))\\ &= \Phi_1(f_1) + i\,\Phi_1(f_2) - \Phi_2(f_2) + i\,\Phi_2(f_1).\end{aligned}$$

Hence, since $\int(\cdot)\,d\mu$ is a linear functional, for every $f \in C(X,\mathbb{C})$,

$$\begin{aligned}\Phi(f) &= \int f_1\,d\nu_1 + i\int f_2\,d\nu_1 - \int f_2\,d\nu_2 + i\int f_1\,d\nu_2\\ &= \int(f_1 + i\,f_2)\,d\nu_1 + i\int(f_1 + i\,f_2)\,d\nu_2\\ &= \int f\,d\nu_1 + i\int f\,d\nu_2 = \int f\,d\eta,\end{aligned}$$

where $\eta = \nu_1 + i\,\nu_2$ is a complex measure on $\mathcal{X}_\mathcal{T}$, which is unique since the finite signed measures ν_1 and ν_2 on $\mathcal{X}_\mathcal{T}$ are unique. Finally, to verify that

$$\|\Phi\| = |\eta|(X)$$

we proceed as follows. First recall that since $C(X) \subseteq B(X)$ whenever X is compact, we get $\sup|f| < \infty$ for every $f \in C(X) = C(X,\mathbb{C})$, and also

$$|\Phi(f)| = \left|\int f\,d\eta\right| \leq \int|f|\,d|\eta| \leq \sup|f|\,|\eta|(X)$$

for every $f \in C(X) = C(X, \mathbb{C})$ (as we saw in Remark 10.6), which implies

$$\|\Phi\| = \sup_{f \in C(X),\, |f| \leq 1} |\Phi(f)| = \sup_{f \in C(X),\, |f| \leq 1} \left| \int f\, d\eta \right| \leq |\eta|(X).$$

Next we show the reverse inequality, so that

$$\sup_{f \in C(X),\, |f| \leq 1} \left| \int f\, d\eta \right| = |\eta|(X).$$

In fact, we have already seen in Section 10.3 that the above identity holds when the supremum is taken over all complex-valued *integrable* functions f on X such that $|f| \leq 1$. We now show that this actually happens (i.e., the above identity holds) even when the supremum is taken over the subcollection of all complex-valued *continuous* functions f on X such that $|f| \leq 1$. Indeed, since X is a compact Hausdorff space, we can apply Lemma 12.2 and Corollary 12.3. By Lemma 12.2 it follows that *if* $\{U_i\}$ *is any finite open covering of* X*, then there is a finite compact covering* $\{K_i\}$ *of* X *such that* $K_i \subseteq U_i$ *and, for each index* i*, there exists a function* $f_i \in C_c(X) = C(X)$ *such that* $\chi_{K_i} \leq f_i \leq \chi_{U_i}$. Let $\{A_1^+, A_1^-\}$ and $\{A_2^+, A_2^-\}$ be arbitrary Hahn decompositions of X with respect to the signed measures ν_1 and ν_2, and consider the collection $\mathcal{A}_{1,2}^{+-} = \{A_1^+ \cap A_2^+,\, A_1^+ \cap A_2^-,\, A_1^- \cap A_2^+,\, A_1^- \cap A_2^-\}$, which also is a covering of X. An argument similar to that in Remark 10.3 shows that for every $\mathcal{X}_\mathcal{T}$-measurable covering of X there is a $\mathcal{X}_\mathcal{T}$-measurable covering of X consisting of subsets of the four sets in $\mathcal{A}_{1,2}^{+-}$. Thus take a finite open covering $\{\mathcal{U}_i\} \subseteq \mathcal{A}_{1,2}^{+-}$ of X. Hence (cf. Remark 10.5),

$$|\eta(K_i)| \leq \left| \int f_i\, d\eta \right| \leq |\eta(U_i)|,$$

and so

$$\sum\nolimits_i |\eta(K_i)| \leq \sum\nolimits_i \left| \int f_i\, d\eta \right| \leq \sum\nolimits_i |\eta(U_i)|.$$

Observe that η is quasiregular in the sense that $\eta = \nu_1 + i\,\nu_2$ where ν_1 and ν_2 are quasiregular (finite) signed measure (i.e., they are differences of quasiregular finite positive measures, which are Borel measures μ on $\mathcal{X}_\mathcal{T}$ such that $\mu(U) = \sup\{\mu(K)\colon K \subseteq U,\ K \text{ compact}\}$ for open sets U). Then

$$\sup \sum\nolimits_i |\eta(K_i)| = \sup \sum\nolimits_i \left| \int f_i\, d\eta \right| = \sup \sum\nolimits_i |\eta(U_i)|,$$

where the supremum is taken over all finite open coverings $\{U_i\} \subseteq \mathcal{A}_{1,2}^{+-}$ of X, and so over their corresponding compact coverings $\{K_i\}$ with $K_i \subseteq U_i$, and over all finite collections $\{f_i\}$ of continuous functions on X such that

$\chi_{K_i} \le f_i \le \chi_{U_i}$. Since $|\eta|(X) = \sup \sum_j |\eta(E_j)|$, where the supremum is taken over all finite $\mathcal{X}_\mathcal{T}$-measurable partitions $\{E_i\}$ of X (cf. Section 10.3), we may infer that

$$\sup \sum\nolimits_i \left| \int f_i \, d\eta \right| = |\eta|(X).$$

Now, arguing as in Corollary 12.3, we may take $\{f_i\}$ such that $\sum_i f_i = 1$. Indeed, set $f = \sum_i f_i$, which is a real-valued continuous function on X such that $f \ge 1$. By setting $f' = f_i/f$ we get a finite collection $\{f_i'\}$ of functions in $C(X)$ such that $\chi_{K_i} \le f_i' \le \chi_{U_i}$ and $\sum_i f_i' = 1$. Such a system of functions is referred to as a *partition of the unity* on X corresponding to an open covering $\{U_i\}$ of X. Thus, with $\sum_i f_i = 1 \in C(X)$ (and so $\left|\sum_i f_i\right| = 1$) and $0 \le f_i \le \chi_{U_i}$ for each index i, it is readily verified (taking again the supremum over all finite open coverings $\{U_i\} \subseteq \mathcal{A}^{+-}_{1,2}$ of X) that

$$\sup \sum\nolimits_i \left| \int f_i \, d\eta \right| = \sup \sum\nolimits_i \left| \int_{U_i} f_i \, d\eta \right| \le \sup_{f \in C(X),\, |f| \le 1} \left| \int f \, d\eta \right|.$$

Thus we get the reverse inequality,

$$|\eta|(X) = \sup \sum\nolimits_i \left| \int f_i \, d\eta \right| \le \sup_{f \in C(X),\, |f| \le 1} \left| \int f \, d\eta \right|.$$

Therefore,

$$\|\Phi\| = \sup_{f \in C(X),\, |f| \le 1} |\Phi(f)| = \sup_{f \in C(X),\, |f| \le 1} \left| \int f \, d\eta \right| = |\eta|(X). \qquad \square$$

12.4 Additional Propositions

Take the Banach space $L^p = L^p(\mu) = L^p(X, \mu) = L^p(X, \mathcal{X}, \mu)$ of all (equivalence classes of) $\mathbb{F}$-valued functions $f\colon X \to \mathbb{F}$ on a nonempty set X that are p-integrable if $p \ge 1$, or essentially bounded if $p = \infty$, where $\mu\colon \mathcal{X} \to \mathbb{R}$ is a positive measure on a σ-algebra $\mathcal{X}$ of subsets of X (Chapter 5). If $p > 1$, then the Hölder conjugate $q > 1$ of p is the solution to the equation $\frac{1}{p} + \frac{1}{q} = 1$; for $p = 1$, set $q = \infty$ (and vice versa). In Section 7.2, after proving the Radon–Nikodým Theorem, we remarked that "an application of the Radon–Nikodým Theorem is the Riesz Representation Theorem". The referred version of the Riesz Representation Theorem reads as follows.

Proposition 12.A. *If $p \ge 1$ and $\Phi\colon L^p \to \mathbb{F}$ is a bounded linear functional, then there is a unique $g \in L^q$, where q is the Hölder conjugate of p, such that*

$$\Phi(f) = \int f \overline{g} \, d\mu \quad \text{for every} \quad f \in L^p.$$

Moreover, $\|\Phi\| = \|g\|_q$. If $p = 1$ (so that $q = \infty$) then the (positive) measure μ is supposed to be σ-finite.

A particularly important especial case is that of $p = q = 2$. As we have also noticed in Section 7.2, such a particular case can be proved without using the Radon–Nikodým Theorem. Indeed, recall that a *Hilbert space* is a Banach space (i.e., a complete normed spaces, as in Chapter 4) whose norm is induced by an inner product. In other words, let $\mathcal{L}$ be an arbitrary (abstract) linear space over a field $\mathbb{F}$. A functional $\langle\;;\;\rangle\colon\mathcal{L}\times\mathcal{L}\to\mathbb{F}$ is an *inner product* on $\mathcal{L}$ if the following conditions hold for all vectors f, g, and h in $\mathcal{L}$ and all scalars α in $\mathbb{F}$: (i) $\langle f+g\,;h\rangle = \langle f\,;h\rangle + \langle g\,;h\rangle$, (ii) $\langle \alpha f\,;g\rangle = \alpha\langle f\,;g\rangle$, (iii) $\langle f\,;g\rangle = \overline{\langle g\,;f\rangle}$, (iv) $\langle f\,;f\rangle \ge 0$, (v) $\langle f\,;f\rangle = 0$ only if $f = 0$. An *inner product space* $(\mathcal{L}, \langle\;;\;\rangle)$ is a linear space $\mathcal{L}$ equipped with an inner product $\langle\;;\;\rangle$. If $\mathcal{L}$ is a real or complex linear space, so that $\mathbb{F} = \mathbb{R}$ or $\mathbb{F} = \mathbb{C}$, equipped with an inner product on it, then it is referred to as a *real* or *complex inner product space*, respectively. Every inner product induces a norm. The norm $\|\;\|\colon\mathcal{L}\to\mathbb{R}$ induced by an inner product $\langle\;;\;\rangle$ is defined by $\|f\|^2 = \langle f\,;f\rangle$ for every $f \in \mathcal{L}$. A *Hilbert space* $\mathcal{H}$ is an inner product space $(\mathcal{L}, \langle\;;\;\rangle)$, which is complete in the sense that the normed space $(\mathcal{L}, \|\cdot\|)$, whose norm is induced by the inner product $\langle\;;\;\rangle$, is complete. The Riesz Representation Theorem in an abstract Hilbert space $\mathcal{H}$ reads as follows.

Proposition 12.B. *Let $\mathcal{H}$ be an arbitrary Hilbert space. For every bounded linear functional $\Phi\colon\mathcal{H}\to\mathbb{F}$, there exists a unique vector $g \in \mathcal{H}$ such that*

$$\Phi(f) = \langle f\,;g\rangle \quad \text{for every} \quad f \in \mathcal{H}.$$

Moreover, $\|\Phi\| = \|g\|$. Such a unique vector g in $\mathcal{H}$ is called the Riesz representation *of the functional Φ.*

The space L^2 is a concrete Hilbert space. Among the Banach spaces L^p, for every $p \ge 1$ and for $p = \infty$, the only one which is a Hilbert space is L^2 (only $\|\;\|_2$ is induced by an inner product). The inner product in L^2 is

$$\langle f\,;g\rangle = \int f\overline{g}\,d\mu \quad \text{for every} \quad f, g \in \mathcal{H}.$$

Thus the Riesz Representation Theorem in L^2 can be independently obtained either by Proposition 12.A or by Proposition 12.B. In fact, we can get it through Proposition 12.B, and use it to yield another proof of the Radon–Nikodým Theorem (different from our proof of Theorem 7.8), and the Radon–Nikodým Theorem is, in turn, used to prove Proposition 12.A (see e.g., [36, Remark 6.17] or [4, Exercise 8.V]).

The next result extends Theorem 12.5 from the linear space $C_c(X)$ of all continuous functions with compact support to the linear space $C_0(X)$ of all continuous functions that vanish at infinity. A function $f\colon X\to\mathbb{F}$ on a topological space X is said to *vanish at infinity* if for every $\varepsilon > 0$ there is a compact set $K_\varepsilon \subset X$ such that $|f(x)| < \varepsilon$ whenever $x \in X\backslash K_\varepsilon$. Recall that a *regular* (*quasiregular*) *signed measure* $\nu = \nu^+ - \nu^-$ is one such that the finite positive measures ν^+ and ν^- are regular (quasiregular); a *regular* (*quasiregular*) *complex measure* $\eta = \nu_1 + i\,\nu_2$ is one such that the (finite) signed measures ν_1 and ν_2 are regular (quasiregular) — cf. Proof of Theorem 12.7.

Proposition 12.C. *Let X be a locally compact Hausdorff space. If Φ is an $\mathbb{F}$-valued bounded linear functional on $C_0(X)$, then there is a unique quasiregular $\mathbb{F}$-valued measure η on a Borel σ-algebra $\mathcal{X}_{\mathcal{T}}$ of subsets of X such that*

$$\Phi(f) = \int f\,d\eta \quad \text{for every} \quad f \in C_0(X).$$

Moreover, $\|\Phi\| = |\eta|(X)$.

Let X be a topological space. Observe that $C_c(X) \subseteq C_0(X) \subseteq B(X) \subseteq L^\infty(X,\mu)$. Also, $C_c(X) \subseteq L^1(X,\mu)$ if $L^1(X,\mathcal{X}_{\mathcal{T}},\mu)$ is given in terms of a positive Borel measure μ on $\mathcal{X}_{\mathcal{T}}$, as we had noticed at the opening of Section 12.3. Thus (Problems 5.8 and 5.12), $C_c(X) \subseteq L^p(X,\mu)$ for every $p \geq 1$ if $L^p(X,\mathcal{X}_{\mathcal{T}},\mu)$ is given in terms of a Borel measure μ on $\mathcal{X}_{\mathcal{T}}$. Note that the inclusions in $L^p(X,\mu)$ or $L^\infty(X,\mu)$ are interpreted in terms of the equivalence classes of each representative in $C_c(X)$ or $B(X)$. The next proposition shows a denseness result of fundamental importance for the linear spaces upon which the Riesz Representation Theorems were built, viz., $C_c(X)$, $C_0(X)$, and $L^p(X,\mu)$ (recall: $C_c(X) = C_0(X) = C(X)$ if X is compact).

Proposition 12.D. *Let X be a locally compact Hausdorff space. If μ is a positive quasiregular Borel measure on a Borel σ-algebra $\mathcal{X}_{\mathcal{T}}$ of subsets of X, then the linear space $C_c(X)$ is dense in the Banach space $(L^p(X,\mu), \|\,\|_p)$ for every real $p \geq 1$ and for $p = \infty$.*

Notes: These are classical and crucial results in functional analysis, which extend the original form of the Riesz Representation Theorem (Theorem 12.5) in many directions. Proposition 12.A is the standard form of it in the concrete Banach spaces L^p (for proofs using somewhat different approaches see, e.g., [4, Theorems 8.14 and 8.15], [35, Theorems 11.29 and 11.30], or [36, Theorem 6.16]). Proposition 12.B is the traditional version for abstract Hilbert spaces, which does not depend on any measure-theoretical concept (see, e.g., [26, Theorem 5.62]). For the extension to the concrete linear space $C_0(X)$ in Proposition 12.C see, e.g., [36, Theorem 6.19], and for the pivotal denseness result of Proposition 12.D see, e.g., [6, Theorem 29.41].

Suggested Reading

Bauer [6], Brown and Pearcy [8], Conway [11], Halmos [18], Kingman and Taylor [23], Royden [35], Rudin [36].

13
Invariant Measures

13.1 Topological Groups

A *binary operation* on a set X is a mapping $\star\colon X\times X \to X$ of the Cartesian product $X\times X$ into X. It is usual to write $z = x \star y$ instead of $z = \star(x, y)$ to indicate that z in X is the value of $\star$ at the point (x, y) in $X\times X$. In this context it is convenient to interpret the binary operation $\star$ multiplicatively, so that $x \star y$ is interpreted as the product of x and y, and it is written in a simplified form as $x\,y$. If a binary operation on X has the property that

$$x\,(y\,z) = (x\,y)\,z$$

for every x, y, and z in X, then it is said to be *associative*. So we can drop the parentheses and write $x\,y\,z$. If there exists an element e in X such that

$$x\,e = e\,x = x$$

for every $x \in X$, then e is said to be the *identity element* (or the *neutral element*) with respect to the binary operation on X. It is easy to show that if a binary operation has a neutral element e, then e is unique. If an associative binary operation on X has an identity element e in X, and if for some $x \in X$ there exists $x^{-1} \in X$ such that

$$x\,x^{-1} = x^{-1}\,x = e,$$

then x^{-1} is called the *inverse* of x with respect to the underlying binary operation. It is also easy to show that if the inverse of x exists with respect

C.S. Kubrusly, *Essentials of Measure Theory*,
DOI 10.1007/978-3-319-22506-7_13

to an associative binary operation, then it is unique. Note that $(x\,y)^{-1} = y^{-1}x^{-1}$ (indeed, $x\,y\,y^{-1}x^{-1} = e$).

A *group* is a nonempty set X on which is defined a binary operation that is associative, has an identity element $e \in X$, and every x in X has an inverse in X. In general, the binary operation is not commutative in the sense that there may exist a pair of elements $x, y \in X$ for which $x\,y \neq y\,x$. If the binary operation on a group X is such that $x\,y = y\,x$ for every $x, y \in X$, then X is said to be a *commutative group* (or an *Abelian group*). A nonempty subset M of a group X is a *subgroup* of X if M is itself a group; that is, if $e \in M$, and x^{-1} and $x\,y$ lie in M for every x and y in M; equivalently, if $x\,y^{-1} \in M$ whenever x and y are in M. Suppose A and B are arbitrary subsets of a group X. Let A^{-1} and AB stand for the subsets of X consisting of all elements of the form x^{-1} and $x\,y$, respectively, for $x \in A$ and $y \in B$. Thus a nonempty subset M of a group X is a subgroup of X if and only if $MM^{-1} \subseteq M$. Note that $A \cap A^{-1} = (A \cap A^{-1})^{-1}$ for every $A \subseteq X$.

We write A^2 for AA. Observe that $(AA)A = A(AA)$, and so we write A^3 for $AAA = (AA)A = A(AA)$. Generalizing, we write A^n for the multiplication of A with itself n times, $A \ldots A \subseteq X$, for every positive integer n. If $e \in A$, then $\{A^n\}$ is an increasing sequence (with respect to the inclusion ordering) since, in this case, $A^n = A^n\{e\} \subseteq A^n A = A^{n+1}$ for every positive integer n. For each $x \in X$, let xA and Ax denote $\{x\}A$ and $A\{x\}$, which are referred to as *left translation* and *right translation* of the set $A \subseteq X$, respectively. If M is a subgroup of X, then the sets xM and Mx are referred to as *left coset* and *right coset* of M. An *invariant subgroup* M of X is one for which $xM = Mx$ for every $x \in X$. A *homomorphism* is a map $\Phi\colon X \to Y$ of a group X into a group Y such that

$$\Phi(x\,y) = \Phi(x)\,\Phi(y) \quad \text{for every} \quad x, y \in X.$$

A *topological group* is a group X that is also a topological space for which the map $(x, y) \mapsto x\,y^{-1}$ is continuous from $X \times X$ (equipped with the usual product topology) to X or, equivalently, for which the multiplication map $(x, y) \mapsto x\,y$ from $X \times X$ to X and the inversion map $x \mapsto x^{-1}$ from X to X are both continuous. Take an arbitrary $x_0 \in X$. The maps $y \mapsto x_0\,y$ and $y \mapsto y\,x_0$ from X to X are referred to as the *left multiplication* by x_0 and *right multiplication* by x_0, respectively. These are sections of the multiplication map $(x, y) \mapsto x\,y$ (see Section 9.2). Since a restriction of a continuous map is continuous, it follows that a section of a continuous function is continuous, and so the right and left multiplication maps are both continuous. It is readily verified that the map $y \mapsto x_0^{-1}y$ from X to X is the inverse of the left multiplication $y \mapsto x_0\,y$, and the map $y \mapsto y\,x_0^{-1}$ from X to X is the inverse of the right multiplication $y \mapsto y\,x_0$, which are again continuous (since these inverses are precisely the left and right multiplications by x_0^{-1}).

A *symmetric neighborhood* in a topological group is a neighborhood N of the identity e such that $N = N^{-1}$. Topological qualifications are naturally attributed to topological groups. We refer to *compact*, or *locally compact*, or *Hausdorff groups* if, as topological spaces, they are compact, or locally compact, or Hausdorff. A map between topological spaces is continuous if the inverse image of open (closed) sets is open (closed). A *homeomorphism* $\Psi: X \to Y$ of a topological space X onto a topological space Y is an invertible map that is continuous and has a continuous inverse. So a homeomorphism between topological spaces maps open (closed) sets into open (closed) sets, and the inverse image of open (closed) sets is an open (closed) set. That is, $\Psi(A)$ is open (closed) in Y if and only if $A = \Psi^{-1}(\Psi(A))$ is open (closed) in X. Observe that the left and right multiplication maps are homeomorphism of a topological group X onto itself (and so is the inversion map).

Next we exhibit four classical examples of topological groups.

(1) Let $\mathbb{F}$ denote either the set of all real or complex numbers, and consider the set $\mathbb{F}\backslash\{0\}$ of all nonzero real or complex numbers. Both $\mathbb{F}$ and $\mathbb{F}\backslash\{0\}$ are commutative locally compact Hausdorff groups, when equipped with the usual metric topology, where $\mathbb{F}\backslash\{0\}$ is equipped with the binary operation defined by the usual multiplication in $\mathbb{F}$ with identify $e = 1$ (a multiplicative group), and $\mathbb{F}$ is equipped with the binary operation defined by the usual addition in $\mathbb{F}$ with identify $e = 0$ (an additive group).

(2) The unit circle about the origin $\mathbb{T} = \{z \in \mathbb{C}: |z| = 1\}$ in the complex plane $\mathbb{C}$, equipped with the usual metric topology, is a commutative compact Hausdorff (multiplicative) group with identify $e = 1$.

(3) Let $GL(n)$ be the collection of all (real or complex) $n \times n$ invertible matrices equipped with the ordinary matrix operations and the uniform topology induced by the usual topology of the normed space $\mathbb{F}^n$. This is a noncommutative locally compact Hausdorff (multiplicative) group with identity $e = I$, where I denotes the identity $n \times n$ matrix.

(4) Consider the collection of all unitary operators on a Hilbert space $\mathcal{H}$. That is, consider the collection of all operators U in the Banach space $\mathcal{B}[\mathcal{H}]$ of all bounded linear operators of $\mathcal{H}$ into itself, equipped with the uniform topology, such that $U^*U = UU^* = I$, where I is the identity operator and U^* denotes the adjoint of U. This is a noncommutative Hausdorff (multiplicative) group, which is not locally compact if $\mathcal{H}$ is infinite-dimensional, with identity $e = I$. Here multiplication means composition. For continuity of the inversion and multiplication maps, see, e.g., [20, Problems 100 and 111].

The following lemma collects standard results on topological groups that will be required in the sequel. The result in item (c) is central.

Lemma 13.1. *Let X be a topological group.*

(a) *If K and C are compact sets in X, then K^{-1} and CK are compact.*

(b) *Take $x \in X$ and $U \subseteq X$ arbitrary. The set U is open if and only if xU is open if and only if Ux is open if and only if U^{-1} is open.*

(c) *If G is an open (bounded) neighborhood of the identity e, then there is a decreasing sequence $\{U_n\}$ of open (bounded) symmetric neighborhoods U_n of e such that, for each integer $n \geq 1$, $U_n^k \subseteq G$ for every $1 \leq k \leq n+1$.*

(d) *If K is compact and G is open (bounded) in X such that $K \subseteq G$, then there is an open (bounded) symmetric neighborhood U of e such that $KU \subseteq G$, $UK \subseteq G$. If X is locally compact and Hausdorff, $UKU \subseteq G$.*

(e) *If X is locally compact and Hausdorff, and if C and K are disjoint compact subsets of X, then there exists an open bounded symmetric neighborhood U of the identity e such that each of UC, CU and UCU is disjoint with each of UK, KU and UKU.*

(f) *X is Hausdorff if and only if $\{e\}$ is a closed set.*

Proof. Let X be a topological group.

(a) Recall: continuous images of compact sets are compact (cf. Claim in the Proof of Lemma 12.1). Since inversion is continuous, it follows that K^{-1} is compact if K is. Since the Cartesian product of compact sets is compact, it follows that $C \times K$ is compact in $X \times X$ whenever C and K are compact in X; since multiplication is continuous, it follows that CK is compact.

(b) Since the left and right multiplication maps are homeomorphism, xU is open if and only if U is, and Ux is open if and only if U is. Since inversion also is a homeomorphism, U^{-1} is open if and only if U is.

(c) Since $G = eG$, it follows that G lies in the range of the multiplication map. Since G is open in X, and since the multiplication map from $X \times X$ to X is continuous (in particular, it is continuous at (e, e)), the inverse image of G under this map is open in $X \times X$, and contains the point (e, e) because G contains e. Thus there exist $A \subseteq X$ and $B \subseteq X$ such that $A \times B$ is open in $X \times X$, $(e, e) \in A \times B$, and $AB = G$. Since $e \in A \cap B$ and $AB = G$, it follows that $A \subseteq G$ and $B \subseteq G$. Therefore, since $A \times B$ is open in $X \times X$, there exist open neighborhoods of e, say, $W_1 \subseteq A \subseteq G$ and $W_2 \subseteq B \subseteq G$, such that $W_1 W_2 \subseteq AB = G$. Hence the set $W = W_1 \cap W_2 \subseteq G$ is open in X (intersection of open sets), contains e, and $W^2 = WW \subseteq W_1 W_2 \subseteq G$. Since W^{-1} is open (according to (b)) and contains e, it follows that $U_1 =$

$W \cap W^{-1} \subseteq W \subseteq G$ is a symmetric ($U_1 = U_1^{-1}$) open neighborhood of e (intersection of open sets containing e), and $U_1^2 = U_1 U_1 \subseteq WW = W^2 \subseteq G$.

Outcome 1: *For every open neighborhood G of the identity e there is a symmetric open neighborhood U_1 of e such that $U_1 \subseteq G$ and $U_1^2 \subseteq G$.*

Thus, applying the above result to U_1 itself (which is an open neighborhood of e), it follows that there exists a symmetric open neighborhood U_2 of e such that $U_2 \subseteq U_1 \subseteq G$ and $U_2^2 \subseteq U_1 \subseteq G$. Therefore, since $e \in U_2$,

$$U_2^3 = U_2 U_2 U_2 \subseteq U_2 U_2 U_2 U_2 = U_2^4 = (U_2^2)(U_2^2) \subseteq U_1 U_1 = U_1^2 \subseteq G.$$

Outcome 2: *For every open neighborhood G of the identity e, there is a symmetric open neighborhood U_2 of e such that $U_2 \subseteq G$, $U_2^2 \subseteq G$, and $U_2^3 \subseteq G$.*

Again, applying the above result to U_2 (which is an open neighborhood of e), it follows that there exists a symmetric open neighborhood U_3 of e such that $U_3 \subseteq U_2 \subseteq G$, $U_3^2 \subseteq U_2 \subseteq G$, and $U_3^3 \subseteq U_2 \subseteq G$. Since $e \in U_3$, we get

$$U_3^4 \subseteq U_3^5 = (U_3^2)(U_3^3) \subseteq U_2 U_2 = U_2^2 \subseteq G.$$

Outcome 3: *For every open neighborhood G of the identity e, there is a symmetric open neighborhood U_3 of e such that $U_3^k \subseteq G$ for every $1 \le k \le 4$.*

Proceeding this way, if G is an open neighborhood of e, we eventually get down to a pair of symmetric open neighborhoods U_n and U_{n+1} of e such that (i) $U_{n+1} \subseteq U_n \subseteq G$, (ii) $U_n^k \subseteq G$ for $1 \le k \le n+1$, and (iii) $U_{n+1}^j \subseteq G$ for $1 \le j \le n+2$, for each integer $n \ge 1$. Thus $\{U_n\}$ is a decreasing sequence of open symmetric neighborhoods of e such that $U_n^k \subseteq G$ for every k such that $1 \le k \le n+1$. Note that $U \subseteq G$ is bounded whenever G is.

(d) Take an arbitrary $x \in K \subseteq G \subseteq X$, where K is compact and G is open (bounded). Thus Gx^{-1} and $x^{-1}G$ are open (bounded) neighborhoods of e. Then, by (c), there are open (bounded) symmetric neighborhoods V_x and W_x of e such that $V_x^2 \subseteq Gx^{-1}$ and $W_x^2 \subseteq x^{-1}G$. Since $e \in V_x \cap W_x$, we get that $K \subseteq \bigcup_{x \in K} V_x x$ and $K \subseteq \bigcup_{x \in K} x\,W_x$. Since K is compact, there exist finite open subcoverings such that $K \subseteq \bigcup_{i=1}^n V_{x_i} x_i$ and $K \subseteq \bigcup_{i=1}^m x_i W_{x_i}$ with x_i in K. Set $V = \bigcap_{i=1}^n V_{x_i}$ and $W = \bigcap_{i=1}^m W_{x_i}$, which are again open (bounded) symmetric neighborhoods of e (because each V_{x_i} and W_{x_i} are). Since $x \in K$, it follows that $x \in V_{x_i} x_i$ and $x \in x_i W_{x_i}$ for some x_i, and so

$$Vx \subseteq V_{x_i} x \subseteq V_{x_i} V_{x_i} x_i = V_{x_i}^2 x_i \subseteq G x_i^{-1} x_i = G \quad \text{for every} \quad x \in K,$$

$$xW \subseteq x W_{x_i} \subseteq x_i W_{x_i} W_{x_i} = x_i W_{x_i}^2 \subseteq x_i^{-1} x_i G = G \quad \text{for every} \quad x \in K.$$

Thus $VK \subseteq G$ and $KW \subseteq G$. If X is locally compact and Hausdorff, then (by Theorem 11.3) there is an open neighborhood N of e and a compact C such that $\{e\} \subseteq N \subseteq C \subseteq V$. Set $V' = N \cap N^{-1}$, an open (cf. (b)) symmetric neighborhood of e such that $V' \subseteq C \subseteq V$. Hence $CK \subseteq VK \subseteq G$. Recall that CK is compact by (a). Then the preceding result ensures that there is an open (bounded) symmetric neighborhood W' of e such that $CKW' \subseteq G$. Set $U = V' \cap V \cap W \cap W'$, an open (bounded) symmetric neighborhood of e. So $UK \subseteq VK \subseteq G$, $KU \subseteq KW \subseteq G$, and $UKU \subseteq V'KW' \subseteq CKW' \subseteq G$.

(e) Suppose X is Hausdorff. If C and K are compact, then C and K are closed (Lemma 11.2(g)), and so $X\backslash C$ and $X\backslash K$ are open. If $C \cap K = \varnothing$, then $C \subseteq X\backslash K$ and $K \subseteq X\backslash C$. Since X is Hausdorff, it follows by Lemma 11.2(f) that for every $x \in X\backslash C$ there is an open set G_1 and an open neighborhood N_x of x such that $C \subseteq G_1$ and $G_1 \cap N_x = \varnothing$. In particular, this holds for every $x \in K \subset X\backslash C$, and so there is an open set $G_2 = \bigcup_{x \in K} N_x$ so that $K \subseteq G_2$ and $G_1 \cap G_2 = \varnothing$. If the Hausdorff X is also locally compact, then we can take G_1 and G_2 bounded by Theorem 11.3. Thus, by item (d), there are open bounded symmetric neighborhoods U_1 and U_2 of e such that $U_1C \subseteq G_1$, $CU_1 \subseteq G_1$, $U_1CU_1 \subseteq G_1$, $U_2K \subseteq G_2$, $KU_2 \subseteq G_2$, $U_2KU_2 \subseteq G_2$. Set $U = U_1 \cap U_2$, an open bounded symmetric neighborhood of e such that $UC \subseteq U_1C \subseteq G_1$, $CU \subseteq CU_1 \subseteq G_1$, $UK \subseteq U_2K \subseteq G_2$, $KU \subseteq KU_2 \subseteq G_2$, $UCU \subseteq U_1CU_1 \subseteq G_1$, and $UKU \subseteq U_2KU_2 \subseteq G_2$. Since $G_1 \cap G_2 = \varnothing$, each of UC, CU, and UCU is disjoint with each of UK, KU and UKU.

(f) Lemma 11.2(e) ensures that if a topological space X is Hausdorff, then $\{e\}$ is closed. Conversely, suppose X is a topological group, and suppose $\{e\}$ is closed. Consider the left multiplication map, so that $\{x\} = x\{e\}$ is closed for every $x \in X$ (reason: the left multiplication map is a homeomorphism so that it maps open sets into open sets and, dually, closed sets into closed sets). Suppose $x \neq e$. Since $\{x\}$ is closed, its complement $X\backslash\{x\}$ is open, so that there exists an open neighborhood G of e included in $X\backslash\{x\}$, and so $x \notin G$. By item (c) there exists an open symmetric neighborhood U of e such that $U^2 \subseteq G$. This ensures that

$$U \cap x\,U = \varnothing.$$

Indeed, if $y \in U \cap x\,U$, then $y = x\,z$ for some $z \in U$, which implies that $x = y\,z^{-1} \in UU = U^2 \subseteq G$ (because U is symmetric; i.e., $U = U^{-1}$), which is a contradiction ($x \notin G$). Hence $U \cap x\,U$ is empty. Thus, since $e \in U$ and $x \in x\,U$, it follows that there exist neighborhoods N_x of x and N_e of e such that $N_x \cap N_e = \varnothing$. Take any distinct points x and y in X. Since $x \neq y$, we get $x^{-1}y \neq e$, and so there exist neighborhoods $N_{x^{-1}y}$ of $x^{-1}y$ and N_e of e such that $N_{x^{-1}y} \cap N_e = \varnothing$, which implies that $xN_{x^{-1}y} \cap xN_e = \varnothing$. Since $y \in xN_{x^{-1}y}$ and $x \in xN_e$, it follows that X is Hausdorff. □

13.2 Haar Measure

Let X be a locally compact Hausdorff group, and let $\mathcal{A}$ be a Borel σ-algebra of subsets of X (i.e., a σ-algebra that includes $\mathcal{X}_{\mathcal{T}}$). A measure μ on $\mathcal{A}$ is *left invariant* or *right invariant* if $\mu(xE) = \mu(E)$ or $\mu(Ex) = \mu(E)$ for every E in $\mathcal{A}$ and every $x \in X$, respectively. It is called an *invariant measure* if it is left invariant. In other words, a measure on a Borel σ-algebra of subsets of a topological group is left invariant or right invariant if it is invariant under left or right translations; for commutative groups a measure is left invariant if and only if it is right invariant (called a *translation invariant* measure, or a measure satisfying the *translation invariance* property).

A left (right) Haar measure is a left (right) invariant nonzero positive Borel measure on a Borel σ-algebra of subsets of a locally compact Hausdorff group. A *Haar measure* is a left Haar measure.

Remarks on Haar Measure: To begin with, observe that there is an asymmetry in our definition of invariant and Haar measures. Left and right translations are naturally distinct but symmetrical properties in a (noncommutative) group. So why favoring "left" when defining Haar measure? The point is that in this context, as we might have already guessed when considering the proofs of items (d) and (e) in Lemma 13.1, the left-right symmetry of the invariance property preserves the pertinent algebraic (group) and topological properties. The reason for this is that the inversion map $x \mapsto x^{-1}$ on X interchanges left and right (i.e., $(xA)^{-1} = A^{-1}x^{-1}$) and, being a homeomorphism, also preserves topological properties. In other words, a result on left translation naturally implies and is implied by a corresponding result on right translation. In particular if μ is a left Haar measure on a Borel σ-algebra, then the set function λ on the same Borel σ-algebra obtained by the composition of inversion and μ (i.e., $\lambda(E) = \mu(E^{-1})$) is a right Haar measure (and vice versa). Another point that is worth noticing is that under the inner regularity assumption the "nonzero" condition in the definition of a Haar measure is equivalent to saying that it assigns a positive value to every nonempty open set. This is shown in the next result.

Lemma 13.2. *Let $\mu\colon \mathcal{A} \to \overline{\mathbb{R}}$ be a positive measure on a Borel σ-algebra of subsets of a locally compact Hausdorff group. Suppose μ is left invariant.*

(a) *If $\mu(U) = 0$ for some open $U \neq \varnothing$, then $\mu(K) = 0$ for every compact K.*

(b) *If $\mu(K) > 0$ for some compact K, then $\mu(U) > 0$ for every open $U \neq \varnothing$.*

If, in addition, μ is an inner regular Borel measure, then

(c) *$\mu \neq 0$ if and only if $\mu(U) > 0$ for every nonempty open set $U \in \mathcal{A}$,*

and, in this case (i.e., *if $\mu \neq 0$*), *then*

(d) *$0 < \int f\,d\mu < \infty$ for every $f \in C_c(X)$ such that $0 \leq f \neq 0$ (for every nonzero nonnegative continuous functions on X with compact support).*

Proof. If K and U are nonempty subsets of X, then for every $k \in K$ and every $u_0 \in U$ there exists $x = ku_0^{-1} \in X$ such that $k = x\,u_0$, and so $\bigcup_{x\in X} xU$ is a covering of K. (If U contains the identity e, then we can take $u_0 = e$ and, in this case, $\bigcup_{x\in K} xU$ is a covering of K.) Thus, if $\mu(U) = 0$ for some open set $U \neq \varnothing$, and if $K \neq \varnothing$ is any compact set, then $\bigcup_{x\in X} xU$ is an open covering of K, and therefore there exists a finite set of points in X, $\{x_i\}_{i=1}^n$, such that $K \subseteq \bigcup_{i=1}^n x_iU$. Since μ is left invariant,

$$\mu(K) \leq \sum_{i=1}^n \mu(x_iU) = \sum_{i=1}^n \mu(U) = n\mu(U) = 0.$$

This proves (a), which is equivalently stated as in (b): if $\mu(K) > 0$ for some compact K, then $\mu(U) > 0$ for every open $U \neq \varnothing$. Suppose, in addition, that μ is an inner regular Borel measure. Since $\mu(U) = 0$ for some open $U \neq \varnothing$ implies $\mu(K) = 0$ for every compact K, and recalling that an inner regular measure is one for which $\mu(E) = \sup\{\mu(K)\colon K \subseteq E,\ K \text{ compact}\}$ for every Borel set E, it follows that $\mu(U) = 0$ for some nonempty open set U implies $\mu(E) = 0$ for every $E \in \mathcal{A}$; that is, $\mu = 0$. In other words, *if every compact set has measure zero, and if the measure is inner regular, then $\mu = 0$.* Thus $\mu \neq 0$ implies $\mu(U) > 0$ for every open $U \neq \varnothing$, proving (c) — the converse is trivial. For (d) note that if $0 \leq f \neq 0$ in $C_c(X)$, then there is an $\varepsilon > 0$ and an open $\varnothing \neq U \subseteq [f]^\circ$ such that $\varepsilon\chi_U \leq f$ and so, according to (c), $0 < \mu(U) = \int \chi_U\,d\mu \leq \int f\,d\mu \leq \mu([f]) < \infty$, since μ is Borel. □

A program to build a Haar measure on a locally compact Hausdorff group is similar to that of building a Borel measure in a locally compact Hausdorff space as in Section 11.3, where it was shown that an inner content on a topology generates a quasiregular outer measure (Lemma 11.8) that is finite at bounded sets, which in turn generates a Borel measure (Theorem 11.7). Now we define the notion of outer content on the collection of all compact sets, which is dual to the notion of inner content on a topology as in Section 11.3, and show in Lemma 13.3 how an outer content on the compact sets generates a quasiregular outer measure, which is the dual of Lemma 11.8.

Let X be a topological space, and let $\mathcal{K}$ denote the class of all compact subsets of X. A *content on the compact sets* is a real-valued set function $\lambda\#: \mathcal{K} \to \mathbb{R}$ on $\mathcal{K}$ such that for arbitrary compact sets K, C, and K_i in $\mathcal{K}$,

(1′) $\lambda\#(\varnothing) = 0$,

(2′) $\lambda\#(K) \geq 0$,

(3′) $\lambda\#(K) < \infty$,

(4′) $\lambda\#(C) \leq \lambda\#(K)$ whenever $C \subseteq K$,

(5′) $\lambda\#(C \cup K) = \lambda\#(C) + \lambda\#(K)$ whenever $C \cap K = \varnothing$,

(6′) $\lambda\#\left(\bigcup_i K_i\right) \leq \sum_i \lambda\#(K_i)$ for finite families $\{K_i\}$.

It is an *outer content* if, in addition,

(7′) $\lambda\#(K) = \inf\{\lambda\#(C)\colon K \subseteq C^\circ,\ C \in \mathcal{K}\}$.

Lemma 13.3. *Let $\lambda\#: \mathcal{K} \to \mathbb{R}$ be an outer content. If X is a locally compact Hausdorff space, then the set function $\mu^*: \wp(X) \to \overline{\mathbb{R}}$ given by*

$$\mu^*(S) = \sup\left\{\lambda\#(K)\colon K \subseteq S,\ K \text{ compact}\right\} \quad \text{for every} \quad S \in \wp(X)$$

is a quasiregular outer measure such that $\mu^*(S) < \infty$ for every bounded S.

Proof. Properties (1′), (2′), and (4′) in the definition of the content $\lambda\#$ imply the properties (a), (b), and (c) in the definition of an outer measure μ^* (see Section 11.3). Property (6′) in the definition of the content $\lambda\#$ implies that

$$\mu^*\left(\bigcup_i S_i\right) \leq \sup_{K_i \subseteq S_i} \lambda\#\left(\bigcup_i K_i\right) \leq \sup_{K_i \subseteq S_i} \sum_i \lambda\#(K_i) \leq \sum_i \mu^*(S_i),$$

for every finite family $\{S_i\}$ of sets in $\wp(X)$, which leads to property (d) in the definition of an outer measure $\mu*$.

Thus μ^* is an outer measure.

Property (4′) in the definition of the content $\lambda\#$ ensures that

$$\mu^*(K) = \lambda\#(K) \quad \text{for every compact set } K.$$

Thus property (iii) in the definition of a quasiregular outer measure (Section 11.3) holds by the definition of μ^*. Moreover, the above identity also shows that property (7′) in the definition of an outer content ensures that

$$\mu^*(K) = \inf\left\{\mu^*(C)\colon K \subseteq C^\circ,\ C \in \mathcal{K}\right\}$$

for every compact set K which, by the definition of μ^*, implies that

$$\mu^*(S) = \sup_{K\subseteq S} \mu^*(K) = \sup_{K\subseteq S} \inf_{K\subseteq C^\circ} \mu^*(C^\circ) = \inf_{S\subseteq C^\circ} \mu(C^\circ)$$

for every bounded set S, which leads to property (i) in the definition of a quasiregular outer measure. Property (5′) in the definition of an outer content implies property (ii) in the definition of a quasiregular outer measure, according to the definition of μ^*:

$$\mu^*(G\cup U) = \sup_{C\cup K\subseteq G\cup U} \lambda\#(C\cup K) = \sup_{C\subseteq G,\, K\subseteq U} \lambda\#(C)+\lambda\#(K) = \mu^*(G)+\mu^*(U)$$

whenever $G\cap U = \varnothing$. (Indeed, according to the definition of μ^*, property (ii) of a quasiregular outer measure becomes, in this case, a consequence of the condition $\mu^*(C\cup K) = \mu^*(C) + \mu^*(K)$ whenever $C\cap K = \varnothing$ — property (5′), since this μ^* coincides with $\lambda\#$ on $\mathcal{K}$ so that $\mu^*(S) = \sup_{K\subseteq S} \mu^*(K)$.)

Thus the outer measure μ^* is quasiregular.

Finally, property (3′) in the definition of the content $\lambda\#$ and property (c) in the definition of an outer measure μ^* imply that

$$\mu^*(S) < \infty \text{ for every bounded set } S \in \wp(X)$$

(i.e., for every $S \subseteq K$ for some $K \in \mathcal{K}$). □

13.3 Construction of Haar Measures

In a locally compact Hausdorff group X, a bounded set means a relatively compact set (i.e., one whose closure is compact) and so a compact set is precisely a closed and bounded set (see Remarks on Boundedness in Section 11.1). Let $B \subseteq X$ be bounded (i.e., B^- is compact), and let $A \subseteq X$ have a nonempty interior (i.e., the open set A° is nonempty). Since $\bigcup_{x\in X} xA^\circ$ is an open covering of B^-, there is a finite set $\{x_i\}_{i=1}^n \subseteq X$ such that

$$B \subseteq B^- \subseteq \bigcup_{i=1}^n x_i A^\circ \subseteq \bigcup_{i=1}^n x_i A.$$

The *covering number* $[B\!:\!A]$ of a bounded set B by a set A with nonempty interior is the least number of translates of A required to cover B. That is,

$$[B\!:\!A] = \min\big\{n \in \mathbb{N}\colon B \subseteq \textstyle\bigcup_{i=1}^n x_i A\big\} > 0$$

whenever $B \neq \varnothing$. Actually, $[\varnothing\!:\!A] = 0$. Observe that

$$[B\!:\!A] = \min\big\{[C\!:\!A]\colon B \subseteq C^\circ,\ C \text{ bounded}\big\} \in \mathbb{N}.$$

Also, the covering number is translation invariant: for every $x \in X$,

$$[xB:A] = [B:A] = [B:xA].$$

If $D \in X$ is bounded with nonempty interior (D^- compact and $D^\circ \neq \varnothing$), then

$$[B:A] \leq [B:D]\,[D:A]$$

(reason: cover B with $[B:D]$ translates of D and cover each translate of D with $[D:A]$ translates of A), and

$$[D:D] = 1.$$

If B is a bounded set and U is a nonempty open set, then the covering number $[B:U]$ can be thought of as a comparison between the sizes of B and U. This is our starting point for constructing a Haar measure. For a given B, the covering number $[B:U]$ gets larger as U gets smaller. To control this growing process we consider instead the *covering ratio* $[B:U]/[D:U]$ for some reference bounded set D with nonempty interior, and define somehow a limit of this covering ratio as U gets smaller and smaller. This will lead us to a left invariant nonzero positive inner regular Borel measure.

From now on we assume the following setup. Let X be a locally compact Hausdorff group and consider the inclusions

$$\mathcal{T} \subseteq \mathcal{X}_{\mathcal{T}} \subseteq \mathcal{A} \subseteq \wp(X) \quad \text{and} \quad \mathcal{K} \subseteq (\mathcal{B} \cap \mathcal{X}_{\mathcal{T}}),$$

where $\mathcal{T}$ is a topology on X, $\mathcal{X}_{\mathcal{T}}$ is the Borel σ-algebra of subsets of X generated by $\mathcal{T}$, $\mathcal{A}$ is any Borel σ-algebra subsets of X, $\wp(X)$ is the power set of X, $\mathcal{K} \subseteq \mathcal{X}_{\mathcal{T}}$ is the class of all compact subsets of X, and $\mathcal{B} \subseteq \wp(X)$ is the class of all bounded subsets of X. Let D be a bounded set with nonempty interior (i.e., $D \in \mathcal{B}$ and $D^\circ \neq \varnothing$). Take an arbitrary nonempty open set $U \in \mathcal{T}$ and consider the rational-valued function $\lambda_U: \mathcal{B} \to \mathbb{Q}$ defined by

$$\lambda_U(B) = \tfrac{[B:U]}{[D:U]} \quad \text{for every} \quad B \in \mathcal{B}.$$

Lemma 13.4. *The covering ratio function $\lambda_U: \mathcal{B} \to \mathbb{Q}$ has the following properties. For arbitrary bounded sets B, B_1, B_2, and B_i in $\mathcal{B}$,*

(a) $\lambda_U(\varnothing) = 0$,

(b) $0 < \lambda_U(B)$ *whenever* $B \neq \varnothing$,

(c) $0 \leq \lambda_U(B) \leq [B:D] < \infty$,

(d) $\lambda_U(B_1) \leq \lambda_U(B_2)$ *whenever* $B_1 \subseteq B_2$,

(e) $\lambda_U(B_1 \cup B_2) = \lambda_U(B_1) + \lambda_U(B_2)$ *whenever* $B_1U^{-1} \cap B_2U^{-1} = \varnothing$,

(f) $\lambda_U\big(\bigcup_i B_i\big) \le \sum_i \lambda_U(B_i)$ *for finite families* $\{B_i\}$,

(g) $\lambda_U(B) = \min\{\lambda_U(C)\colon B \subseteq C^\circ,\ C \in \mathcal{B}\} \in \mathbb{Q}$,

(h) $\lambda_U(xB) = \lambda_U(B)$ *for every* $x \in X$,

(i) $\lambda_U(D) = 1$,

(j) $U \mapsto \lambda_U(B)$ *is a bounded map for each* B.

(k) *If* K_1 *and* K_2 *are disjoint compact sets, then there is an open bounded symmetric neighborhood* U *of the identity* e *such that*

$$\lambda_G(K_1 \cup K_2) = \lambda_G(K_1) + \lambda_G(K_2)$$

for every open neighborhood G *of* e *such that* $G \subseteq U$.

Proof. Properties (a) to (f) are readily verified except, perhaps, for (e). To show (e) take $x \in X$ and observe that if $B_1 \cap xU \neq \varnothing$, then $x \in B_1U^{-1}$. (Indeed, if $b = xu$, then $bu^{-1} = x$, so that $x \in B_1U^{-1}$.) Similarly, if $B_2 \cap xU \neq \varnothing$, then $x \in B_2U^{-1}$. Thus, if $B_1U^{-1} \cap B_2U^{-1} = \varnothing$, then there is no x such that $B_1 \cap xU \neq \varnothing$ and $B_2 \cap xU \neq \varnothing$; that is, no left translate of U meets both B_1 and B_2, and so $[B_1 \cup B_1 : U] = [B_1 : U] + [B_2 : U]$, proving property (e). Property (e) leads to property (k). Indeed, if the K_1 and K_2 are disjoint compact sets, then there exists an open bounded symmetric neighborhood N of the identity e such that $K_1N \cap K_2N = \varnothing$ by Lemma 13.1(e). Set $U = N^{-1}$, again an open bounded symmetric neighborhood of e. Let G be an arbitrary open (bounded) neighborhood of e such that $G \subseteq U$. Set $N' = G^{-1}$, an open (bounded) neighborhood of e. Since $N'^{-1} = G \subseteq U = N^{-1}$, it follows that $N' \subseteq N$, and so $K_1N' \cap K_2N' = \varnothing$, equivalently, $K_1G^{-1} \cap K_2G^{-1} = \varnothing$. Thus $\lambda_G(K_1 \cup K_2) = \lambda_G(K_1) + \lambda_G(K_2)$ by (e), proving property (k). Properties (g) and (h) follow from the facts that $[B : U] = \min\{[C : U]\colon B \subseteq C^\circ,\ C \in \mathcal{B}\}$ and $[xB : U] = [B : U]$, (i) is trivial, and (j) follows from (c) as a consequence of $[B : U] \le [B : D]\,[D : U]$. □

Observe from properties (a) to (f) in the preceding lemma that λ_U (actually, the restriction of it to $\mathcal{K}$) is nearly a content; it fails to be a content on $\mathcal{K}$ just because additivity in property (e) may not hold for every pair of disjoint compact sets — property (k) is an attempt to establish additivity that will actually succeed in the next lemma.

The next lemma plays a central role in proving the existence of Haar measures. There are distinct ways of approaching its proof. Some authors use the Hahn–Banach Theorem (on extension of bounded linear functionals), others use the Arzelà–Ascoli Theorem (on compact subsets of continuous functions), and some use the Tychonoff Theorem, which says that *the Cartesian product* $\prod_{\gamma\in\Gamma} X_\gamma$ *of compact sets* X_γ *is compact* (in the product topology — see e.g., [21, Theorem 5.13]).

Lemma 13.5. *The covering ratio function* $\lambda_U\colon\mathcal{B}\to\mathbb{Q}$ *generates a nonzero left invariant outer content* $\lambda\#\colon\mathcal{K}\to\mathbb{R}$ *on* $\mathcal{K}$.

Proof. Let $\mathcal{U}$ denote the collection of all open neighborhoods of the identity $e\in X$. Fix an arbitrary compact set $D\in\mathcal{K}\subseteq\mathcal{B}$ with nonempty interior, and consider the covering ratio function $\lambda_U\colon\mathcal{B}\to\mathbb{R}$ for an arbitrary $U\in\mathcal{U}$. Take the restriction of each λ_U to $\mathcal{K}$, denoted again by $\lambda_U\colon\mathcal{K}\to\mathbb{R}$, so that $\lambda_U(K)=[K\colon U]/[D\colon U]$ for every $K\in\mathcal{K}$. Let

$$\Lambda=\bigcup_{U\in\mathcal{U}}\{\lambda_U\}$$

be the collection of the covering ratio functions $\lambda_U\colon\mathcal{K}\to\mathbb{R}$ restricted to $\mathcal{K}$ for all $U\in\mathcal{U}$. Take the Cartesian product

$$\Pi=\prod_{K\in\mathcal{K}}[0,[K\colon D]]$$

of the closed intervals $[0,[K\colon D]]$. Since closed (and bounded) intervals are compact sets in $\mathbb{R}$, the Tychonoff Theorem says that the Cartesian product $\prod_{K\in\mathcal{K}}[0,[K\colon D]]$ is compact in the product topology. But the elements of the Cartesian product $\prod_{K\in\mathcal{K}}[0,[K\colon D]]$ are interpreted as real-valued functions on $\mathcal{K}$. Indeed, by the definition of Cartesian product, $\prod_{K\in\mathcal{K}}[0,[K\colon D]]$ consists of all indexed families $\{\varphi(K)\}_{K\in\mathcal{K}}$ such that $\varphi(K)\in[0,[K\colon D]]$ for each $K\in\mathcal{K}$; equivalently, of all real-valued functions $\varphi\colon\mathcal{K}\to\mathbb{R}$ on $\mathcal{K}$ such that $0\le\varphi(K)\le[K\colon D]$ for every $K\in\mathcal{K}$. Since $0\le\lambda_U(K)\le[K\colon D]<\infty$ by Lemma 13.4(c), the function $\lambda_U\colon\mathcal{K}\to\mathbb{R}$ lies in $\prod_{K\in\mathcal{K}}[0,[K\colon D]]$ for each $U\in\mathcal{U}$. That is, $\Lambda\subseteq\Pi$. For each $U\in\mathcal{U}$ consider the set of functions

$$\Lambda(U)=\{\lambda_G\in\Lambda\colon G\subseteq U\}\subseteq\Pi.$$

Let $\{U_i\}_{i=1}^n$ be an arbitrary finite family of open neighborhoods of e (i.e., each U_i lies in $\mathcal{U}$). Observe that $\bigcap_{i=1}^n U_i$ is again a set in $\mathcal{U}$, and also that

$$\varnothing\neq\Lambda\Big(\bigcap_{i=1}^n U_i\Big)\subseteq\bigcap_{i=1}^n\Lambda(U_i).$$

(Reason: $\bigcap_{i=1}^n U_i\subseteq U_j$ so that $\Lambda(\bigcap_{i=1}^n U_i)\subseteq\Lambda(U_j)$ for all $U_j\in\{U_i\}_{i=1}^n$ — cf. Lemma 13.4(d) — and $\Lambda(U)\neq\varnothing$ for every $U\in\mathcal{U}$ because $\lambda_U\in\Lambda(U)$.) Hence the family $\{\Lambda(U)\subseteq\Pi\colon U\in\mathcal{U}\}$ of subsets of $\Lambda\subseteq\Pi$ has the finite intersection property (i.e., every finite subcollection of it has a nonempty finite intersection). In particular, the family $\{\Lambda(U)^- \subseteq \Pi : U \in \mathcal{U}\}$ of closed subsets of the compact space Π has the finite intersection property (since $\varnothing \neq \bigcap_{i=1}^n \Lambda(U_i) \subseteq \bigcap_{i=1}^n \Lambda(U_i)^-$). Therefore, Proposition 11.A ensures that $\{\Lambda(U)^- \subseteq \Pi : U \in \mathcal{U}\}$ has a nonempty intersection. Hence there exists

$$\lambda^\# \in \bigcap_{U \in \mathcal{U}} \Lambda^-(U) = \bigcap_{U \in \mathcal{U}} \{\lambda_G \in \Lambda : G \subseteq U\}^- \subseteq \Pi,$$

where $\lambda^\#$ is a real-valued function on $\mathcal{K}$, as is every element of the set $\Pi = \prod_{K \in \mathcal{K}}[0, [K : D]]$. Thus, for an arbitrary $K \in \mathcal{K}$, the value $\lambda^\#(K) \in \mathbb{R}$ is approached by taking a $\lambda_U(K) = [K : U]/[D : U] \in \mathbb{Q}$ for some $U \in \mathcal{U}$, and letting U get smaller and smaller in the above sense. We show that such a function $\lambda^\# : \mathcal{K} \to \mathbb{R}$ is a nonzero left invariant outer content on $\mathcal{K}$.

(i) First note that properties (a) to (c) in Lemma 13.4 trivially imply properties $(1')$ to $(3')$ in the definition of a content on $\mathcal{K}$.

(ii) To verify property $(4')$ in the definition of a content on $\mathcal{K}$ proceed as follows. For each $K \in \mathcal{K}$ take the projection $\Phi_K : \Pi \to [0, [K : D]]$ defined by $\Phi_K(\varphi) = \varphi(K)$ for every function $\varphi : \mathcal{K} \to [K : D]$ in $\Pi = \prod_{K \in \mathcal{K}}[0, [K : D]]$. The product topology on $\prod_{K \in \mathcal{K}}[0, [K : D]]$ makes the projections Φ_K continuous (see e.g., [21, p. 90]). Take arbitrary sets $K_1, K_2 \in \mathcal{K}$. Thus the difference $\Phi_{K_2 - K_1} = \Phi_{K_2} - \Phi_{K_1} : \Pi \to \mathbb{R}$ is continuous. Consider the set

$$\begin{aligned}\Sigma_{K_2 - K_1} &= \{\varphi \in \Pi : \varphi(K_1) \leq \varphi(K_2)\} \\ &= \{\varphi \in \Pi : 0 \leq \Phi_{K_2 - K_1}(\varphi)\} = \Phi_{K_2 - K_1}^{-1}([0, \infty)) \subseteq \Pi,\end{aligned}$$

which is closed (since $\Phi_{K_2 - K_1}$ is continuous, so that the inverse image of a closed set is closed). If $K_1 \subseteq K_2$, then $\lambda_U(K_1) \leq \lambda_U(K_2)$ for every $U \in \mathcal{U}$ by Lemma 13.4(d), so that $\lambda_U \in \Sigma_{K_2 - K_1}$ for every $U \in \mathcal{U}$, and hence we get $\Lambda(U) \subseteq \Sigma_{K_2 - K_1}$ for every $U \in \mathcal{U}$. Thus $\lambda^\# \in \bigcap_{U \in \mathcal{U}} \Lambda(U)^- \subseteq \Sigma_{K_2 - K_1}$ (since $\Sigma_{K_2 - K_1}$ is closed). Then $\lambda^\#(K_1) \leq \lambda^\#(K_2)$. That is, property (d) in Lemma 13.4 implies property $(4')$ in the definition of a content on $\mathcal{K}$.

(iii) Take $K_1, K_2 \in \mathcal{K}$. Since Φ_K is continuous for each $K \in \mathcal{K}$, it follows that $\Phi_{K_2 + K_1} = \Phi_{K_2} + \Phi_{K_1} - \Phi_{K_1 \cup K_2} : \Pi \to \mathbb{R}$ is continuous. Consider the set

$$\begin{aligned}\Sigma_{K_2 + K_1} &= \{\varphi \in \Pi : \varphi(K_1 \cup K_2) = \varphi(K_1) + \varphi(K_2)\} \\ &= \{\varphi \in \Pi : \Phi_{K_2 + K_1}(\varphi) = 0\} = \Phi_{K_2 + K_1}^{-1}(\{0\}) \subseteq \Pi,\end{aligned}$$

which is closed (since $\Phi_{K_2 + K_1}$ is continuous and every singleton is closed in $\mathbb{R}$). If $K_1 \cap K_2 = \varnothing$, then there exist $U_0 \in \mathcal{U}$ such that $\lambda_G(K_1 \cup K_2) = \lambda_G(K_1) + \lambda_G(K_2)$ for every $G \in \mathcal{U}$ such that $G \subseteq U_0$ by Lemma 13.4(k).

That is, there exists $U_0 \in \mathcal{U}$ such that $\Lambda(U_0) \subseteq \Sigma_{K_1+K_2}$. Therefore, since $\Sigma_{K_2+K_1}$ is closed, it follows that $\lambda^\# \in \bigcap_{U\in\mathcal{U}} \Lambda(U)^- \subseteq \Lambda(U_0)^- \subseteq \Sigma_{K_2+K_1}$. Then $\lambda^\#(K_1 \cup K_2) = \lambda^\#(K_1) + \lambda^\#(K_2)$. That is, property (k) in Lemma 13.4 implies property (5′) in the definition of a content on $\mathcal{K}$.

(iv) Property (6′) of a content on $\mathcal{K}$ results from property (f) in Lemma 13.4 by the same argument of (ii) with the map $\Phi_{K_2+K_1}$ of (iii). Indeed, the set

$$\begin{aligned}\Sigma'_{K_2+K_1} &= \{\varphi \in \Pi \colon \varphi(K_1 \cup K_2) \le \varphi(K_1) + \varphi(K_2)\}\\ &= \{\varphi \in \Pi \colon 0 \le \Phi_{K_2+K_1}(\varphi)\} = \Phi^{-1}_{K_2+K_1}([0,\infty)) \subseteq \Pi\end{aligned}$$

is closed. Since $\lambda_U(K_1 \cup K_2) \le \lambda_U(K_1) + \lambda(K_2)$ for every $U \in \mathcal{U}$ by Lemma 13.4(f), $\lambda_U \in \Sigma'_{K_2+K_1}$, and hence $\Lambda(U) \subseteq \Sigma'_{K_2+K_1}$ for every $U \in \mathcal{U}$. Thus $\lambda^\# \in \bigcap_{U\in\mathcal{U}} \Lambda(U)^- \subseteq \Sigma'_{K_2+K_1}$, and so $\lambda^\#(K_1 \cup K_2) \le \lambda^\#(K_1) + \lambda^\#(K_2)$, which extends by induction to finite unions of compact sets. Then property (f) in Lemma 13.4 implies property (6′) in the definition of a content on $\mathcal{K}$.

(v) Property (g) in Lemma 13.4 leads to Property (7′) of an outer content on $\mathcal{K}$. In fact, take $K \in \mathcal{K}$ arbitrary. Let C be any set in $\mathcal{K}$ such that $K \subseteq C^\circ$. Recall that Φ_K and Φ_C are continuous, and so $\{\Phi_{C-K} = \Phi_C - \Phi_K\}_{K\subseteq C^\circ}$ is a family of continuous maps. Take the set $\Sigma_{C-K} = \{\varphi \in \Pi \colon 0 \le \Phi_{C-K}(\varphi)\}$ of item (ii), which is closed, and consider the set

$$\begin{aligned}\Sigma_K &= \{\varphi \in \Pi \colon \varphi(K) = \inf_{K\subseteq C^\circ} \varphi(C)\} = \{\varphi \in \Pi \colon \Phi_K(\varphi) = \inf_{K\subseteq C^\circ} \Phi_C(\varphi)\}\\ &= \{\varphi \in \Pi \colon \inf_{K\subseteq C^\circ} \Phi_{C-K}(\varphi) = 0\} = \inf_{K\subseteq C^\circ} \Sigma_{C-K} = \bigcap_{K\subseteq C^\circ} \Sigma_{C-K} \subseteq \Pi,\end{aligned}$$

which is closed as well (intersection of closed sets). Thus, as before, since $\lambda_U(K) = \inf_{K\subseteq C^\circ} \lambda_U(C)$ by Lemma 13.4(g), so that $\lambda_U \in \Sigma_K$, for every $U \in \mathcal{U}$, we get $\lambda^\# \in \bigcap_{U\in\mathcal{U}} \Lambda(U)^- \subseteq \Sigma_K$ (because Σ_K is closed). Therefore, $\lambda^\#(K) = \inf_{K\subseteq C^\circ} \lambda^\#(C)$. Then property (g) in Lemma 13.4 ensures property (7′) in the definition of an outer content on $\mathcal{K}$.

Hence, according to (i)-(v), the function $\lambda^\# \colon \mathcal{K} \to \mathbb{R}$ is an outer content on $\mathcal{K}$. That $\lambda^\#$ is nonzero follows from the fact that $\lambda^\#(D) = 1$. Indeed the same continuity argument ensures that since Φ_D is continuous, the set

$$\Sigma_D = \{\varphi \in \Pi \colon \varphi(D) = 1\} = \{\varphi \in \Pi \colon \Phi_D(\varphi) = 1\} = \Phi_D^{-1}(\{1\}) \subseteq \Pi$$

is closed. Since $\lambda_U \in \Sigma_D$ by Lemma 13.4(i), we get that $\Lambda(U)^- \subseteq \Sigma_D^- \subseteq \Sigma_D$, for every $U \in \mathcal{U}$, and so $\lambda^\# \in \Sigma_D$. Finally, take $x \in X$, $K \in \mathcal{K}$, and consider the continuous map $\Phi_{xK-K} = \Phi_{xK} - \Phi_K \colon \Pi \to \mathbb{R}$. Then the set

$$\Sigma_{xK-K} = \{\varphi \in \Pi \colon \varphi(xK) = \varphi(K)\} = \{\varphi \in \Pi \colon \Phi_{xK} = 0\} = \Phi^{-1}_{xK}(\{0\}) \subseteq \Pi$$

is closed, and so, according to Lemma 13.4(h), the same argument ensures that $\lambda\# \in \Sigma_{xK}$, which means that $\lambda\#$ is left invariant. □

The preceding lemma allows us to apply Theorem 11.7 to ensure the existence of a Haar measure on a Borel σ-algebra of subsets of an arbitrary locally compact Hausdorff group.

Theorem 13.6. *There exists a quasiregular Haar measure on a Borel σ-algebra of subsets of every locally compact Hausdorff group.*

Proof. We will prove the following statement that leads to the above one.

A left invariant outer content $\lambda\#: \mathcal{K} \to \mathbb{R}$ on $\mathcal{K}$ (generated by a covering ratio function $\lambda_U: \mathcal{B} \to \mathbb{Q}$) generates a quasiregular outer measure $\mu^*: \wp(X) \to \overline{\mathbb{R}}$, finite on $\mathcal{B}$, which in turn generates a quasiregular complete Haar measure (i.e., a left invariant nonzero positive quasiregular complete Borel measure) $\lambda^*: \mathcal{A}^* \to \overline{\mathbb{R}}$ *on a Borel σ-algebra $\mathcal{A}^*$ of subsets of a locally compact Hausdorff group X. Its restriction $\lambda: \mathcal{X}_{\mathcal{T}} \to \overline{\mathbb{R}}$ to the Borel σ-algebra $\mathcal{X}_{\mathcal{T}}$ generated by $\mathcal{T}$ is again a quasiregular Haar measure although λ may not be compete.*

Indeed, consider the covering ratio function

$$\lambda_U: \mathcal{B} \to \mathbb{Q}$$

defined by $\lambda_U(B) = [B:U]/[D:U]$ for every $B \in \mathcal{B}$, which is nonzero and left invariant by Lemma 13.4(h,i). This λ_U generates a nonzero left invariant outer content on $\mathcal{K}$ (whose existence was proved in Lemma 13.5),

$$\lambda\#: \mathcal{K} \to \mathbb{R},$$

which in turn generates a quasiregular outer measure

$$\mu^*: \wp(X) \to \overline{\mathbb{R}}$$

defined by $\mu^*(S) = \sup\{\lambda\#(K): K \subseteq S,\ K \in \mathcal{K}\}$ for every $S \in \wp(X)$, which is finite on $\mathcal{B}$, according to Lemma 13.3. Therefore, by Theorem 11.7, there exists a quasiregular complete Borel measure

$$\lambda^* = \mu^*|_{\mathcal{A}^*}: \mathcal{A}^* \to \overline{\mathbb{R}}$$

on a σ-algebra $\mathcal{A}^*$ that includes the Borel σ-algebra $\mathcal{X}_{\mathcal{T}}$. It remains to verify that the (positive) Borel measure λ^* is nonzero and left invariant.

(a)] The measure λ^* is nonzero. In fact, $\lambda\#$ is nonzero by Lemma 13.5, and hence μ^* is nonzero according to its very definition in Lemma 13.3. Thus $\mu^*(X) > 0$. Then, since $X \in \mathcal{A}^*$, we get $\lambda^*(X) = \mu^*|_{\mathcal{A}^*}(X) = \mu^*(X) > 0$.
(b) The measure λ^* is left invariant. Indeed, since $\lambda\#$ is left invariant by

Lemma 13.5 (i.e., $\lambda\#(xK) = \lambda\#(K)$ for every $x \in X$ and every $K \in \mathcal{K}$), it follows by the definition of μ^* in Lemma 13.3 that

$$\begin{aligned}\mu^*(x^{-1}S) &= \sup_{K\subseteq x^{-1}S} \lambda\#(K) = \sup_{xK\subseteq S} \lambda\#(K)\\ &= \sup_{xK\subseteq S} \lambda\#(xK) = \sup_{K'\subseteq S} \lambda\#(K') = \mu^*(S)\end{aligned}$$

for every $x \in X$ and $S \in \wp(X)$, so that μ^* is left invariant, and so is the restriction $\lambda^* = \mu^*|_{\mathcal{A}^*}$ of it to $\mathcal{A}^*$ so that for every $x \in X$ and $E \in \mathcal{A}^*$,

$$\lambda^*(xE) = \mu^*(xE) = \mu^*(E) = \lambda^*(E).$$

Again, as in the proof of Theorem 11.7, the above properties of the quasiregular complete Haar measure λ^*, except for completeness, are transferred to all restrictions of it to any Borel σ-algebra of subsets of X included in $\mathcal{A}^*$; in particular, to its restriction $\lambda = \lambda^*|_{\mathcal{X}_\mathcal{T}}$ to the smallest Borel σ-algebra $\mathcal{X}_\mathcal{T}$ of subsets of X, so that λ is a quasiregular Haar measure on $\mathcal{X}_\mathcal{T}$. □

It is clear that a Haar measure on a Borel σ-algebra of subsets of a given locally compact Hausdorff group X is not unique. Reason: μ is Haar if and only if $\gamma\,\mu$ is Haar for every positive number γ. However, this is essentially the only way they can differ.

Theorem 13.7. *If λ and μ are quasiregular Haar measure on the same Borel σ-algebra of subsets of a given locally compact Hausdorff group, then there exists a positive constant γ such that*

$$\lambda = \gamma\,\mu.$$

Proof. Let μ and ν be quasiregular Haar measures on a Borel σ-algebra $\mathcal{A}$ of subsets of a locally compact Hausdorff group X. Take the product $\mu\times\nu$ on $\mathcal{A}\times\mathcal{A}$ as in Theorem 9.5. Recall that continuous functions with compact support are integrable with respect to Borel measures. Consider the set

$$C_c(X)^+ = \{g \in C_c(X,\mathbb{R})\colon\, g(X) > 0\}$$

consisting of all nonzero and nonnegative continuous real-valued functions $g\colon X \to \mathbb{R}$ on X with compact support (see Section 12.1). Take an arbitrary μ-integrable Borel function $f \in \mathcal{L}(X,\mathcal{A},\mu)$, and an arbitrary function $g \in C_c(X)^+$, so that $\int f\,d\mu < \infty$ and $0 < \int g\,d\nu < \infty$ by Lemma 13.2(d). Consider the function $h\colon X\times X \to \mathbb{R}$ defined by

$$h(x,y) = \frac{f(x)g(yx)}{\int g(zx)\,d\nu(z)} \quad \text{for every} \quad x,y \in X,$$

which is again a $\mu\times\nu$-integrable Borel function: $h \in \mathcal{L}(X\times X, \mathcal{A}\times\mathcal{A}, \mu\times\nu)$ (cf. Problem 9.20). Now recall that the Fubini Theorem (Theorem 9.9) allows us to interchange the order of the iterated integrals (cf. Problem 9.20 again). The translation invariance property in Proposition 13.E says that we can swap x with zx in the argument of integrable functions when integrating them. (In particular, we can replace x with $y^{-1}x$ if the argument is x, and y with xy if the argument is y, and so we can replace $y^{-1}x = (y)^{-1}x$ with $(xy)^{-1}x = y^{-1}x^{-1}x = y^{-1}$ if the argument is y.) Therefore,

$$\begin{aligned}\int\Big(\int h(x,y)\,d\nu(y)\Big)\,d\mu(x) &= \int\Big(\int h(x,y)\,d\mu(x)\Big)\,d\nu(y)\\ &= \int\Big(\int h(y^{-1}x,y)\,d\mu(x)\Big)\,d\nu(y) = \int\Big(\int h(y^{-1}x,y)\,d\nu(y)\Big)\,d\mu(x)\\ &= \int\Big(\int h(y^{-1},xy)\,d\nu(y)\Big)\,d\mu(x)\end{aligned}$$

(where we reverse the integration order, then we replace x with $y^{-1}x$, reverse the integration order again, and replace $y^{-1}x$ with y^{-1} and y with xy). Since

$$h(y^{-1},xy) = \frac{f(y^{-1})g(x)}{\int g(zy^{-1})\,d\nu(z)} \quad \text{for every} \quad x,y \in X,$$

it then follows that

$$\begin{aligned}\int f(x)\,d\mu(x) &= \int f(x)\left(\frac{\int g(yx)\,d\nu(y)}{\int g(zx)\,d\nu(z)}\right)d\mu(x) = \int\left(\int\frac{f(x)g(yx)}{\int g(zx)\,d\nu(z)}\,d\nu(y)\right)d\mu(x)\\ &= \int\Big(\int h(x,y)\,d\nu(y)\Big)\,d\mu(x) = \int\Big(\int h(y^{-1},xy)\,d\nu(y)\Big)\,d\mu(x)\\ &= \int\left(\int\frac{f(y^{-1})g(x)}{\int g(zy^{-1})\,d\nu(z)}\,d\nu(y)\right)d\mu(x) = \int\left(\int\frac{f(y^{-1})\,d\nu(y)}{\int g(zy^{-1})\,d\nu(z)}\right)g(x)\,d\mu(x).\end{aligned}$$

Set $\alpha = \alpha(f,g,\nu) = \big(\int f(y^{-1})\,d\nu(y)\big)/\big(\int g(zy^{-1})\,d\nu(z)\big)$, which is a real number that does not depend on μ such that $\int f\,d\mu = \alpha\int g\,d\mu$. Thus, if λ also is a quasiregular Haar measure on the same Borel σ-algebra $\mathcal{A}$ of subsets of X, then $\int f\,d\lambda = \alpha\int g\,d\lambda$ whenever $f \in \mathcal{L}(X,\mathcal{A},\lambda)$. Hence,

$$\frac{\int f\,d\lambda}{\int g\,d\lambda} = \frac{\int f\,d\mu}{\int g\,d\mu}, \quad \text{which implies} \quad \int f\,d\lambda = \left(\frac{\int g\,d\lambda}{\int g\,d\mu}\right)\int f\,d\mu,$$

for every $f \in \mathcal{L}(X,\mathcal{A},\mu)\cap\mathcal{L}(X,\mathcal{A},\lambda)$, for an arbitrary $g \in C_c(X)^+$. Then there is a positive constant $\gamma = \int g\,d\lambda/\int g\,d\mu$ such that $\int f\,d\lambda = \gamma\int f\,d\mu$ for every $f \in \mathcal{L}(X,\mathcal{A},\mu)\cap\mathcal{L}(X,\mathcal{A},\lambda)$. Therefore, since μ and λ are Borel measures on the Borel σ-algebra $\mathcal{A}$, it follows that for every compact set K, its characteristic function χ_K lies in $\mathcal{L}(X,\mathcal{A},\mu)\cap\mathcal{L}(X,\mathcal{A},\lambda)$, and so

$$\lambda(K) = \int_K d\lambda = \int \chi_K \, d\lambda = \gamma \int \chi_K \, d\mu = \gamma \int_K d\mu = \gamma \, \mu(K)$$

for every compact set K (all of them lie in $\mathcal{A}$). Since the Borel measures λ and μ on the Borel σ-algebra $\mathcal{A}$ are quasiregular, the argument that closes the proof of Theorem 12.5 (which uses Lemma 12.2) ensures that if the above identity holds for every compact set K, then $\lambda = \gamma \, \mu$. $\square$

13.4 Additional Propositions

Recall that a *discrete space* is a topological space X whose topology $\mathcal{T}$ is the largest topology on X; that is, whose topology coincides with the power set of X, so that every subset of X is open (and closed). Such a topology $\mathcal{T} = \wp(X)$ is called the *discrete topology.*

Proposition 13.A. *If μ is a Haar measure on a Borel σ-algebra of subsets of a locally compact Hausdorff group X, then X is discrete if and only if $\mu(\{x\}) \neq 0$ for at least one point $x \in X$.*

Proposition 13.B. *Let $\mathcal{A}$ be a Borel σ-algebra of subsets of a locally compact Hausdorff group X. If a Haar measure on $\mathcal{A}$ is finite, then X is compact.*

Proposition 13.C. *If μ is a Haar measure on a Borel σ-algebra of subsets of a locally compact Hausdorff group X, then the following assertions are pairwise equivalent.*

(a) *The measure μ is σ-finite.*

(b) *The space X is σ-compact.*

(c) *Every disjoint family of nonempty open sets is countable.*

(d) *For every nonempty open set U, there exists a sequence $\{x_n\}$ of points in X such that the family $\{x_n U\}$ covers X* (i.e., $X = \bigcup_n x_n U$).

Proposition 13.D. *If μ is a Haar measure on a Borel σ-algebra of subsets of a locally compact Hausdorff group X, and if f and g are real-valued continuous functions on X, then $f = g$ everywhere if and only if $f = g$ μ-almost everywhere.*

Proposition 13.E. *Let X be a locally compact Hausdorff group. For each real-valued function $f\colon X \to \mathbb{R}$ on X and each $y \in X$ let $f_y\colon X \to \mathbb{R}$ be defined by $f_y(x) = f(y^{-1}x)$ for every $x \in X$. Let $\mu\colon \mathcal{A} \to \overline{\mathbb{R}}$ be a nonzero pos-*

itive quasiregular Borel measure on a Borel σ-algebra $\mathcal{A}$ of subsets of X. The measure μ is Haar (i.e., *it is left invariant*) *if and only if*

$$\int f_y\,d\mu = \int f\,d\mu,$$

where $f_y \in \mathcal{L}(X,\mathcal{A},\mu)$, for every $y \in X$ and every $f \in \mathcal{L}(X,\mathcal{A},\mu)$; that is, for every μ-integrable Borel function $f\colon X \to \mathbb{R}$.

A *Haar integral* is an integral with respect to a Haar measure. The above property, namely, $f_y \in \mathcal{L}(X,\mathcal{A},\mu)$ *whenever* $f \in \mathcal{L}(X,\mathcal{A},\mu)$ *and the integrals coincide*, is referred to as the *translation invariance property* for the Haar integral (which, according to Proposition 13.E, characterizes the Haar measure). Recall again that continuous functions with compact support are integrable with respect to Borel measures. A positive linear functional Φ on the linear space $C_c(X,\mathbb{R})$ (where X is a locally compact Hausdorff group) with the above translation invariance property is also referred to as a *Haar integral*. Example: if $X = \mathbb{R}$, which is a locally compact Hausdorff additive group, then Proposition 13.E ensures that one can make the substitution $x \to x - y$ under the integral sign.

Proposition 13.F. *Lebesgue measure is invariant under addition on $\mathbb{R}$ — either viewed as a measure on the Borel algebra $\mathfrak{R}$ generated by the usual topology of $\mathbb{R}$, or as a measure on its completion $\mathfrak{I}^*$, the Lebesgue algebra. Since it is a regular Borel measure* (see Proposition 11.H), *it follows that it is a regular Haar measure* (see the remarks that close Section 8.3).

Proposition 13.G. *Consider the Lebesgue measure $\mu\colon \mathfrak{R} \to \overline{\mathbb{R}}$ on the Borel σ-algebra $\mathfrak{R}$ of subsets of $\mathbb{R}$ generated by the usual topology of $\mathbb{R}$. Take the set $\mathbb{R}^+ = (0,\infty)$ of all the positive real numbers, which is a locally compact Hausdorff multiplicative group. The set function $\lambda\colon \mathfrak{R}^+ \to \overline{\mathbb{R}}$ on the Borel σ-algebra $\mathfrak{R}^+ = \mathfrak{R} \cap \wp(\mathbb{R}^+)$ of subsets of $\mathbb{R}^+$ defined by*

$$\lambda(E') = \int_{E'} \tfrac{1}{x}\,d\mu(x) \quad \text{for every} \quad E' \in \mathfrak{R}^+$$

is a Haar measure. (Reason: $\int \frac{dx}{x} = \log(x)$ and $\log(\beta) - \log(\alpha) = \log(\frac{\beta}{\alpha})$.)

The notion of absolute continuity on the same σ-algebra (cf. Definition 7.6) is naturally extended to a couple of measures, one of them acting on a sub-σ-algebra of the σ-algebra upon which the other measure acts. Indeed, if $\mu\colon \mathcal{X} \to \overline{\mathbb{R}}$ is a measure on a σ-algebra $\mathcal{X}$, and $\lambda\colon \mathcal{E} \to \overline{\mathbb{R}}$ is a measure on the σ-algebra $\mathcal{E} = \mathcal{X} \cap \wp(E)$, for some $E \in \mathcal{X}$, then λ is *absolutely continuous* with respect to μ (same notation: $\lambda \ll \mu$) if, for an arbitrary $E' \in \mathcal{E} \subseteq \mathcal{X}$,

$$\mu(E') = 0 \quad \text{implies} \quad \lambda(E') = 0$$

(i.e., $\lambda(E') = 0$ for every $E' \in \mathcal{E} \subseteq \mathcal{X}$ such that $\mu(E') = 0$). Now consider the setup of the previous proposition, and observe that $\Re^+ = \Re \cap \wp(\mathbb{R}^+)$ is a sub-σ-algebra of the σ-algebra $\Re$. We know from Propositions 3.5(c) and 3.7(b) that the Haar measure λ on $\Re^+$ is absolutely continuous with respect to Lebesgue measure μ on $\Re$, and therefore (recall the Radon–Nikodým Theorem — Theorem 7.8) the function $f\colon \mathbb{R}^+ \to \mathbb{R}$ defined by $f(x) = \frac{1}{x}$ for every $x \in \mathbb{R}^+$ is the Radon–Nikodým derivative of λ with respect to μ. Now consider the sub-σ-algebra $\Re_0 = \Re \cap \wp(\mathbb{R}_0)$ of the σ-algebra $\Re$, where $\mathbb{R}_0 = \mathbb{R}\backslash\{0\}$ is the set of all nonzero real numbers, which is again a locally compact Hausdorff multiplicative group.

Proposition 13.H. *The same Haar measure λ on the σ-algebra $\Re_0$ of subsets of the multiplicative group $\mathbb{R}_0$ is absolutely continuous with respect to Lebesgue measure μ on the σ-algebra $\Re$ of subsets of the additive group $\mathbb{R}$.*

Observe that the Radon–Nikodým derivative of λ with respect to μ is the function $f\colon \mathbb{R}_0 \to \mathbb{R}$ given by $f(x) = \frac{1}{|x|}$ for every $x \in \mathbb{R}_0$. (Proposition 3.G — note that $\mathbb{R}_0 = \mathbb{R}^+ \cup (\mathbb{R}^+)^{-1}$ and so $\int_{\mathbb{R}_0} \frac{1}{|x|}\, d\lambda = \int_{\mathbb{R}^+} \frac{1}{x}\, d\lambda - \int_{(\mathbb{R}^+)^{-1}} \frac{1}{x}\, d\lambda$).

Notes: Translation invariance was first mentioned at the end of Chapter 8, and it was fundamental to build nonmeasurable sets there. The propositions in this section complement the results on invariant measures discussed along the chapter. For Propositions 13.A, 13.B and 13.C see, e.g., [18, Problems 5.6, 5.8, 5.9], and for Proposition 13.D see, e.g., [10, Exercise 9.2.3]. Proposition 13.E plays an important role in the proof of Theorem 13.7 (see, for instance, [7, Theorem 79.1]). We have already met those properties in the previous propositions when we looked at Lebesgue measure. In fact, Proposition 13.F says that Lebesgue measure is a prototype of a regular Haar measure (see, e.g., [35, Proposition 14.24] or [6, Corollary 8.2]). Propositions 13.G and 13.H show that Haar measures on the Borel sets generated by the multiplicative groups $\mathbb{R}^+$ and $\mathbb{R}_0$ are absolutely continuous with respect to Lebesgue measure on the Borel sets generated by the additive group $\mathbb{R}$ (see e.g., [10, Exercise 9.2.3] and [18, Problem 60.1]).

Suggested Reading

Berberian [7], Cohn [10], Halmos [18], Nachbin [31], Royden [35]

References

1. S. Abbott, *Understanding Analysis*, Springer, New York, 2001.

2. E. Asplund and L. Bungart, *A First Course in Integration*, Holt, Rinehart and Winston, New York, 1966.

3. R.G. Bartle, *The Elements of Real Analysis*, 2nd ed., Wiley, New York, 1976.

4. R.G. Bartle, *The Elements of Integration and Lebesgue Measure*, Wiley, New York, 1995; enlarged 2nd ed. of *The Elements of Integration*, Wiley, New York, 1966.

5. R.G. Bartle, *A Modern Theory of Integration*, Graduate Studies in Mathematics, Vol. 32, Amer. Math. Soc., Providence, 2001.

6. H. Bauer, *Measure and Integration Theory*, Walter de Gruyter, Berlin, 2001.

7. S.K. Berberian, *Measure and Integration*, Chelsea, New York, 1965.

8. A. Brown and C. Pearcy, *Introduction to Operator Theory I: Elements of Functional Analysis*, Springer, New York, 1977.

9. A. Brown and C. Pearcy, *An Introduction to Analysis*, Springer, New York, 1995.

C.S. Kubrusly, *Essentials of Measure Theory*,
DOI 10.1007/978-3-319-22506-7

10. D.L. Cohn, *Measure Theory*, 2nd ed., Birkhäuser-Springer, New York, 2013.

11. J.B. Conway, *A Course in Abstract Analysis*, Graduate Studies in Mathematics, Vol. 141, Amer. Math. Soc., Providence, 2012.

12. J. Dugundji, *Topology*, Allyn & Bacon, Boston, 1966.

13. N. Dunford and J.T. Schwartz, *Linear Operators – Part I: General Theory*, Interscience, New York, 1958.

14. K.J. Falconer, *The Geometry of Fractal Sets*, Cambridge University Press, Cambridge, 1986.

15. K.J. Falconer, *Fractal Geometry: Mathematical Foundations and Applications*, 3rd ed., Wiley, Chichester, 2014.

16. G.B. Folland, *Real Analysis: Modern Techniques and their Applications*, 2nd ed., Wiley, New York, 1999.

17. C. Goffman and G. Pedrick, *First Course in Functional Analysis*, Prentice-Hall, Englewood Cliffs, 1965.

18. P.R. Halmos, *Measure Theory*, Van Nostrand, New York, 1950; reprinted: Springer, New York, 1974.

19. P.R. Halmos, *Naive Set Theory*, Van Nostrand, New York, 1960; reprinted: Springer, New York, 1974.

20. P.R. Halmos, *A Hilbert Space Problem Book*, Van Nostrand, New York, 1967; 2nd ed., Springer, New York, 1982.

21. J.L. Kelley, *General Topology*, Van Nostrand, New York, 1955; reprinted: Springer, New York, 1975.

22. J.L. Kelley and T.P. Srinivasan, *Measure and Integral – Volume 1*, Springer, New York, 1988.

23. J.F.C. Kingman and S.J. Taylor, *Introduction to Measure and Probability*, Cambridge University Press, Cambridge, 1966.

24. A.N. Kolmogorov and S.V. Fomin, *Introductory Real Analysis*, Prentice-Hall, Englewood Cliffs, 1970.

25. C.S. Kubrusly, *Measure Theory: A First Course*, Academic Press-Elsevier, San Diego, 2007.

26. C.S. Kubrusly, *The Elements of Operator Theory*, Birkhäuser-Springer, New York, 2011; enlarged 2nd ed. of *Elements of Operator Theory*, Birkhäuser, Boston, 2001.

27. C.S. Kubrusly, *Spectral Theory of Operators on Hilbert Spaces*, Birkhäuser-Springer, New York, 2012.

28. S. Lang, *Real and Functional Analysis*, 3rd ed., Springer, New York, 1993.

29. J.N. McDonald and N.A. Weiss, *A Course in Real Analysis*, Academic Press, San Diego, 1999.

30. M.E. Munroe, *Measure and Integration*, 2nd ed., Addison-Wesley, Reading, 1971.

31. L. Nachbin, *The Haar Integral*, Van Nostrand, New York, 1965.

32. M. Reed and B. Simon, *Methods of Modern Mathematical Physics I: Functional Analysis*, 2nd ed., Academic Press, New York, 1980.

33. F. Riesz and B. Sz.-Nagy, *Functional Analysis*, Frederick Ungar, New York, 1955.

34. C.A. Rogers, *The Hausdorff Measure*, Cambridge University Press, Cambridge, 1970.

35. H.L. Royden, *Real Analysis*, 3rd ed., Macmillan, New York, 1988.

36. W. Rudin, *Real and Complex Analysis*, 3rd ed., McGraw-Hill, New York, 1987.

37. H. Sagan, *Space-Filling Curves*, Springer, New York, 1994.

38. G.E. Shilov and B.L. Gurevich, *Integral, Measure and Derivative: A Unified Approach*, Prentice-Hall, Englewood Cliffs, 1966.

39. P. Suppes, *Axiomatic Set Theory*, Van Nostrand, New York, 1960.

40. R.L. Vaught, *Set Theory: An Introduction*, 2nd ed., Birkhäuser, Boston, 1995.

41. A.J. Weir, *Lebesgue Integration and Measure*, Cambridge University Press, Cambridge, 1973.

42. A.J. Weir, *General Integration and Measure*, Cambridge University Press, Cambridge, 1974.

Index

C.S. Kubrusly, *Essentials of Measure Theory*,
DOI 10.1007/978-3-319-22506-7

Printed in the United States
By Bookmasters